CHEMICAL SPILLS AND EMERGENCY MANAGEMENT AT SEA

Chemical Spills and Emergency Management at Sea

Proceedings of the First International Conference
on "Chemical Spills and Emergency Management at Sea",
Amsterdam, The Netherlands, November 15–18, 1988

edited by

P. BOCKHOLTS

and

I. HEIDEBRINK

TNO Department of Industrial Safety, Apeldoorn, The Netherlands

KLUWER ACADEMIC PUBLISHERS
DORDRECHT / BOSTON / LONDON

ISBN-13: 978-94-010-6887-1   e-ISBN-13: 978-94-009-0887-1

DOI: 10.1007/978-94-009-0887-1

Published by Kluwer Academic Publishers,
P.O. Box 17, 3300 AA Dordrecht, The Netherlands.

Kluwer Academic Publishers incorporates
the publishing programmes of
D. Reidel, Martinus Nijhoff, Dr W. Junk and MTP Press.

Sold and distributed in the U.S.A. and Canada
by Kluwer Academic Publishers,
101 Philip Drive, Norwell, MA 02061, U.S.A.

In all other countries, sold and distributed
by Kluwer Academic Publishers Group,
P.O. Box 322, 3300 AH Dordrecht, The Netherlands.

INTRODUCTION

International shipping is of great importance for the transport
of a great many types of cargo. Substances and products considered
dangerous constitute almost 50% of all the payload. It is obvious
that stringent regulations are required in order to minimize the
risks of accidents. These regulations, which are derived from good
practice and which are based on research, have been adopted by a
great number of countries. However, emergencies do occur in spite
of all precautions. Such emergencies require fast and adequate
response in order to confine the consequences for man and his
environment to a minimum. Emergency response has political, legal,
financial and technical aspects. This makes decision making
extremely difficult.
The papers carefully prepared and assembled in this book present
an up-to-date picture of today's achievements, knowledge and
difficulties that are being faced.
It was the intention of Oilchem Recovery Denmark and TNO to bring
the wide scatter of aspects together in a joined perspective. We
also intended to spread the information on latest developments
among the many people who are involved in combating calamities and
in particular in decision making.
Finally, we hope that this conference may help all of us to come
to a safer transport of chemicals and a better aquatic environment.

We thank all the authors for their magnificent contribution.

                                    P. Bockholts, chairman
                                    I. Heidebrink, co-chairman

TABLE OF CONTENTS

PREVENTION/PREPAREDNESS FOR ACCIDENTS

THE PREVENTION OF CHEMICAL SPILL INCIDENTS

O.H.J. Dijxhoorn
International Maritime Organization (IMO)
Maritime Safety Division
4 Albert Embankment
London, SE1 7SR
United Kingdom

ABSTRACT. The issue of chemical spills at sea raises the question as to what international regime exists aimed at the prevention of such incidents. The international body regulating maritime matters is the International Maritime Organization (IMO) and it is necessary to turn towards that organization to find the answer. At the same time it is important to be aware of the fact that the work of IMO and the effectiveness of the instruments it adopts depends wholly on the work of Governments which undertake to implement and if necessary review such instruments.

The prevention of chemical spills is first and foremost a safety matter and IMO is responsible for the adoption of a number of conventions currently in force and directed at greater safety at sea. In addition an important convention for the prevention of pollution of ships is in force. Safety conventions usually concern ships in general but there are a number of IMO conventions which have been directly affected by the development of the chemical trade. This paper restricts itself to discussing the latter.

1.   Introduction

The International Convention for the Prevention of Pollution from Ships, 1973, as modified by the Protocol of 1978 relating thereto, commonly referred to as MARPOL 73/78, defines an incident as an event involving the actual or probable discharge into the sea of a harmful substance, or effluents containing such a substance. Harmful substance in this context means any substance which, if introduced into the sea, is liable to create hazards to human health, to harm living resources and marine life, to damage amenities or to interfere with other legitimate uses of the sea, and includes any substance subject to control by MARPOL 73/78. Chemical spills are addressed in Annex II of that convention which, in its Appendix II, lists the so-called noxious liquid substances that may be carried in bulk by ships and for that reason it would seem obvious to consider Annex II

3

P. Bockholts and I. Heidebrink (eds.), Chemical Spills and Emergency Management at Sea, 3—14.
© 1988 by Kluwer Academic Publishers.

to come within the scope of the Conference on Chemical Spills and Emergency Management at Sea. This is true although one needs to consider more than one convention to seek an answer to the question whether an adequate régime exists to prevent such incidents from occurring.

Chemical spills of course should not happen and conferences such as this one should not be necessary. Unfortunately spills do occur although, considering the magnitude and complexity of chemical trade, infrequently, and this conference will serve a useful purpose in bringing together the necessary expertise and in imparting knowledge to those less experienced. Also the question on what is being done internationally to prevent such incidents from occurring is appropriate to this conference and this paper elaborates on a number of international conventions addressing this problem.

MARPOL 73/78 as indicated above is a convention aimed at the prevention of pollution from ships, but its primary aim is the control of discharges which result from the day-to-day operation of chemical tankers at sea. Marine pollution resulting from massive spills are usually the result of accidents to ships such as strandings, collisions or explosions. First and foremost therefore the prevention of chemical spills which warrant combating is accident prevention which reflects in collision regulations, properly constructed and equipped ships, good seamanship to prevent pollution caused by stranding or wrong cargo/ballast handling and adequate knowledge of chemicals to prevent accidents resulting from the properties of the cargoes. Because ships travel worldwide, the problem should be addressed worldwide, which makes it an issue subject to international conventions. Such conventions have been adopted time and again under the auspices of different maritime nations, but since 1958 the internationally recognized forum for shipping matters has been the International Maritime Organization (IMO).

2.   Method of work of the IMO

When the establishment of a specialized agency of the United Nations concerned solely with maritime affairs was first proposed, the main concern was to evolve international machinery to improve safety at sea.

This was understandable for two main reasons. In the first place, seafaring has always been one of the most dangerous of occupations. In the second place, because of the international nature of the shipping industry, it had long been recognized that action to improve safety in shipping operations would be more effective if carried out at an international level rather than by individual countries acting unilaterally and without co-ordination with others. Although a number of important international agreements had already been adopted, many States agreed that there was a need for a permanent

body which would be able to co-ordinate and promote
further measures on a more continuing basis.

It was against this background that the United Nations Maritime
Conference of 1948 adopted the Convention establishing the
International Maritime Organization (IMO)* as the first ever
international body devoted exclusively to maritime matters.

In the ten-year period between the adoption of the Convention and its
entry into force in 1958, other problems related to safety, but
requiring slightly different emphasis, had attracted international
attention. One of the most important of these was the threat of
marine pollution from ships, particularly pollution by oil carried in
tankers. An international convention on this subject was actually
adopted in 1954, four years before IMO came into existence; and
responsibility for administering and promoting this convention was
assumed by IMO at the inception of its work in January 1959. Thus,
from the very beginning, the improvement of maritime safety and the
prevention of marine pollution have been IMO's most important
objectives.

The governing body of IMO is the Assembly, which meets once every two
years and consists of all the Member States. In the period between
the sessions of the Assembly, a Council exercises the functions of
the Assembly in running the affairs of the Organization. For the
time being, the Council consists of 32 Member Governments elected for
two-year terms by the Assembly.

The Organization's technical work is carried out by a number of
committees, the most senior of which is the Maritime Safety Committee
(MSC). This has a number of sub-committees on: safety of
navigation; radiocommunications; life-saving search and rescue;
standards of training and watchkeeping; carriage of dangerous
goods; ship design and equipment; fire protection; stability and
load lines and fishing vessel safety; containers and cargoes; and
bulk chemicals.

The Sub-Committee on Bulk Chemicals is also a sub-committee of
another technical committee, the Marine Environment Protection
Committee (MEPC), which deals with the Organization's marine
pollution control activities. Because of the legal issues involved
in much of its work, the Organization also has a Legal Committee,
while the Committee on Technical Co-operation co-ordinates and
directs IMO's activities in this area. These committees are all
recognized by the IMO Convention. Finally there is the Facilitation
Committee, which deals with measures to simplify and minimize
documentation in international maritime traffic. This is a subsidiary
body of the Council.

---

*   Until 22 May 1982 the Organization was called the
      Inter-Governmental Maritime Consultative Organization (IMCO).

In order to achieve its objectives IMO has, in the last thirty years, promoted the adoption of conventions and protocols and in addition adopted a large number of codes and recommendations on various matters relating to maritime safety and the prevention of pollution. The initial work on a convention is normally done by a committee which may delegate part of its work to a sub-committee; a draft instrument is then produced which is submitted to a conference to which delegations from all States within the United Nations system - including States which may not be IMO Members - are invited. The conference adopts a final text, which is submitted to Governments for ratification.

3.   The role of Governments

With the successful adoption of the convention, the onus for action moves to Governments. The speed with which the convention enters into force (that is, becomes binding on States which have agreed to be bound by it) depends upon the time taken by Governments to ratify or accept it. The consent to be bound may be expressed by signature, ratification, acceptance, approval or accession, depending on the wish of the States concerned. This procedure is generally referred to as "ratification". IMO treaties enter into force after a specific number of States have ratified them. Most IMO conventions require that a certain proportion of the world's total tonnage be covered before the conventions enter into force.

A Government ratifying a convention or a treaty has to ensure that its own national law conforms with its provisions. This usually involves some form of domestic legislative action.

After the requirements for entry into force of a treaty have been achieved, there is a "period of grace" before it actually comes into force. This period varies from a few months to a year or even two years, and is designed to enable the Governments concerned to take the necessary legislative or administrative measures for implementing the provisions of the convention.

The third stage is implementation, in many ways the most important stage. In most cases, the main responsibility for the enforcement of an international treaty lies on the State under whose flag the ships concerned operate. Basically, each Government is responsible for ensuring that ships which fly its flag conform to the requirements of treaties which it has ratified. However, many IMO treaties also contain provisions permitting or requiring other States, particularly port States, to enforce the requirements of the conventions concerned.

The effectiveness of a convention therefore depends to a considerable extent on the way in which it is enforced by the States entrusted with its implementation. IMO as an Organization has no authority or means to enforce or implement conventions against individual 'ships or States. The Organization's role is to encourage the Governments concerned to take the required measures. Where necessary, the Organization provides technical advice and assistance to Governments which may need such advice and assistance in taking the requisite action.

When discussing the prevention of pollution by chemical spills it is difficult to identify the convention which has the greatest impact in this respect. It may be the convention aimed at the avoidance of collisions or the one assigning loadlines but this paper limits itself to highlighting three conventions which are of particular significance because these conventions have been and still are subject to amendments due to the development of the chemical trade.

4.   SOLAS 74

The first conference organized by IMO in 1960 was, appropriately enough, concerned with safety matters. In 1948 an International Convention on the Safety of Life at Sea had been adopted at a conference convened by the United Kingdom, but developments during the intervening years had made it necessary to bring this up to date without delay.

This Safety of Life at Sea Convention - the SOLAS Convention as it is known for short - became the basic international instrument dealing with matters of maritime safety and in response to new developments it was amended several times. However, because of the rather difficult requirements for bringing amendments into force, none of these amendments actually became binding internationally.

To remedy this situation and effect the needed improvements more speedily, IMO convened a conference, in 1974, to adopt a new International Convention on the Safety of Life at Sea which would incorporate the amendments adopted to the 1960 Convention as well as introduce other necessary improvements.

It was felt that it would be easier to bring the new Convention into force than to secure the acceptances necessary for the amendments to become international law. The new Convention also has a much easier amendment procedure under which amendments will normally enter into force on certain predetermined dates, unless a stipulated number of States Parties to the Convention indicate that they object to them.

The 1974 Convention entered into force on 25 May 1980.

In the meantime a considerable amount of work had been done in updating the Convention. A Protocol adopted in 1978 entered into force in May 1981 and the first of a series of important amendments was adopted in November 1981. The amendments entered into force in September 1984. A second set of amendments was adopted in June 1983 and entered into force on 1 July 1986.

Initially SOLAS 74 introduced regulations for the protection of
tankers in general against fire and explosion, but the convention did
not cater for the many and different hazards posed by the carriage in
bulk of increasingly divers chemicals. It was only with the 1983
amendments to chapter VII of SOLAS that binding requirements were
introduced for the construction and equipment of ships carrying
dangerous chemicals in bulk, For this purpose chapter VII refers to
the International Bulk Chemical Code (IBC Code).

5.   MARPOL 73/78

The International Convention for the Prevention of Pollution from
Ships was adopted in 1973, the Protocol thereto, which contained
important amendments mainly with respect to oil tankers, was adopted
in 1978 and the combined instrument, MARPOL 73/78, entered into force
on 2 October 1983. The amendment procedure is similar to the one of
SOLAS 74 making adjustments in the light of developments
practicable. The Convention addresses several forms of pollution by
means of provisions contained in five technical annexes. Annex I
addresses oil, Annex II noxious liquid substances in bulk, Annex III
harmful substances in packaged forms, Annex IV sewage and Annex V
garbage. Acceptance of MARPOL 73/78 by contracting Governments
implies acceptance of all five annexes although Annexes III to V, the
so-called optional annexes, may be excepted. Since a number of
Contracting Governments excepted those annexes only Annex I and Annex
II have entered into force. Annex V has since obtained sufficient
ratifications and will enter into force in December 1988. The 1978
Protocol made it possible to allow Annex II to enter into force at a
later date than Annex I and as a result Annex II, for which many
technical problems had to be overcome, only went into effect on
7 April 1987.

One of the purposes of Annex II of MARPOL 73/78 is to minimize the
likelihood that noxious substances are accidentally released into the
marine environment from ships transporting these in bulk. Like
chapter VII of SOLAS 74, Annex II of MARPOL 73/78 refers to the
International Code for the Construction and Equipment of Ships
Carrying Dangerous Chemicals in Bulk (IBC Code). Unlike SOLAS 74, it
also makes the IBC Code's forerunner, the Bulk Chemical Code (BCH
Code) mandatory for ships constructed prior to 1 July 1986.

Under Annex II, chemicals that are shipped in bulk are evaluated on
the basis of their pollution hazard, assigned one of four pollution
categories (category A, B, C or D) and included in Appendix II.
Chemicals which pose insufficient pollution hazard to warrant
categorization are included in Appendix III of the Annex.

6.   The development of the chemical tanker

The chemical tanker is basically a development of the last forty
years. The development of the chemical industry in the United States

following the end of World War II led to a demand for ships in which to carry the industry's products. A number of T-2 tankers, mass produced during the war, were converted by installing special tanks, double bottoms and suitable structural and piping arrangements.

For the next decade or so tankers used in the carriage of chemicals were nearly all conversions. As the trade developed these ships became more refined, with the addition of tank linings, cofferdams and other features.

The range of products carried in the first chemical tankers was relatively limited and the products themselves were not technically too demanding. The products also all tended to be owned by one company.

By the 1960s the chemical trade was becoming more complex. The number of substances being transported at sea was increasing rapidly - as was the total tonnage - and the products were technically more complicated. At the same time a growing number of what came to be called 'parcel tankers' were making their appearance. These were tankers designed to carry a range of chemical products for a number of different owners.

These ships were of necessity more complex than the original first-generation chemical tankers and during the early 1960s the first purpose-built chemical tankers made their appearance.

By the mid-1960s the chemical tanker had developed into a ship which was different from any other type, including other tankers. The cargoes carried by chemical tankers were probably potentially the most dangerous substances afloat, to the ships, their crews and to the marine environment. All of this made it imperative for something to be done to ensure that the ships themselves were suitable for the task. The answer was the Bulk Chemical (BCH) Code.

7.   The Bulk Chemical Code

In adopting the BCH Code in resolution A.212(VII) the IMO Assembly urged Governments having fleets of chemical tankers to implement the Code and to inform the Organization accordingly. A number of flag States have implemented the Code through national law. Other States leave the matter to the discretion of shipowners and only issue a code certificate, known as a Certificate of Fitness, upon request. A number of port States also require foreign flag ships to have a Certificate of Fitness as a condition of port entry into their ports.

The action taken by some States of applying the Code to all ships entering their ports gave the Code a de facto convention status.

Considering this ambiguity of the status of the Code to be unsatisfactory the IMO Maritime Safety Committee decided that it should be made mandatory and chose the 1974 SOLAS Convention as the mechanism for doing so.

It was agreed that, before making the Bulk Chemical Code mandatory, it should first be harmonized with the Gas Carrier Code which serves a similar purpose for ships carrying liquefied gases in bulk and which, by virtue of having been developed later, was therefore more detailed and complete.

The revised BCH Code now contains nineteen chapters instead of the original seven and follows closely the layout and text of the Gas Carrier Code, which in the harmonization process was also further improved. To distinguish this mandatory Code from the voluntary Code under resolution A.212(VII), it was agreed that it should be named the International Code for the Construction and Equipment of Ships carrying Dangerous Chemicals in Bulk (International Bulk Chemical Code or IBC Code) and be applicable to ships built on or after the date of their entry into force.

No action was taken under SOLAS to make the existing BCH Code mandatory. It will remain a voluntary Code and apply to ships of any size built before the date of coming into force of the new mandatory code. Under MARPOL 73/78 the BCH Code has become mandatory.

The Maritime Safety Committee at its forty-eighth session adopted the IBC Code with resolutions MSC.4(48)and decided that it should apply to chemical tankers of any size built on or after 1 July 1986.

At the same time the Maritime Safety Committee in order to make the Code mandatory under SOLAS 74 included in the 1983 SOLAS Amendments a revised Chapter VII "Carriage of Dangerous Goods" with two new parts; Part B on construction and equipment of ships carrying dangerous chemicals in bulk and Part C ships carrying liquefied gases in bulk. For ships carrying liquefied gases a similar dual régime exists: The so-called GC Code for ships built before 1 July 1986 with the same status as the BCH Code and the IGC Code, mandatory for ships built on or after 1 July 1986.

According to new Regulation 10 of the amended Chapter VII, new chemical tankers shall comply with the requirements of the IBC Code and shall, in addition to the applicable requirements of Chapter I of the 1974 SOLAS Convention, be surveyed and certified as provided for in that Code. For the purpose of Regulation 10 the requirements of the Code shall be treated as mandatory.

In the Codes, chemicals considered sufficiently dangerous to warrant special precautions are listed in chapter VI (BCH Code) or 17 (IBC Code) and initially these chapters only listed dangerous chemicals.

With the advent of Annex II of MARPOL 73/78, the list of chemicals underwent some changes. In recognition that many of the safety features of the Codes have the effect of preventing accidental marine pollution, regulation 13 of Annex II requires that, as a minimum ships carrying category A, B or C substances must comply with the Codes. All such substances are therefore listed in chapters VI or 17 of the Codes even if not posing safety hazards to man. This has the effect that chemicals may be listed in those chapters on safety grounds (indicated by S), pollution grounds (indicated by P) or both (indicated by S/P).

8.   STCW 78

According to the convention establishing IMO the major purpose of the Organization is 'to provide machinery for co-operation among governments in the field of governmental regulations and practices relating to technical matters of all kinds affecting shipping engaged in international trade; to encourage the general adoption of the highest practicable standards in matters concerning maritime safety, efficiency of navigation and the prevention and control of marine pollution from ships ...'.

Even in 1960 it was recognized that training was vital to safety at sea. Regulation 13 of Chapter V of the International Convention for the Safety of Life at Sea adopted in that year stated that: 'The Contracting Governments undertake, each for its national ships, to maintain or if necessary to adopt measures for the purpose of ensuring that from the point of view of safety of life at sea all ships shall be safely and efficiently manned.'

The 1960 Conference also adopted Recommendation 39 calling upon Contracting Governments to 'ensure that the education and training of masters, officers and seamen in the use of aids to navigation, of life-saving and of authorized devices designed for the prevention, detection and extinction of fires or for preventing or alleviating casualties at sea is sufficiently comprehensive, and also that, by supplementary or refresher courses, or by other appropriate means, such education and training is kept up-to-date and in step with modern technological developments in this field ...'.

The Recommendation also called upon IMO to co-operate with governments and the International Labour Organisation, in achieving these aims.

In 1964 the two Organizations established a Joint Committee on Training and in the same year produced a Document for Guidance containing guidelines on the training and education of masters, officers and seamen. Updated versions of the document were issued by the two Organizations in 1969, 1971, 1975 and 1977.

The Joint Committee was aware in 1964 of the particular problems facing developing countries, for the document states: 'Systematic training is essential in all countries, and it is greatly hoped that

this guidance will achieve a very wide and general measure of
application. It is believed that it will be of particular use to
those countries which are known to be instituting or overhauling
maritime training programmes.'

The document then goes on to outline in detail the sort of subjects
which should be covered at training schools and the knowledge which
seafarers should be required to possess. Sample courses are given in
a number of appendices.

In 1967 the IMO Assembly adopted a recommendation on crew training
which was inspired particularly by a series of fires on board ship
which had resulted in heavy loss of life. It recommended that all
IMO Member States should 'aim at training all its seafarers in fire
prevention and fire fighting to an extent appropriate to their
functions on board ship' ...'.

The Document for Guidance remained IMO's chief instrument on the
subject of training for another ten years, but by the 1970s IMO had
decided that something much more ambitious was needed. In 1971 a
special Sub-Committee on Standards of Training and Watchkeeping was
established as a sub-committee of the Maritime Safety Committee with
the primary task of developing a draft convention on the subject of
training and certification.

The Convention on Standards of Training, Certification and
Watchkeeping for Seafarers, 1978 (STCW 78), is the first attempt to
establish global minimum professional standards for seafarers.
Previously such standards were set by individual governments, usually
without reference to practices in other countries. As a result
standards and procedures varied widely, even though shipping is the
most international of all industries.

The Convention prescribes minimum standards which Contracting Parties
are obliged to meet. In the majority of developed maritime
countries, standards are often higher than those stipulated in the
convention. In some countries, however, standards are not so high
and, by ratifying or accepting the Convention, Governments undertake
to implement and enforce its requirements. The effect of the
Convention's entry into force will therefore be to raise standards in
the world as a whole.

The requirement for entry into force of the Convention was acceptance
by 25 countries whose combined fleets of merchant shipping represent
at least 50 per cent of world tonnage. This target was reached on 27
April 1983 and the Convention entered into force one year later, on
28 April 1984.

One especially important feature of the Convention is that it will
apply to ships of of non-Party States when visiting ports of States
which are Parties to the Convention. Article X requires Parties to

apply the control measures to ships of all flags to the extent
necessary to ensure that no more favourable treatment is given to
ships entitled to fly the flag of a State which is not Party than is
given to ships entitled to fly the flag of a State that is not
Party. As this could lead to difficulties for ships of States which
are not Parties to the Convention it is expected that the number of
Parties will increase considerably in the near future. By January
1988 it had been accepted by 63 countries.

Regulation V/2 of the Convention and resolution 11 adopted at the
conference address the training and qualification of masters,
officers and ratings of chemical tankers.

The amendment procedure is again similar to the one in SOLAS 74.

9. Technical assistance

For those charged with implementing the Convention even the great
detail in the appendices to the convention and the resolutions
adopted at the conference is not enough and there is still a widely
felt need for guidelines on implementation of the standards. Also,
training of seafarers constitutes only one element towards safety.
Of equal if not more importance is the availability of trained shore
staff.

The challenge was met by IMO through a complex set of measures of
which the establishment of the World Maritime University (WMU) at
Malmö in 1983 is perhaps the most outstanding. The WMU is designed
to provide training for senior maritime administrations, teachers and
instructors, examiners, technical port managers, surveyors, accident
investigators and others at similar levels - the very people, in
fact, who are most closely involved in putting IMO's technical
standards into effect. Advanced training of the type provided at the
WMU is almost non-existent in many developing countries and the
university was established by IMO to fill this gap in the most
practical way possible and at the least cost.

Other technical assistance projects in the field of training evolve
around the participation of IMO in a great number of regional or
national seminars on a wide range of maritime subjects. In addition,
there are a number of advisers, some based in the regions and some at
IMO Headquarters, whose task it is to visit countries, on request, to
give advice on a wide range of subjects. Depending on the
requirements of the maritime administration the adviser will be
expected either to identify immediate maritime needs and assist
governments in achieving desired objectives, or deal with specific
requests on concrete problems as a first step towards obtaining
further assistance.

A relative new-comer in the field of technical assistance is the
joint IMO-Norwegian programme of model courses to assist in the

implementation of the STCW Convention and to achieve a more rapid
transfer of information and skills regarding new developments in
maritime technology. The model course project encompasses five
programmes, directed at sea-going staff (basic and advanced),
maritime safety/pollution administrators, port authorities and
shipping company staff. This in addition to the certificate courses
for masters, deck and engineer officers. The project, which is still
under way, has currently completed some fifteen courses which can be
obtained at the Publications Section of the IMO or through bookshops
handling IMO publications.

10. Conclusion

Over the years the international maritime community has pursued the
development of international conventions, codes of practice and
recommendations to enhance safety of life at sea. It was thereby not
only working towards the protection of life at sea, but at the same
time, through decreasing the number of accidents, diminishing the
chance of pollution incidents. Nevertheless accidents and spills at
sea continue and subsequent inquiries have led to the conclusion that
in by far the majority of cases human error rather than technical
failure was the underlying cause.

Part of the answer may be improved training. This applies to
seafarers as well as those ashore dealing with maritime matters.
However, this is more easily stated than done in view of the fact
that technical requirements in respect of ships through sheer
magnitude, are increasingly difficult for many maritime
administrations to follow up. Also it should be realized that IMO
produces increasingly complex operational requirements for seafarers.

Whether improved training alone is sufficient is a question that
still requires an answer. A new challenge to the future work of IMO
which will have to come up with an answer that equates the complexity
of the carriage of cargoes by sea with the correct combination of
technical feasibility and what one can confidently expect seafarers
to accomplish.

Views and opinions expressed in this paper are the responsibility of
the author and do not necessarily reflect those of IMO.

CONTINGENCY PLANNING

TERENCE M. HAYES
Marine Environment Division
International Maritime Organization
4, Albert Embankment
London SE1 7SR
United Kingdom

ABSTRACT. This paper discusses the problems which have been identified in dealing with spillages of hazardous chemicals in bulk and packaged form. The work of the International Maritime Organization (IMO) in developing contingency plans and manuals for prevention of and combating oil spillages is described. The Marine Environment Protection Committee (MEPC) of IMO has assigned priority to the preparation of the Manual on chemical pollution and the paper also describes the current status of this Manual. The elements of a chemical spill contingency plan and factors to be considered are discussed. Both IMO and the Commission for the European Communities have undertaken a number of initiatives in developing data bases and conducting training courses. Both organizations are in the process of developing arrangements to provide emergency assistance to developing countries in the event of a chemical spillage.

Introduction

Whilst the principal concern of coastal States in the past decade has been with oil pollution from tanker accidents and exploration or production well blowouts, there is a growing concern over the possible problems created by an incident involving a chemical tanker. The world chemical tanker fleet has increased dramatically over the past decade to 886 dedicated chemical tankers and 543 oil/chemical tankers. Although there have been relatively few accidents involving chemical tankers and no significant spillages, the risk does exist. The recent incident involving the "Anna Broere" has caused concern to the authorities of the Netherlands and involved a difficult and costly salvage operation. A similar operation also recently took place off the coast of Yugoslavia involving removal of a cargo of vinyl chloride monomer gas from the "Brigitta Montanari".

15

P. Bockholts and I. Heidebrink (eds.), Chemical Spills and Emergency Management at Sea, 15—27.
© *1988 by Kluwer Academic Publishers.*

Transport of chemicals, both in bulk and in packaged forms, is increasing rapidly. Consequences of such transport to the marine environment were foreseen by the International Conference on Marine Pollution, 1973 in adopting Annexes II and III of the International Convention for the Prevention of Pollution from Ships, 1973, as modified by the Protocol of 1978 relating thereto (MARPOL 73/78). Similarly, dangers to the ship and crew and other personnel involved in handling chemicals have long been recognized in the SOLAS Convention and in the development of the International Maritime Dangerous Goods (IMDG) Code, the Code for the Construction and Equipment of Ships Carrying Dangerous Chemicals in Bulk (IBC and BCH Code) 1986 editions and related recommendations.

The design, construction and cargo loading of chemical tankers is such as to minimize the likelihood of spillage of the most hazardous chemicals. However, a severe collision or grounding in adverse weather conditions could result in the release of noxious liquid substances. The increasing quantities of chemicals carried in packaged form by general cargo and container ships can also present particular problems in the event of a casualty. The September 1985 incident in Mogadishu, Somalia, involving the vessel Ariadne highlighted the problems which can be faced by a developing country in responding to a chemical spillage, or threat of same.

Contingency plans for oil spillages

Under its technical assistance programme, the International Maritime Organization (IMO) has provided advisory services to developing countries by missions on prevention of, control of and response to marine pollution. A draft national oil spill contingency plan has also been prepared for consideration by the Government during these missions.

Such contingency plans should be action oriented and cover such aspects as reporting, alerting, assessment, operations, administration, finance, public relations and arrangements with other contiguous States. The plan should assign responsibility for various tasks to the relevant Government and private agencies, identifying trained personnel and equipment resources and means of access to same. Such resources may be specialized pollution combating equipment or multi-purpose such as vessels, aircraft, construction equipment and communications systems.

Many countries have already established disaster preparedness or relief committees and enabling legislation which confers extraordinary powers on either Cabinet or designated ministers. It is considered desirable that Governments should treat a medium to major oil spillage as a man-made disaster, and that their response capability should preferably be centred on an existing organization for response to national disasters. Much of the national legislation relating to emergency co-ordination can readily be modified or amended to include medium to major oil spillages as a national disaster.

The national organizational structure for response to oil spills should include representation from such agencies as fisheries, tourism, port authorities and other sectors of the economy which could be affected by an oil spillage. Those coastal areas and essential installations which should be given priority in protection and clean-up operations should be identified and classified well in advance of an actual spillage. Each country has different concerns; those that are dependent on tourism should give high priority to their beaches; others may be more concerned about the possible effects on the fishery and maritime commerce; whilst water distillation and other plants which use sea water for processing or cooling may need to be accorded high priority.

IMO has prepared a Manual on Oil Pollution which addresses oil pollution problems rather than safety measures. It is a particularly useful guide for Governments of developing countries and for those persons directly associated with the sea transportation and transfer of oil. The Manual is divided into four sections:

- Section I - Prevention (1983 revised edition) contains practical information on means of avoidance or prevention of pollution of the sea by oil, in accordance with the requirements of MARPOL 73/78.
- Section II - Contingency Planning (1988 edition) provides guidance on ways of establishing response organizations and preparing contingency plans, both local and regional.
- Section III - Salvage (1983 edition) provides guidance on effectively mitigating the effects of oil pollution casualties, concentrating particularly on the salvage of vessels and cargoes.
- Section IV - Practical Information on Means of Dealing with Oil Spillages (revised 1980 edition). Oil spills pose numerous risks to the environment and this section of the Manual concerns practical ways of dealing with environmental problems resulting from oil spills; its suggestions should generate unique and innovative methods when required by local circumstances. This Manual is currently under revision and should be reissued towards the end of 1988.

Contingency plans for chemical spillages - initiatives by IMO

As previously mentioned, considerable progress has been made in the development of national contingency plans to deal with marine oil spills, however, much remains to be done in terms of contingency planning for response to spills of noxious liquid substances in bulk and harmful substances in packaged form.

In late 1981, the Marine Environment Protection Committee (MEPC) at its sixteenth session began its deliberations on spillages of substances other than oil and considered a basic text on Spillages of Substances Other than Oil - Hazardous Materials Response. This was amended at subsequent sessions of the Committee with the objective of providing guidance to shore or port personnel dealing with spillages of substances other than oil in locations on board ship and into the sea from both chemical tankers and ships carrying packaged chemicals. It was recognized that ship's crews would be aware of the relevant response methods and in the case of chemical tankers would have undergone specialized training. At its twenty-first session in May 1985, the Committee approved the final draft of the Manual on Chemical Pollution Section 1 - "Problem Assessment and Response Arrangements" and agreed to begin work on Section 2 - "Search and Recovery of Packaged Goods Lost Overboard at Sea". However, after some preliminary consideration, this section was deferred until the twenty-sixth session of the Committee, pending a decision on selection criteria for marine pollutants.

At its twenty-fifth session the Committee agreed on a criteria for selection of marine pollutants for the purpose of Annex III - "Regulations for the Prevention of Pollution by Harmful Substances Carried by Sea in Packaged Forms" of MARPOL 73/78. The criteria for identifying packaged goods as marine pollutants were based on the hazard evaluations carried out by the Group of Experts on the Scientific Aspects of Marine Pollution (GESAMP). The following GESAMP hazard ratings were taken into account:

- substances bioaccumulated to a significant extent and known to produce a hazard to aquatic life or human health;
- substances bioaccumulated with attendant risk to aquatic organisms or to human health, with a short retention time of the order of one week or less;
- substances liable to produce tainting of seafood; and
- substances known to be highly toxic to aquatic life.

At the twenty-sixth session of MEPC (September 1988) a working group will continue deliberations on the draft text for Section 2 of the Manual on Chemical Pollution, based on the material prepared by Member Governments and the Secretariat.

Section 1 of the Manual on Chemical Pollution approved by the twenty-first session of MEPC in 1985, was published and distributed to Member Governments in early 1987 and provides guidance to Governments on ways of assessing hazards associated with spillages of harmful substances other than oil and setting up response organizations.

Planning for response to chemical spillages

At the national level, assessment and response to chemical spills can be integrated into an existing national oil spill contingency plan. However, it should be recognized that national oil spill contingency plans are normally prepared to deal with large oil spills from tankers and/or terminals which require major commitments of resources in terms of personnel, equipment and material, together with the necessary command and administrative structure.

With a few notable exceptions the developing countries do not import or export chemicals in bulk and the majority of the chemical traffic is in packaged form. During my missions to many of these countries, there have been reports of leaking drums, broken packages or mishaps during handling at ports. Most administrations are in possession of the IMO International Maritime Dangerous Goods Code (IMDG Code) and are applying this code in transport, handling and storage of dangerous goods. Spillages from such sources would normally be the responsibility of the port authorities at the local level rather than at the national level. During advisory missions regarding development of contingency plans, the attention of administrations has been drawn to the need for port contingency plans as the majority of marine casualties occur in the approaches to, or in port. Such port contingency plans would deal with response to fires, collisions, groundings, spillage of oil or chemicals, etc. and clearly assign responsibility and authority. In most cases the harbour police and fire departments would mount the initial response. However, there is an identified lack of resources and training in most ports in developing countries.

Reports of spillages and releases of hazardous material, which generally refer to hazardous substances, petroleum, liquefied gases, acutely toxic chemicals and other chemicals, are increasing. With the risk of hazardous material incidents, during the initial stages of such an incident Governments will be forced to rely upon local resources. In order to make optimum use of available human and material resources, the national contingency plan should include a section on response to chemical spillages.

It should be recognized that although there is some similarity in the types of incidents when dealing with oil spillages, additional factors have to be taken into account when chemicals are involved. For example, some chemicals, being highly volatile, will evaporate rapidly over a limited area and may produce a fire hazard that may, in certain cases, extend beyond the area covered by the pool of spilled material. In addition, because many bulk chemicals are toxic to man at low concentrations in air, dangerous conditions may prevail downwind from such a spill to distances greatly in excess of the pool dimensions and the moving cloud may continue to constitute a threat even after the pool of spilled chemical has entirely evaporated into the air, which may take only a few hours.

Less volatile chemicals which float will behave rather differently. The threat in these cases will not be from the atmosphere but from the movement of such slicks on the sea surface, in a manner analogous to weathered crude oil or fuel oil spills. Such spilled material may eventually come ashore. Those chemicals which are not too toxic or corrosive to be handled may be dealt with in the same way as an oil spill. Floating chemicals may disperse or dissolve in the sea giving rise to a toxic body of water, depending on the toxicity of the chemical, its rate of dilution and its subsequent movement. However, as in cases of dispersed oil, responses to such spillages are not likely to take place. Lastly, some chemicals will sink and may or may not dissolve in seawater. Here, depending on the water depth, attempts may be made to recover spillages if the need is judged to be great enough.

Thus, in the case of chemicals, oil spill type response will not always be appropriate. Sometimes the surface slick will evaporate too quickly, but in such cases atmospheric dangers may result in necessitating the issuing of instructions to unauthorized personnel to avoid the area. In other cases solution or rapid dispersion in the sea will greatly reduce the need for response, although avoidance may be appropriate. Finally, recovery from the sea-bed may be possible or desirable and again, avoidance may be necessary.

The possibility of legal proceedings should be borne in mind in all cases of pollution. However, there are at present no schemes for recovery of clean-up expenditures for spills involving substances other than oil. Internationally agreed provisions relating to government intervention on the high seas are contained in the Protocol Relating to Intervention on the High Seas in Cases of Marine Pollution by Substances Other than Oil, 1973.

The chemical spill contingency plan should consider the assessment of hazards associated with a chemical spillage and describe possible response, both of the avoidance and spillage clearance type, and the provision of suitable response organizations and personnel. It should also describe the provision of appropriate training. The diligent application of safety procedures in cargo handling described in a number of other publications should help to reduce the threat of accidents and thus reduce the risk of pollution arising from them.

Identification of hazard and assessment

In the case of an oil spill, an evaluation of the threat posed by the oil spill to aquatic and coastal resources is necessary. An uncontrolled release of chemicals may pose many hazards to personnel in the locality at the time of the release, including those who may be required to respond. Hazards include fire, explosion and toxic effects. In planning the response, it is essential to acquire as much information as possible concerning the identity of the chemical or chemicals spilled, the quantities released and the risk of further spillage. In general, response personnel should not be sent into the scene of a spill until the identity of the chemical or chemicals is known and a hazard assessment has been made.

During and after identification of the hazardous material that has been spilled, continuous assessment of the hazard must take place. The person in charge must be cognizant of the hazards presented by the spill as well as of the safety measures needed to protect both the public and response personnel. Safety should be the primary concern. All other concerns, such as environmental protection and protection of property, are secondary.

In acquiring information about a release, the first point of contact must be the personnel engaged in the operation which gave rise to it. Thus, port authorities or relevant agents of central government must endeavour to obtain as much information as possible from the ship's crew, ship's owner, cargo owner or agents.

If the ship is carrying a single substance then the
identification process is usually rapid, only requiring a
confirmation from the shipping papers or shipping company. If the
ship is carrying a number of different commodities, the ship's crew
will determine from the position of the damage and the cargo loading
pattern which chemical or chemicals are likely to leak or to have
leaked.

If several substances are involved, additional investigation and
caution is needed, particularly in determining if there are any
incompatible chemicals on the same ship. Compatibility charts are
not infallible and an expert should be consulted regarding any
mixture which may be part of the spill. A list of experts in this
field should be established nationally in advance.

Information on the hazards of most substances can be obtained
through published material. A selection of such information sources
is set out in annex 2 to Section 1 of the Manual on Chemical
Pollution. Before accessing such information it is important to
ensure that the name or names being used to describe the spilled
material are known. It is recommended that the information obtained
from such information sources be compared with observations made at
the scene of the incident, to ensure that no mistake has been made
as to the identity of the spilled material.

Measurements of atmospheric concentrations

Whether or not an ability to predict atmospheric concentrations of a
released contaminant exists, it will be necessary to measure these
concentrations wherever and whenever possible.

Portable devices which can measure the concentrations of
hazardous substances and the extent of certain hazards in the
atmosphere surrounding a spill should be part of the response
equipment inventory. Most devices require identification of the
hazardous material in question before reliable concentrations can be
measured. However, there are portable devices that can also assist
in the identification of certain chemicals.

Descriptions of common devices for measuring concentrations of
flammable gases, oxygen and toxic substances are given Section 1 of
the Manual on Chemical Pollution. It should be noted that each
instrument does have limitations caused either by design, quality,
the hazard itself or user error. The manufacturer usually publishes
the specifications for the unit along with an operator's manual.
Methods of operation and limitations of the more common instruments
are also found in annex 4 to Section 1 of the Manual on Chemical
Pollution.

Phases of hazard evaluation

The purpose of the assessment and evaluation process is to collect
sufficient data about the spill situation, so that a decision about

physical response methods can be made.

Throughout the span of an incident, from first notification to final disposition, it is necessary to obtain information for evaluating the impact created by the incident. Evaluating the impact or effects is generally a two-phase process. The first phase is a preliminary evaluation based on information that is initially available or on the first rapid collection of data needed to determine if emergency protective measures are necessary. The second phase is a more methodical process designed to enhance, refine and enlarge the initial data base for decision-making concerning response operations. The latter phase provides more comprehensive information for evaluating environmental hazards associated with the incident.

Response

The plan should describe the various options which are available to deal with spillages from a ship, and spread beyond its original location. As in the case of oil, every effort should be made on board the ship to limit the release of spilled material to the sea by such means as cargo transfer, temporary patches, etc.

Chemicals are normally divided into four groups for response purposes:

1 Substances which form gas and vapour clouds. As a minimum, people in downwind areas should be informed, alerted or possibly evacuated. Access to the casualty and its surrounding area should be restricted. Gas clouds may be explosive or toxic and due regard should be given to the risks of explosion or the dangers to living organisms.
2 Substances which float. These may be classified as either volatile or non-volatile and both may be toxic. Volatile and toxic substances may have to be dealt with according to the procedures for gases and vapours. If the material is non-volatile and safe to approach and handle, it may be handled in a similar manner to oil as described in the IMO Manual on Oil Pollution, Section IV – Practical information on means of dealing with oil spillages.
3 Substances which dissolve or disperse. The hazard presented by these substances will depend on the nature of the chemical and its concentration. The concentration may be expected to decrease with time if sufficient depth and exchange of water is present. Response to such a spillage would consist of restriction of access, cessation of fishing activities within a prescribed safety zone and possible restriction of use of recreational beaches by bathers and pleasure boaters, accompanied by a monitoring programme to determine chemical concentrations in the water column.

4 Substances which sink. Substances which neither dissolve nor undergo a chemical reaction and have a density greater than that of water will sink and settle on the bottom. In areas where depth permits, dredge recovery can be implemented if considered necessary. Benthic species can be severely affected by these chemicals and restrictions may have to be imposed on exploitation of these species by sport and commercial fishermen.

5 Substances which react with air and seawater. A limited number of substances react with air and seawater. These may give rise to special problems according to the nature of the reaction which may result in the generation of harmful fumes.

Control of operations

Whenever work is undertaken at the spill site, it is necessary to give due regard to both safety and personnel protection. Removal of possible sources of ignition, provision of protective clothing and, if necessary, breathing apparatus are essential.

Some guidelines for control of operations are:
- establishing control zones with physical barriers and checkpoints to exclude unauthorized personnel;
- restrictions on use of machinery, vehicles and vessels;
- evacuation of non-essential personnel; and
- establishment of a communications system.

Response organization

As in the case of an oil spill an On-Scene Commander or Co-ordinator (OSC) should be designated who will be responsible for directing the employment of the necessary resources for containment, clean-up, disposal and restoration. His staff will be similar to that for oil spill response but should also include a scientific adviser who should have access to data on types of chemicals, flammability and toxicity and methods of response. He would also be responsible for sample collection, monitoring activities and monitoring of results of field and laboratory tests.

Disposal

During and after a clean-up that results in the collection of hazardous material, the problem of ultimate disposal of the material must be addressed. The various disposal options are described in Section 1 of the Manual on Chemical Pollution.

Discussion

As described earlier, it is considered that the national oil spill
contingency plan can provide a basic framework for a chemical spill
contingency plan, recognizing the need for additional scientific
personnel and equipment.

In the case of chemical spills different response strategies may
be involved than for oil and the "Ariadne" incident highlighted the
problems which may be faced by a developing country in responding to
spillage of chemicals in packaged form into the marine environment.
The specialized equipment which was required in this incident had to
be air-freighted into Somalia, together with technical experts
familiar with the various chemicals.

Whilst most port fire departments have the basic fire-fighting
equipment, together with breathing apparatus, they lack the
technical knowledge regarding correct response methods for chemical
spills and the appropriate protective clothing. As previously
indicated, there have not been many incidents of chemical spillages
occurring in ports and this problem has not been addressed as a
priority item, but with the gradual development of port contingency
plans under the framework of a national contingency plan, response
to chemical spills should receive increasing attention by maritime
and port administrations. Electronic data and information systems
on chemicals are available from a number of Government and private
sources.

It should also be noted that the majority of spills of either
bulk or packaged chemicals occur during transport from the port area
to the consignee, collisions, loss "overboard" or overturning of
trucks and rail cars occur much more frequently than in the port and
its approaches. Again, the municipal police and fire departments
would mount the initial response but the training and equipping of
these shore agencies does not fall within IMO's mandate.

Training and technical co-operation activities

The difficulties faced by developing countries in responding to a
chemical spillage due to lack of appropriate data, trained personnel
and specialized equipment have been recognized by IMO and the
Commission of the European Communities.

INTERNATIONAL MARITIME ORGANIZATION

A course on Sea Transport of Chemicals was held under the
International Maritime Organization/Swedish International
Development Authority (IMO/SIDA) Programme for the Protection of the
Marine Environment at the Revinge Civil Defence Training Centre of
the Swedish National Rescue Administration, Revinge, Sweden from
6 June to 8 July 1988. The course was organized in co-operation
with the World Maritime University (WMU), Malmö, Sweden.

The course, which covered the provisions of SOLAS 74, Chapter VII and MARPOL 73/78, Annexes II and III, was intended to provide training for officers of harbour and port administrations and of maritime administrations from developing countries, with responsibility for ensuring that hazardous substances are transported according to internationally agreed safety and environmental standards. The course contents included:
- The legislative background to the carriage of dangerous goods and harmful substances by ships insofar as this concerns liquid chemicals carried in bulk, including noxious liquid substances and dangerous goods in packaged form, including marine pollutants.
- The practical application of rules and recommendations relative to the stowage, segregation and carriage of those substances including the disposal of residues arising from their carriage.
- Ships' operations in port necessary to prevent pollution of the marine environment.
- Issues related to the ship/shore interface.
- Basics of dealing with accidents involving chemicals.

The course was attended by 15 participants from developing countries.

COMMISSION FOR THE EUROPEAN COMMUNITIES

A number of initiatives have been taken by the Commission for the European Communities including, inter alia:
1 the Seabel Project, consisting of four modules:
   Module 1 - covers the information gathering process. It supplies the user with information collection in a structured way in order to avoid omissions. For this purpose a set of check lists is available. Furthermore, a set of data bases is available in the module in order to research data on chemicals, equipment, etc. The output of the module is a list of all relevant items on the described accident, including all gathered data.
   Module 2 - covers the simulation of the fate and effects of the spilled chemical. Various mathematical models are available in Seabel by which the behaviour of the chemicals can be described.
   Module 3 - merges the estimated effects of the spilled material with location dependent information which is available in a geographic data base.
   Module 4 - handles the decision to be made on emergency response, consisting of two stages. In the first stage all applicable measures are selected and in the second stage the measures are evaluated to select the most appropriate option.
2 Training courses have been arranged through the Centre de Documentation de Recherches et d'expérimentations sur les pollutions accidentelles des eaux (CEDRE), Brest, France, the Regional Oil Combating Centre for the Mediterranean Sea (ROCC), Malta and SOGESTA, Urbino on oil spill combating, and more recently through the Rotterdam International Safety Centre B.V. (RISC) on Hazardous Material Spills in the Maritime Sector.

3 Emergency assistance. Such aid in the event of a serious chemical spillage is possible in the case where a partner of the Commission, entrusted with the execution of the task, accepts the modes which govern the implementation of the operations. Requests for such assistance should be addressed to the Commission by the Governments concerned.

4 Community Task Force. The recently created Task Force is composed essentially of designated governmental experts. In case of an emergency these experts would be called upon to give advice and, above all, attend at the scene of the incident to provide every possible assistance to the authorities concerned. It has been indicated in a recent paper that the services of this Task Force could be made available on request to developing countries facing emergency situations.

Whilst the foregoing programmes are having a limited impact in providing the required training, the lack of specialized equipment in developing countries remains a very real problem. During the "Ariadne" incident the Vice Minister of Maritime and Ports, whilst appreciative of the response to his requests for assistance by provision of advisers and experts, indicated that what was really needed was monitoring and analysis equipment, protective clothing and respirators.

Although these were eventually provided by the salvors, there were considerable delays due to complex negotiations with the insurers. During this time the Somali authorities were unable to initiate any positive action, due to the aforementioned lack of resources.

It is hoped that, based on experience with the Community Task Force, a response package can be developed which would include the necessary equipment to enable immediate remedial action to be taken in the event of a chemical spill.

References

1 Manual on Oil Pollution, Section II - Contingency Planning (1988 edition). IMO, London 1988
2 Manual on Chemical Pollution, Section 1 - Problem Assessment and Response Arrangements. IMO, London 1987
3 International Maritime Dangerous Goods Code (IMDG) (1988 edition) including amendment 24 of 1986. IMO, London 1988
4 Emergency Procedures for Ship's carrying dangerous goods (EmS) (1985 edition). IMO, London 1985
5 Recommendations on the Safe Transport, Handling and Storage of Dangerous Substances in Port Areas (1983 edition). IMO, London 1983
6 Medical First Aid Guide for use in Accidents Involving Dangerous Goods (MFAG) (1985 edition). IMO, London 1985
7 International Tanker Owners Pollution Federation Ltd. Technical Information Paper No.9 - Contingency Planning for Oil Spills. ITOPF, London 1985

8 <u>Transport Canada Dangerous Goods Guide to Initial Emergency Response</u>. Transport Canada, Ottawa 1986
9 <u>US Department of Transportation, Emergency Response Guidebook</u>. US Department of Transportation, Washington D.C. 1984
10 <u>US National Response Team Hazardous Materials Emergency Planning Guide</u>. Washington D.C. 1987
11 TNO Division of Technology for Society 'FACTS databank for accidents with hazardous materials'
12 P. Bockholts, 'Seabel Hazard Identification and Decision Support System for Emergency Response of Chemical Spills at Sea'

RISK ANALYSIS

Risk Analysis for Marine Chemical Spills – A Survey

J.R. Taylor
ITSA
Jernbanegade 52A
DK-4000 Roskilde, Denmark

There are two primary situations in which risk analysis is necessary for marine chemical spills. One is prior to any accident,
when the potential range of accidents is enormous. The other
occurs after an accident has happened, when the range of potential
consequences must be determined.

Risk analysis during the planning stage involves several steps.

- Determination of traffic intensity, including chemical
  transport traffic intensity.

- Determination of collision, grounding, and foundering probability.

- Determination of release rates.

- Determination of immediate release source terms.

- Calculation of spread and dispersion patterns.

- Calculation of consequences.

Determination of traffic intensity

Perhaps surprisingly, determining actual traffic intensities and
densities one of the more difficult tasks in marine risk analysis.
For an isolated country with little neigbourhood traffic outside
its own parts, traffic intersity can be determined from part
records. For straits, radar observations can be used.

However, in order to be able to calculate collision probabilities,

31

P. Bockholts and I. Heidebrink (eds.), Chemical Spills and Emergency Management at Sea, 31–43.
© 1988 by Kluwer Academic Publishers.

it is necessary to know the width of the sailing route, and actual
sailing patterns. While sea charts may be used to some extent,
arial reconnaisance is the only reliable method.

Determination of the volume of chemical transport activity is even
more of a problem. Specific large scale transports are easily
recognised and recorded, for example ammonia, LPG, liquid methane,
and bulk chemicals. Discussions with shippers can often give a
good picture. (see eg Stenström). Shipping of packaged chemicals
is much harder to record, not least because formulations and pro-
prietary mixtures are often involved. General statistical experience is at present the best guide to the level of risk here,
except in the few cases. where detailed studies have been made.

Probability of accidents

Given the intensity of traffic, the simplest aproach to determining accident probability is to use average statistics concerning
the number of collisions, grounding etc, per nautical mile sailed,
this approach has been found to give a good first approcimation to
accident risk. Unfortunately as within a factor of 2, the accident
rate for larger areas, it does not indicate in detail where
accidents are likily to happen. This presents a problem when risks
to special fishing grounds or tourist areas are a major problem.

A more precise estimation of risk can be obtained by using various
traffic density models. These are mostly based on the work of
Fujii and of Mc. Duff.

The simplest of these models concerns the risk of grounding
(Fig.1). The probability is proportional to the traffic intensity,
the size of the grounding obstacle, the velocity of the traffic.
In Fujii's model, there is a basic assumption that traffic is
bidirectional. Often, situations arise in which traffic passes in
several directions, as for example in crossing shipping lanes. The
models do not apply properely in these situations. An extension to
the case of uniform directional probability is given in fig. 2.

For a proper evaluation of risk in critical situations, a model of
the navigation process itself is necessary. These make it possible
to take into account the effect of manning, visibility, navigation
aids etc. The background for these models is the use of navigation
simulators.

In order to be able to cover a complete range of situations, two
and for some situations three ship simulations are requered. The
main problem here is to obtain a sufficiently complete range of
navigational background, and flexible simulation systems are
required in order to allow proper accident rate estimate. The work
here is highly experimental. Initial results give a reasonably
good agreement with traffic density models, and an enormous
improvement in insight into the accident process.

Accident damage

Once an accident occurs, the <u>extent</u> of the accident becomes
important. Even for a ship which has sunk, the quantity of material released depends on leakage paths and hole sizes. Generally,
the larger accidents involve holing of the transport vessels, (or
for deck cargo, holing of the transport containers).
The distribution of hole sizes can be obtained statistically from
accident records, in the case of conventional ships. In the case
of special transport ships, with double hulls, the size of hole
must be calculated. The usual method involves calcultation of
deformation energy of the hull for a typical ship, and comparison
of this with the energy of collision. This energy can be determined approximately from navigation models, or more directly, from
reports of grounding and collision incidents.

Fortunately, the actual size of holing can often be estimated from
actual experience, a fact which makes it possible to check the
theoretical calculations.

A special case of hazard concerns fire in the tanks or hold of
chemical transport. Hare, the pattern of risk can be exceedingly
complex, and the resulting threat difficult to evaluate. This is
especiaaly a problem for <u>actual</u> risk situations, where questions
of proper estinguishing methods, likelihood of explosion, and
protection of emergency personnel become acute.

Some modelling work has been done in this direction, particularly
as a result of the studies of the bombing runs made on the Torrey
Canyon. Studies of fires in tanks and buildings on land supplement
these studies.

Release of Toxic Substances

For the purposes of risk analysis, texic substances can be divided
into seven main groups (fuel oils have not been considered here)

- liquified toxic gases

- flammable toxic gases (LPG, Methane)

- Soluble toxic liquids

- insoluble, floating toxic liquids

- dense toxic liquids

- packaged toxic liquids

- packaged toxic solids.

The spread of liquefied gases on water has been relatively well studied, with several large scale experimental series, both on land and water. Releases on water have the special feature that heating is relatively rapid. The rate of boiling depends on heat transfer from water. In the case of ammonia, heat is also derived from the absorption reaction with ammonia.

Gas dispersion is by now fairly well undertstood, there being several popular mathematical models, with good agreement with experimental results, (see review by Havens 1986). In general, the acute hazard distances are at most a few kilometers, even for relatively large releases, making the calculations primarily relevant for narrow straits, harbours and estuaries. The most interesting cases are generally for heay gases, (which most often excludes methane except in the case of jet releases which result in mixing with air (cold air is dense).

A relatively overlooked case concerning combustion of toxic chemicals, producing toxic smoke. Here, the limited acces of air to holds can create conditions for partial combustion.

For substances which float, a number of standard models exist for spread (see expeciaaly the work of Blokker and of Fay), for evaporation (especially Sutton, for for dispersion (eg, Anderson). A typical set of models is given in fig. 4. These models have to some extent been verified on offshore trials and from actual spill incidents. Conpared with oil spills, release analysis are much more critical - the effects of hydrolysis are important, and for some substances, the relevant concentrations are very low. Existing models have so far not been well tested for these applications.

For substances which sink existing models are <u>tentative</u>. Fig. 5 gives the structure of one model developed by the author. This gives the spread of a stream of liquid, falling through the water column. The result is a collection of pools of liquid on the sea bed, which will typically be moved by sea currents. The transfer of toxin into the water column then depends on the mass transfer accross the liqued/water interface and the dispersion by current turbulence at the sea bed. The background for these models is standard process engineering mass transfer theory. However, there are some considerable unknowns, including especially the form of the sea bed and the dependence of constants in the dispersion equations on current. Order of magnitude agreement with experience is all that can at present be achieved.

Overall effect

In order to determine the overall effect of toxic releases, it is not sufficient just to determine the location of the accident itself. The most likely transportation of substances is required. Here, tide, currents, and wind directions are needed. For risk analysis for planning purposes, introduced as a pattern of winds and current period of a year, and tidal currents over a period of a day. For planning under accident conditions, the actual weather and current conditions at the time are important.

A special problem in this respect is the pattern and velocities of bottom currents, which are seldom known, and must often be inferred.

Given reasonable current data, however, the probability or frequency of impact can be calculated in a reasonably straight forward way – the simplest method is to use Monte Carlo simulation (Fig. 6).

The human impact of toxic releases can be determined relatively straight forwardly, toxic limits generally being known to within a factor of 3 at worst. The uncertainties involved do not represent so large a problem. Human effects will primarily be the result of fires, exposions or of toxic gases in coastal areas.

By contrast, the effect of toxic substances on commercial fisheries are extremely difficult to analyse. A simplifying approach is to determine the area over which tainting can be expected. It is a reality that such an area will be closed to commercial exploitation during the period of contamination – even if the contamination is biologically irrelevant. Few companies would risk marketing fish from known contaminated area, for fear of the effect sales from other areas. The main question then is – how long will the contamination last? At present this information is available only for very few chemicals.

References

1. Fujii Y. et al Traffic Capacity. Journal of navigation Vol.24 No 4 p 453–552 and The probability of Stranding, Journal of Navigation, Vol 27, No 2 p 339–243

2. Macduff, T. The Probability of Vessel Collision. Ocean Industry, Sept 1974

3. Havens, J. Dispersal of Heavy Gases, A Review, in 5th Int Symposium on Loss Prevention and Safety Promotion in the Process Industries, Cannes, 1986

4. Bender K, and Taylor J.R. Oil Spill Contingency Planning in Thailand. Int Oil Spill Conf, Baltiomore 1986.

5. Stenström, B. Risk Analysis for Spills of Chemicals by Tankers in the Baltic Sea. Seminar on Risk Assessment for Spills of Oil and Other Harmful Substabnces, Copenhagen, 1988.

Ship Collision Model

$Fc = N . Fu . 1/pi . L . l . sin(theta/2) / d^2$

N = Ships passing per year

Fu = Fujii constant, $10^-3.7$

L = Length of channel or sailing route

l = length of ship

theta = angle of collision

d = mean distance between ships

(Fujii, Macduff)

itsa

Model for Grounding

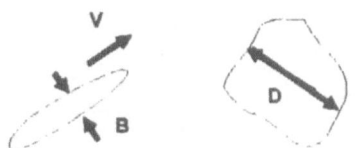

$$Fg = N \cdot Fu \cdot (D + B) \cdot rh \cdot V$$

N = Number of ships per year

Fu = Fujii number, $10^{-3.7}$

D = Diameter of object

B = Width of Ship

rh = Density of shipping, per unit area

V = Typical navigation speed

itsa

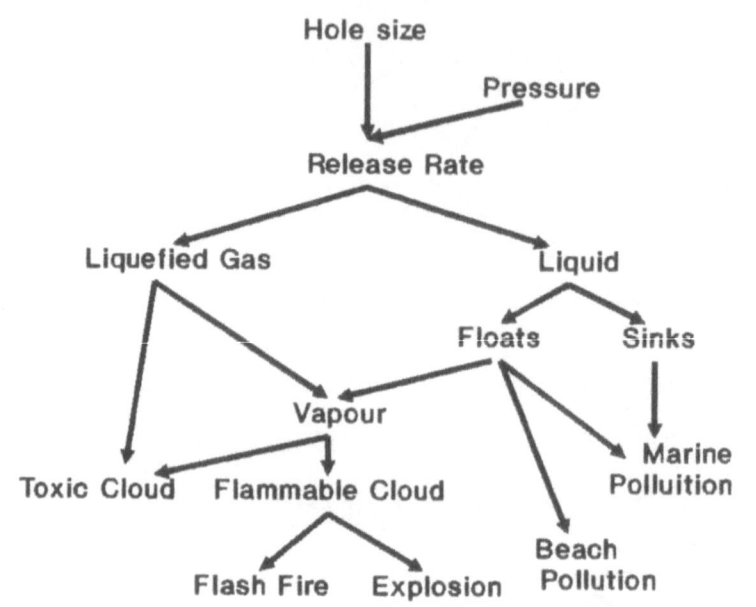

Release Model

Hole size

Pressure

Release Rate

Liquefied Gas

Liquid

Floats

Sinks

Vapour

Marine Polluition

Toxic Cloud

Flammable Cloud

Flash Fire

Explosion

Beach Pollution

itsa

Spreading of Liquid

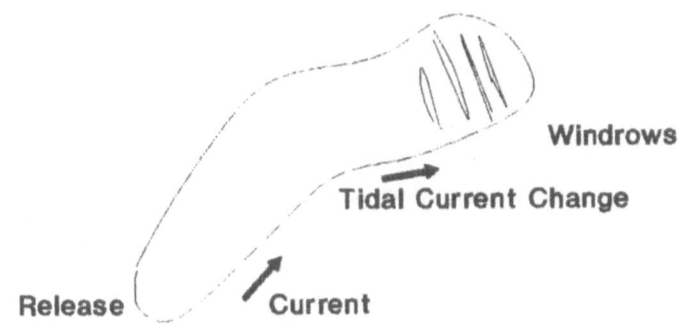

Windrows

Tidal Current Change

Release Current

Gravity -Viscous Regime

$$I = (g' V^2 t^{1.5} / v^{.5})^{.166}$$

Surface Tension - Viscous Regime

$$I = (sigma^2 . t^3 /(rh^2 v))^{.25}$$

sigma = surface tension   rh = density
V = volume of oil   I = diameter of slick
v = viscoscity   g' = gravity · relative density of oil
                                                        (Fay)

itsa

Evaporation of Liquid

Oil

$$Qdot = (.0074 + .00187 \cdot U) \ P \ M \ Q \ /(R \ T \ H \ rh)$$

(Audunson)

Ammonia

$$R = 500 \ Q^{.375}$$

$$Tevap = .675 \ (V/G \ Ydot^2)^{0.25}$$

(USCG)

Volatile Liquid

$$Mdot = K \ Mw \ (rhw - rha) + Pw/Pa \ M$$

(Bird et al)

tsa

Subsea Dispersion

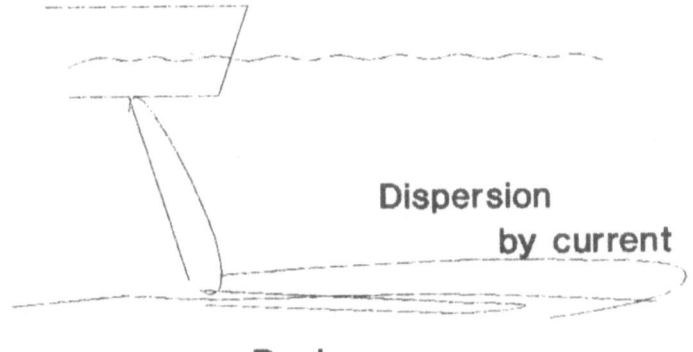

Dispersion
by current

Pools

Drop size

$$V_p = \pi \, d_p{}^3 \, /6$$
$$= 20 * mu * Q * d0 \, /(d_p{}^2 * g * deltarh)$$

(Scheel and Meister)

Transfer into Water Column

$$M = Km \ A$$

itsa

Subsea Dispersion

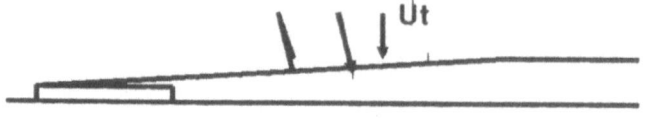

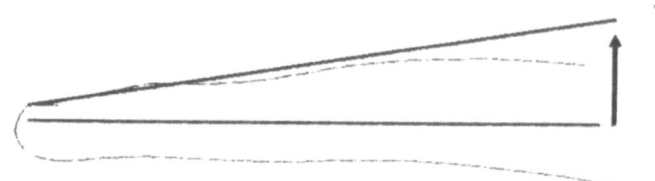

Ut = K U / Ri
H = K2 x

C = Mdot / 2 * H * Y * U

Ydot = 1.4 * sigmay

siagmay = a x^b

itsa

Spread Simulation

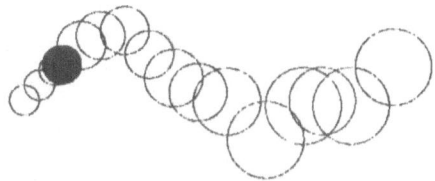

Spill trajectory
Speed is a function of
current and wind
Size is given by spread models
Lifetime is given by
dispersion and evaporation models

For monte carlo simulation
typically 1000 calculations
are performed for each position

itsa

TRANSPORT STRUCTURES OF CHEMICAL AND GASTANKER CARGOES

R.L. Tollenaar
Maritime Economic Research Centre
P.O. Box 1555, 3000 BN Rotterdam
The Netherlands

INTRODUCTION

It is an honour to address a distinguished audience on such an important topic. It is up to me to indicate the relative importance of
seaborne transport for those products, which, when not dealt with
properly, spills in the marine environment may cause havock. At MERC
we have dealt with chemical logistics for a long time, our main task
is to monitor the transport demand for all commodities. We undertake
research within all maritime sectors and this from a variety of
angles, from the market point of view, in relation to port developments, to assess newbuilding demands for ships, etc.

LAY OUT OF THE PAPER

I aim to start with a presentation of the total maritime transport
structure of all IBC and Annex 11 commodities. Apart from the present
trade structure the slides present the 1975 situation. The second
part of my paper deals with the tonnages involved in the carriage of
the substances. In this section I will to present some data on
deployment, on number of shipments, parcelsizes and related issues.
These afterall have great relevance for the spill potential. The
final part of my presentation is the easiest, since it deals with the
future. Here you are invited to join me and my cristal ball.

PRODUCT CONTEXT

I'am not a technician and will therefore not dwell upon all kinds of
technical aspects. Still regulations are regulations and for that one
person under you not familiar with phrases such as IBC and Annex 11 I
would like to point out that the IBC as well as the MARPOL Annex 11

P. Bockholts and I. Heidebrink (eds.), Chemical Spills and Emergency Management at Sea, 45—58.
© 1988 by Kluwer Academic Publishers.

Code relate to cargoes and embrace all commodities which are either included for safety reasons or are hazardous for the environment. First aspects are covered through constructional demands for the vessels involved in the carriage of the products through the IBC/BCH codes, often referred to as IMO classes.

The second through the MARPOL Annex 11 and Gastanker Codes. You will not be surprised to note that there is a near complete overlap in the cargo coverage from the two aspects. For further details on these matters I refer to later sessions.

1. TRADES

1.1. Summary of main trade findings

Summarizing the coverage of products included in my figures we note all liquid organic and inorganic chemicals, the animal and vegetable oils and fats and the molasses. A further breakdown will be presented in the course of my speech. The total maritime transport volume we are faced with amounts to no less than 150 mln tonnes. A first identification indicates: (1986)

product/category	deepsea	regional	national	total volumes
propane/butane and mixtures (LPG)	12	9	3	24 mln tonnes
ammonia in all its liquid forms	2	3	–	5 mln tonnes
liquified chemical gases	2.5	3.5	–	6 mln tonnes
liquid organic chemicals	16	12	4	32 mln tonnes
liquid inorganic chemicals	7	4	1	12 mln tonnes
animal/vegetable oils and fats	8	6	2	16 mln tonnes
molasses	4	1	–	5 mln tonnes
Total	51.5	38.5	10	100 mln tonnes

In 1975 the 100 mln tonnes package aggregated a volume of 55 mln tonnes.

Additionally we may add <u>10</u> mln tonnes of other Annex 11 substances such as pygas, lubs, white spirits, wine and consumer alcohols. Also 52 mln m3, or <u>38</u> mln tonnes of maritime LNG transport can be mentioned and, to complete the picture 20 mln tonnes of Annex 11 product transport on the continental inland waterways. This constitutes the broad transport context for this conference. I will now deal with the subsectors, starting with gases.

Liquified gastransportation

- In the gassector we see LNG transport conducted by only 70 large vessels. Over 75% of the capacity is engaged in Japanese trades. Earlier anticipated fast trade developments never materialised.
- LPG –also an energy product- requires smaller refrigerated tonnages. Japan is the major importer, the Middle East the prime exporter.
- Ammonia for fertilizer and chemical industries originates from LNG producers such as the USSR and Trinidad, small vessels are common here.
- The chemical gas sector poses the biggest problem, here we find a large variety of often difficult products such as Vinyl Chloride Monomer, Propylene Oxide but also easier ethylene, butadiene and propylene. Over 500 small pressurised or refrigerated vessels operate in this sector.
  The trade is widely dispersed, shipments and lotsizes extremely small.

Liquid organic chemicals

These commodities are the most relevant ones within the context of this conference. There are ten thousands of known chemical products, thousands are being transported of which roughly a thousand liquid, some 600 of which are carried in bulk, half of these however, in extreme small lotsizes by tank lorries or in tank containers. Within the context of our 32 mln tonnes of maritime transport a mere 12/15 commodities account for over 80% of the volumes. Most important are: methanol, ethanol, benzene, toluene, xylene, styrene, the glycols, EDC, AN, cyclohexane, ethyl benzene, etc. Another 15 products cover another 10%. Here we find phenol, acetone, propanol, butanol, TML, MEK, MtBE, acetic acid, etc.
This means that some 600 different substances account for only 3 mln tonnes of seatransport, an average of around 5.000 tonnes per product.

Since half of the more voluminous ones are also carried in bulk by chemical tankers we are faced with a 300 products with extreme small transport volumes.
The significance of this divers trade picture in terms of hazards of chemical spills, will undoubtedly be discussed in other contexts on this conference.

Liquid inorganic chemicals
These trades are relatively simple, in all we deal with some 5 mln tonnes of phosphoric acid (mainly from North Africa and the US) and of caustic soda solution, 1.5 mln tonnes of sulfuric acid and some minor volumes of liquid phosphorus, or of boric or hydrochloric acids. Most heavy acids are carried in large volumes by dedicated vessel types.

Animal and vegetable products
These commodities are only found in the MARPOL context, the main liquids are the vegetable palm and soybean oils which originate repectively from the Far East and from the Americas and the animal fish oils and tallow. Additionally we find here the unlisted molasses, an important Indian export commodity. Because of contamination hazards, cleaning problems and heating requirements the stuff represents the chemical tanker operators nightmare and is often considered to be more difficult to transport than most hazardous chemicals.

Other products
The other products are relatively new entries in the Annex 11. Basically we are talking about former Annex 1 mineral oilproducts with a high chemical content, such as pyrolysis gasoline and lubricating oils. Statistically the trade in these commodities is difficult to identify.

1.2. Inland waterway transportation

The commodity structure on the European inland waterways is very much alike the one as described sofar for the maritime sector. The regulations however, differ. After all Marpol (and IBC) only deal with seagoing transport sectors.

2. SHIPS

2.1. Gas tankerfleets and their deployment structures

The size and basic employment structure of the LNG fleet was already indicated. Due to a persistant overcapacity caused by wrong forecasts from people like me only 50 vessels in the 125.000 m3 class are employed, 32 of these to Japan. All others operate regionally, from N. Africa to (mainly South) European destinations.
The LP/chemical gasfleet comprises 650 vessels with an aggregate 7.6 mln m3, 70 large vessels >50.000 m3 transport LPG between Middle East and East Asian ports and Japan and between North Africa and Europe, no less than 500 ships measure less than 10.000 m3 and operate in regional distributive trades in Europe, the Far East or in the Caribbean.

2.2. Chemical tankerfleets and their employment structures

The total amount of chemical tankertonnage – close to 900 vessels- stands at around 12.5 mln dwt, 300 of these with 1.5 mln dwt operate regionally or nationally in Europe (75%), the Far East (15%) or in the Carribean. The deepsea sector – nearly 600 vessels with 10.5 mln dwt – operate worldwide, with the smaller ones participating in shorter hauls also.

2.3. Short sea

The small vessel category contains all 'multi cargo' units, designed for the carriage of IMO 2 products. Generally these vessels carry 1 to 3 parcels of organic commodities and switch regularly between products. A small number of vessels operate dedicated in some higher rated specialities or in oils.

2.4. Deepsea

Operations of larger vessels for longer hauls can be categorised as follows:

```
1.  large sophisticated chemical tankers 160 vessels with 3.8 mln dwt
2.  small sophisticated chemical tankers 100 vessels with 1.0 mln dwt
3.  'easy' IMO 3 cargo  chemical tankers 200 vessels with 3.3 mln dwt
4.  oil / chemical tankers                20 vessels with 0.7 mln dwt
5.  dedicated 'one cargo' chem.tankers    65 vessels with 1.2 mln dwt
6.  IMO 3/vegetable oil and fats tankers  35 vessels with 1.0 mln dwt
--------------------------------------------------------------------
                                         580 vessels with 11  mln dwt
```

- Part of the category 1 vessels carry on average up to 15 diffe-rent commodities in around 40 separate tanks. These vessels operate as 'liquid general cargo vessels' on transatlantic, transpacific or round the world voyages. Generally the vessels are never empty, being engaged in part loading or discharging operations. All products and all parcel sizes are accepted. Other 'real' chemical tankers may concentrate on larger difficult parcels. European owners dominate these sectors.
- The smaller category 2 vessels are more limited in their cargo scope. Full cargoes of vegetable oils or caustic soda next to more complicated organic chemical commodity structures are no exception. Japanese owners with speculatively built vessels dominate this sector.
- Large multi shiptype companies such as Maersk, Nedlloyd, Petro-bras, Canadian Pacific, MISC etc. and oil companies such as Petroleos Mexicanos, Shell or Marpetrol operate the category 3 & 4 shiptypes. Category 3 mainly in large parcels easy chemicals such as benzene or toluene, combining at the most 3 or 4 parcels in one trip or carrying a full load of vegetable oils, category 4 vessels with more emphasis on mineral oil commodities such as naphtha and fuel oils.
- Commodity producers - phoshoric acid, methanol, ethanol, phos-phorus, sulfuric acid etc, operate their own specialised tonnages (category 5).
  Several NIC's or LDC's are found here, Saudi Arabia, Morocco, etc.
- Vegetable oil tankers are owned by Greek, Hong Kong or Bermuda interests, most of these vessels may also operate in clean petroleum products so that the vessels are fitted according to both Annex 1 and all standards. Multi product transportation is rare here. Usually easy chemicals are going up to the Far East, vegetable oils coming down.

2.5. Number of shipments and parcel sizes

Theoretically the combined trade and fleet data presented sofar allow
for the identification of the number of sailings. By allocating trade
portions to shiptypes and sizes and assuming a load level the number
of sailings is determined. The resulting total shipments must be
subdivided by the number of parcels. These ultimate parameters should
be indicative for the determination of the potential hazards of
spillage by utilising and projecting sample results of known accidents. This however, is not my task to deal with.

Excluding the straightforward LNG sector the average shortsea LP
gastanker measures 2100 m3, the deepsea vessel 43.000 m3. The average
chemical short sea shipsize can be set at 5000 dwt, the deep sea
vessels at nearly 19.000 dwt. On the basis of a realistic all over
loadlevel of 60% the number of worldwide voyages can now be calculated from the trade tabulation. Having applied the load factor the
voyages incorporate all empty and/or loaded trips and do generally
include of course more than one portcall on each end.

short sea:
LPG ships: 18.5 mln T /28 mln m3, av.shipsize 2100 m3 - req. no.trips
22.100
chemical tankers: 30 mln tonnes, av.shipsize 5000 dwt - req. no.trips
10.000

deep sea:
LPG ships: 16.5 mln T/27 mln m3, av.shipsize 43000 m3 - req. no.
trips 1.000
chemical tankers: 35 mln tonnes, av.shipsize 19.000 dwt - req.no.
trips 3.000.

Division of the number of trips by the number of available vessels
provides the number of average loaded voyages made by vessel per
annum. (43 and 35 for the shortsea, 6.5 for both deepsea ship
categories). Of course these figures are only indications, in reality
most vessels are hardly ever empty.

The data can also be presented by area; having determined that short
sea Europe is responsible for 22 mln tonnes of the volumes, 15 mln
tonnes of chemical - 7 mln tonnes (10.5 mln m3) of gastanker cargoes
the number of trips here stands at respectively: 5000 chemical tanker
and 8300 gastanker sailings, i.e Europe is responsible for respectively 50 and 37% of all short sea trips undertaken in the world.
Roughly 55% of these European trips take place in Atlantic Europe,
25% in the Mediterranean and 20% between the two sailing areas. The
same exercise can be performed for the other areas.

Parcelsizes

A final word on parcel and tank sizes. A crude oil tanker may carry ten thousands of tonnes in non segregated tanks, a producttanker still thousands of tonnes. The tanksize of chemical and gas tankers is limited by the parcelsizes - afterall they have to cater for hundreds of different products - and tanksizes may range as a consequence from 150 tonnes to a maximum of around 1000 tonnes.

A parcel often demands more than 1 tank, sometimes special small decktanks are required for a parcel, not to mention the carriage of chemical tank containers by regular container vessels. Easy commodities such as many the hydrocarbons and most inorganic products are often transported in volumes of several thousand tonnes, many specialities such as TDI in extreme small quantities of a few hundred tonnes or less. The oils and fats and eventual mineral oil cargoes on the other hand are often carried in full ship loads.

In sophisticated gas and chemical vessels all tanks are fully segregated, simpeler ones still have several separations. Due to this construction and to the increasingly common double skin structure spill hazards are minimised as far as possible. I would like to point out that the safety record of these vessels relatively speaking is the best in the maritime world.

3. FUTURE DEVELOPMENTS

3.1. General

Up to the first oilcrisis of the early seventies trade growth rates were in excess of 20% per annum, hereafter they declined rapidly even to a single figure level in the eighties. Two separate developments caused this, namely:
- The product substitution effect was gradually coming to an end,
- The steep increases in oil/naphtha prices over the period.
After all, the cost of the raw material -naphtha- may represent more than half of the cost price of base chemicals such as the olefins and the hydrocarbons however, down to one tenth for higher products such as acrylonitrile.
Notwithstanding the recent drop in oilprices double figure growth-rates are presently only apparent in high speciality sectors, such as for electronic-, diagnostic- and oleochemicals and for advanced polymer composites.

3.2. Forecasting methodology

The liquid or liquified commodities we deal with are eventually converted into thousands of end products, from artificial snow to windscreens, from plastic bags to snowboots and everything inbetween. Generally we allocate the commodities to 14 different base industries and determine consequently:
- the growth prospects of that industry or a part thereof as a whole,
- the growth prospects of our substance within that particullar branch.

Furthermore, we analyse regional production/consumption patterns in order to be able to allocate transport capacity to the anticipated cargo flows.

Industries indentified within the 14 include:

- agricultural sector,incl.fertilizers and others (pesticides, insecticides);
- automotive industry, incl. automobile parts and rubber produce,
- fuel and related sectors, incl. fuel additives such as TML and MtBE;
- domestic and consumer good sectors, iron & steel, paint, cosmetics, etc.

Within the methodology products are allocated to final users and not to the often intermediate chemical reprocessor. Sometimes a liquid serves more purposes and is to be allocated proportionately to more user industries.

forecasts:
Our forecasts aim at the year 2000 and are based on the earlier presented 100 mln tonnes seaborne trade. Trade volumes nearly doubled over our 1975/86 period, the deepsea trade growth stood at an annual 8%, with a 5.5% increase p.a. for the short sea, positive figures compared to the much lower general trade developments. For both trading sectors organic products were responsible for the highest growths, followed by inorganics with the veg oil & fats development in line with population growths of just over 2.5% p.a.
Within our forecast scenario the selected trades are expected to nearly double again by the year 2000, representing an all over growth of 5% p.a. Within the limited time of my presence here it it impossible to indicate all the parameters which have been taken into consideration in forecasting.

The main argumentation is briefly:

- LPG shipments are dominated by the Japanese demand and double by 2000;
- Ammonia production from LNG is planned long ahead, trade and transport, requirements for this fertilizers product will grow only moderately;
- Chemical gas transport, mainly ethylene, propylene, butadiene and VCM will increase to only 8.5 mln tonnes. Trade possibilities face restrictions;
- The organic chemical section remains the most promising one with growths of 7% for the deepsea, a little less for the short sea areas;
- Inorganics are connected to fertilizer and alumina refining developments. Deepsea expectations are more positive than the prospects for short trades;
- The oils & fats are consumer products and depend on population growths, with the same applying for the animal food product molasses.

Summarising the anticipated development for the year 2000 we note:

product category	deepsea	regional	national	total	volumes
propane/butane and mixtures (LPG)	25	19	6	50	mln tonnes
ammonia in all its liquid forms	3.5	4.5	–	8	mln tonnes
liquefied chemical gases	4	4.5	–	8.5	mln tonnes
liquid organic chemicals	41	25	8	74	mln tonnes
liquid inorganic chemicals	12	7	2	21	mln tonnes
animal/vegetable oils and fats	12	10	3	25	mln tonnes
molasses	7.5	1	–	8.5	mln tonnes
Total	105	71	19	195	mln tonnes

Analysing the outcome one might nearly conclude that the trade doubles every 12 years. With this information as input it is possible to assess the future market situation. Based on a status quo market structure it is possible to determine the number of trips, the amount of required ship capacity, or to perform other exercises, which go beyond the context of this conference.

Constraints from environmental points of view may cause structural changes away from a status quo, away from the carriage of dangerous intermediates to the transport of less hazardous endproducts. Afterall there is a growing awareness of the true nature of some of the commodities. From my own experience 1 recall the time that benzene was considered to be fit to swim in, when I handled acrylonitrile with open tanktops. Now we know different.

Still we must remember that the bases for this acrylonitrile as an example is oxygen, propylene and ammonia, end-products are fibres for the production of nitrile rubber, carpets, shirts, and many other less harmfull materials. Can we not limit the transportation of harmfull substances by the increasing conversion of those products to basic endproducts on the same location prior to transportation. It is, however, just a thought from a layman.

PRODUCT GROUP	MLN TONNES	
	1975	1986
LP & CHEMICAL GAS	16	35
ORGANIC CHEMICALS	15	32
INORGANIC CHEMICALS	8	12
ANIMAL/VEGETABLE PRODUCTS	15	21
TOTAL MAIN CARGOES	55	100

DISTRIBUTION OF 100 MLN TONNES OF CHEMICALS, LP GASES AND OILS & FATS

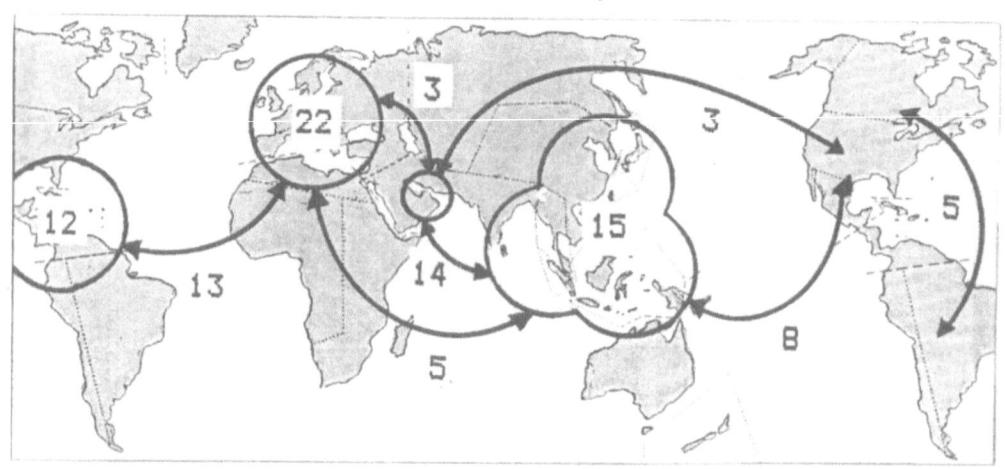

LP & CHEMICAL GAS 1986: 35 MLN TONNES

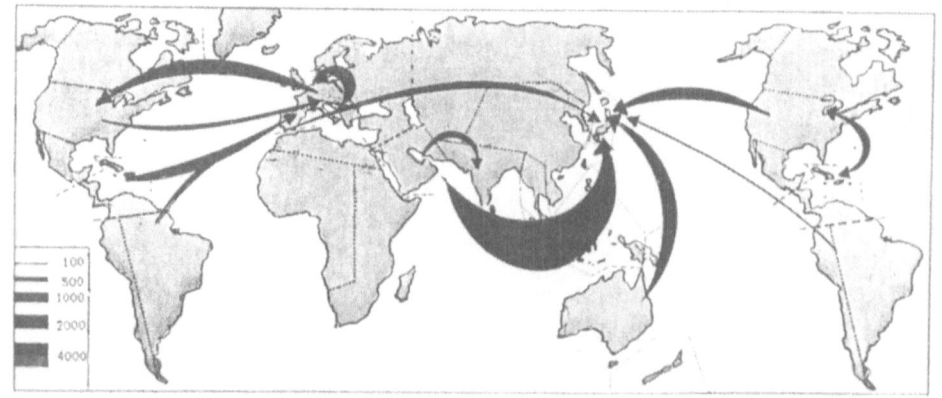

57

OILS/FATS 1975: 12 MLN TONNES

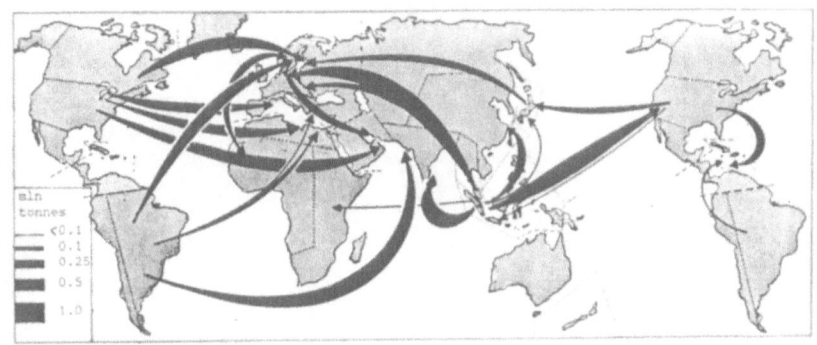

OILS/FATS 1986: 16 MLN TONNES

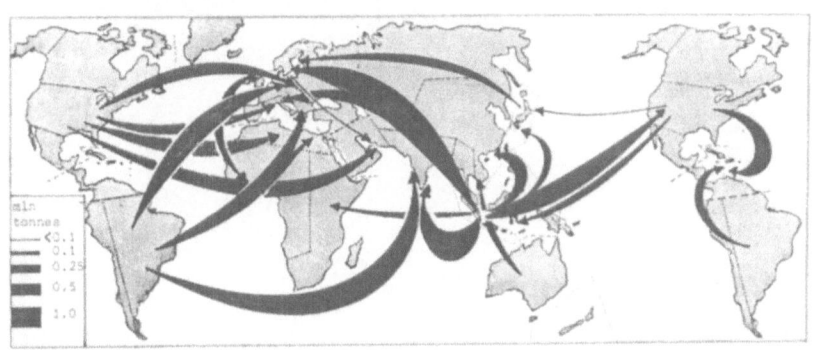

MOLASSES 1986: 5 MLN TONNES

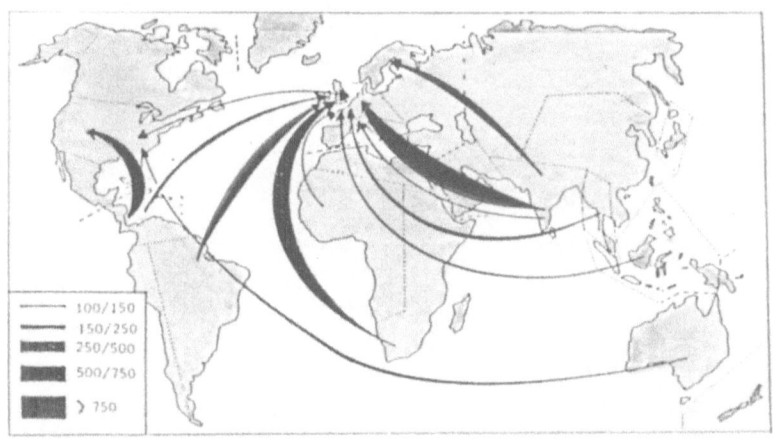

ORGANICS
1975
15 mln T

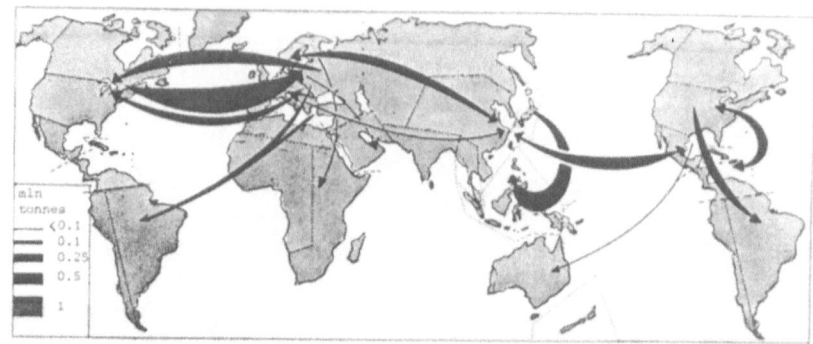

ORGANICS
1986
32 mln T

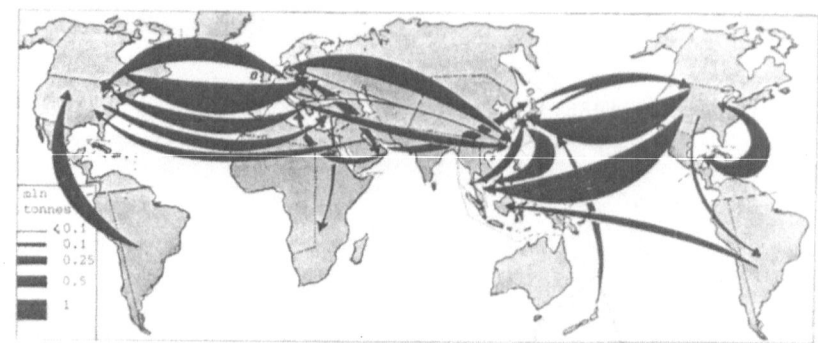

INORGANICS
1975
8 mln T

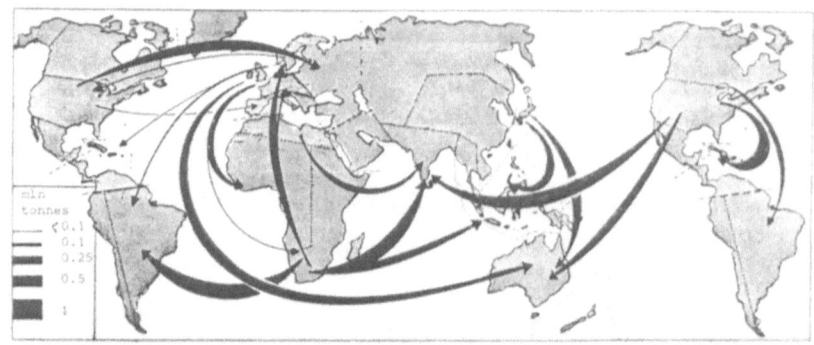

INORGANICS
1986
12 mln T

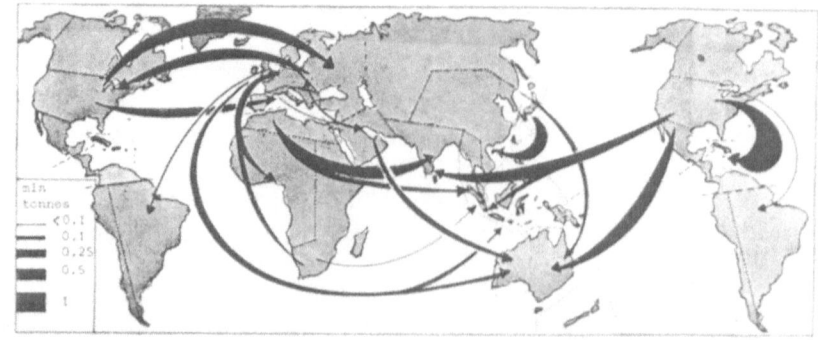

THE USE OF MATHEMATICAL MODELS IN ACCIDENT CONSEQUENCE ASSESSMENT
FOR TRANSPORT OF HAZARDOUS MATERIALS AT SEA

Ir. C.J.H. van den Bosch
Netherlands Organization for Applied Scientific Research TNO
Division of Technology for Society
Department of Industrial Safety
P.O. Box 342, 7300 AH Apeldoorn, The Netherlands

ABSTRACT

The transport of hazardous material at sea constitutes certain risks to
the environment. In estimating the consequence of accidents,
mathematical models are necessary to describe the behaviour of the
alien hazardous material in the marine environment.
Models to predict the threat that chemicals present after an accidental
release at sea, concern a standard sequence of calculations: the
release of the material into the environment, the dispersion of the
material in the environmental compartments, and the resulting
consequences to the environment.
After a general introduction about mathematical modelling, an overview
of relevant existing models, their applications and limitations, will
be given.
The existing mathematical models predicting physical effects on the
marine environment, are complex dedicated models.
In accident consequence assessment for transport of hazardous materials
at sea, and the evaluation of emergency situations, it is a necessity
to computerize the available relevant mathematical models and to
incorporate them into one computersystem.

Introduction

Chemicals are commonly used products in our society. Applications are
widely spread resulting in an extensive transport; throughout the world
large quantities of chemicals are shipped over the oceans, seas and
other waterways.
The transport of (bulk-shipped) hazardous material at sea constitutes a
certain risk to the environment. In the past accidents at sea have
happened causing pollution of seas and oceans, of coastal areas and/or
were a threat to the health of people.
In estimating the consequences of potential and actual accidents,
predictive calculations are necessary to describe the behaviour of the
alien hazardous material in the marine environment.

P. Bockholts and I. Heidebrink (eds.), Chemical Spills and Emergency Management at Sea, 59–71.
© 1988 by Kluwer Academic Publishers.

60

In general a model can be defined as a representation of the essential
aspects of a chosen part of the world, called the "system", presenting
knowledge of that system in a usable form.
In many cases modelling starts with the application of basic physical
laws, like mass and energy/heat balances.
The relevant equations are combined to describe the system under study
in the best possible way. To validate the (concept) model the predicted
results should always be compared with experimental data.
Mathematical models consist of a series of interrelated formulas that
are able to calculate numerical values of the desired output variables
of interest. The models can be extremely complicated, but implemented
in computer programs the calculations can be done in a fairly short
time and the results may be displayed graphically.
The benefit of models is the possibility to use them as a tool to
decide upon alternative management strategies. Models create the
opportunity to simulate experiments, which could never be done in
reality, and show without risk the consequences of certain decisions
and policies or the consequences of hypothetical large accidents.
Models are necessary in evaluating emergency conditions involving the
accidental discharge of hazardous chemicals into navigable waters and
for contingency planning (preplanned assessments and responses, drills,
etc.).

In using models, it should be realized that model predictions are
always associated with uncertainties caused by the deficiency of the
system's description. Not seldom simplifications of applied theories
have been made for several reasons, like the availibilty of limited
(experimental) data, computational approximations, etc.
Models to predict the threat that chemicals present after release in an
accident at sea, concern a standard sequence of calculations: the
release of the material into the environment, the dispersion of the
material in the environmental compartments and the resulting
consequences to the environment. A description of relevant models will
be given in this order.

Classification of bulk-shipped chemicals

Consideration of the physical properties of bulk-shipped chemicals,
like density, solubility and vapour pressure, leads to a behaviour
classification in terms of evaporators, floaters, dissolvers/mixers and
sinkers.
This classification leads to a need for a relatively small number of
general applicable models for: spreading on the sea surface, dilution
in the air, transport on the sea-surface, and in the water column [5].

Table 1: Classification of liquids based on physical properties.

CLASS	Sub class	Example
EVAPORATORS	riser remainder dissolver	hydrogen butane ammonia
FLOATERS	evaporator real floator dissolver evaporator & dissolver	toluene styrene benzene hexanol
MIXERS/DISSOLVERS	evaporator real dissolver	methanol ethylacrylate
SINKERS	insoluble dissolver	tetrachloroethylene carbondisulphide

The release of material into the environment

The release of hazardous chemicals from ruptured vessels directly into the atmosphere or into the marine environment can be described by models which are developed for normal industrial equipment and sites, like: outflow of liquids and gases, two-phase flow, turbulent free jet [7]. Some types of accidental releases at sea require specific models, like underwater releases, and release onto the water surface.
The main interesting characteristics of the release to be predicted by model calculations, are the fraction of the total released chemical that disperse into the atmosphere, the fraction that remains in or on the water body, and the fraction that sinks to the bottom.
To use the models that estimate the physical effects of the release of chemicals into the environment, the following types of information must be obtained [4]:
- discharge chemicals characteristics (identity, storage pressure, temperature, quantity);
- discharge conditions (tank size, location of discharge opening relative to water's surface and to tank level, the rate of discharge;
- the location where the discharge of the spill occurs;
- the time when the spill has occurred or the discharge has begun;
- marine conditions (current speed, water depth at spill site, spill geography).

The dispersion of the material in the atmosphere

The dispersion of flammable/explosive or toxic gases in the atmosphere is of limited importance in the estimation of the consequences of releases of hazardous materials at sea. The relatively rapid dilution of the gas in the atmosphere cause the expected consequences to be local. Nevertheless the dispersion of toxic gases could be relevant in estimating consequences of releases in estuaries and coastal areas. The models that describe the atmospheric dispersion are the Gaussian plume model [7, 14] and the so-called heavy gas models [15], which could be used under certain conditions.

The dispersion of the material in the marine enviroment (short-term) [9]

In the case of the pollution of (floating) hydrocarbons (oil), it is possible to predict the movement of the pollutant by relatively simple models. Problematic may be the estimation of the source terms, the real-time acquisition of the relevant information on the actual movement of the spilled material, the meteorological conditions in the area and of good local wheather forecasts.
More complex is the prediction of the movement of water-miscible chemical pollutants. A wide range of multi-layer or tri-dimensional models is available, which perform sufficiently accurate predictions of the transport of the materials when the water movement is mainly caused by tidal currents. The results are becoming poor when other mechanisms play the main role, like:
- the turbulence which caused a superimposed dispersion process;
- the influence of the wind.
The missing dispersion data could be obtained by:
- large scale experiments;
- monitoring concentrations in the area of accidental release of material.

The movement of seawater is caused by different forces: tide, general currents, swell currents, and the wind.
The behaviour of the pollution and the interaction with the sea-surface confine wheatering, surface-activity, transformation of the chemical, evaporation, self-destruction and fixation by deposition on the seabed, and adsorbtion by flowing particles.
Difficulties in the description of tidal currents arise due to the irregular coastal boundaries and the different depths.
Three-dimensional modelling can be performed at present, but a major draw-back is the long processing times on main frame computers. This means that the models are very expensive to use.
Further more only currents are calculated and not the important dispersion processes caused by the turbulence in the upper layers, which play an important role. In the so-called multi-layer models the problems solved, are in fact two-dimensional; the depth can vary, and also the exchange between the layers.

Models to describe currents in estuaries and bays [10]

For spills in estuaries and bays the boundaries should be explicitly
included in the description and the various flows computed as part of
the model. The Dynamic Estuary Model (DEM) is structured conceptually
for a specific system as a single-dimensioned network approximation of
a vertically mixed system of interconnected channels and embayments.
First the set of motion equations written for the channels of the
network and a set of continuity equations for the embayments, must be
solved. This means the computation of the tidal flows and water surface
elevation. The main restriction of DEM is its limitations to estuaries
wherein vertical stratification is virtually absent and tidal
velocities can be handled in a depth-integrated manner.

Using the calculated hydrodynamic data the transport of specified
constituents into and out of each embayment due to convection and
dispersion, can be determined. The computations result in the temporal
or intratidal variations of concentration and the tidally averged
concentrations at the center of each embayment.
The basic approach for applications of the DEM to a specific estuary
involves the discretization of the system into a series of volumetric
embayments called "nodes" connected by a system of flow elements called
"links". The dimensions of the embayments (volume, surface, diameter,
water surface elevations) and the patern by which they are connected,
the dimensions of the "links" (length, depth, radius), are
characteristic for the estuary.

A significant amount of effort is needed for each use of the estuary
model. For the evaluation of rapid responses the DEM can only be
applied unless a library of models has already been assembled for all
estuaries and bays of potential interest.

An alternative reading of the same principle of dividing the water body
into segments, are the models based upon a derivation of partially
differential equations for the non-steady motions of the compressible
fluids [12].
This derivation uses vertical integration of the equations of motion
and the continuity in Eulerian coordinates. In the so-called finite
difference representation of the hydrodynamic equations, the flow field
is divided into a large number of elements, the computational grid.
Regardless of the size and speed of the available computer it always
appeared that the simulations of the currents are extensive; the
computer time required for a simulation is often expressed in hours.

Models to describe water movement in open sea [11]

The factors influencing the displacement of the water surface at sea
are wind effects, tidal current effects and wave effects.
The deviation of the direction into which the spill moves due to the
Eckman wind drift (10 degrees to the right of the wind direction,

caused by the inertia of the underlying water column) and the Coriolis force (10 degrees to the right caused by the rotation of the earth) can be neglected taking into account the overall accuracy of the computations and input data.
The drift of the sea due to the wind can be estimated by assuming that the wind induced velocity of the water surface is about 2-3 % of the wind velocity at a height of 10 m.
The tidal currents of seawater are strongly time- and location dependent. The direction of the current at one point at the North sea only depends on the hightide cycle (of about 12.5 hours). The magnitude of speed of the current is also governed by the spingtide-cycle (period of about 28 days).
Besides the daily varying tidal currents there exists a averaged net current, flowing in northern direction (0.04-0.2 knots).
The current field of the North Sea can be described by a pattern of reference points for which the direction and magnitude of the current speed is known as a function of time after high water at Dover. The current speed on a location can be calculated by interpolation.

The Mediterranean Sea is with a few exceptions a non-tidal sea, which receives only about one-third of the amount of water lost by evaporation from the rivers that flows into it. Consequently there is a continuous inflow of surface water from the Atlantic Ocean. After passing through the Strait of Gibraltar, the main body of the incoming surface water flows eastward along the north coast of Africa. This current is the most constant component of the circulation of the Mediterranean Sea. It is most powerful in the summer, when the evaporation is at maximum. The fundamental surface circulation of the Mediterrean Sea consists of separate clockwise movements in the basins. Apart from these continuous currents (0.4-0.8 knots, in June), currents will be governed by the wind.

The movement of slicks [13]

An important class of models describes the dynamics of thick slick (gravity-viscous spreading phase) since thick slicks contains nearly all spilled chemicals and represents the most prolonged hazard.
The spreading of floating insoluble or slighty soluble and slightly volatile materials, is comparable to the spreading of crude oil. The main difference is that in general the spreading and evaporation of the spilled chemical are faster and no non-volatile residuals remains after the dilution process; most chemicals are dispersed into the environment within a few hours. In general, the mass and heat transfer processes associated with chemical spills in the environment will be turbulent.
The primary source of information applicable to the present problem for spills on water is the literature on air-sea interactions.
The spreading of oil is governed by many factors: the gravitation, the net surface tension, inertia, viscosity. Wave effects have a negligible effect on the speed of displacement of oil.

There are many models that attempt to predict both the mass of oil
remaining in the slick and the chemical composition and physical
properties of the slick, as a function of time and environmental
parameters, f.i. temperature and wind speed.
They are developed for use in assessing the impact of spilled oil in
marine environments. When the model requires that the oil to be
wheatered, is characterized in terms of distallation cuts, than it is
not oil-specific but can be applied to the variety of crude and refined
petroleum products for which distallation data is available.

The following wheather processes should be taken into account:
evaporation, dispersion (oil into water), mousse (water into oil) and
spreading. These processes are used to predict the mass balance and
composition of oil remaining in the slick as a function of time and
environmental parameters. Dissolution of oil into the water column may
be neglected because this wheatering process is not significant with
respect to the overall material balance of the oil slick.
It is generally accepted to assume the oil well mixed, but it has been
observated that this might not be the case.
As the oil wheathers its viscosity increases resulting in a slab-like
oil phase. The mass transfer within the oil will change drastically in
going from a well-mixed to a slab-like phase. The dispersion process is
a function of the oil viscosity; oil viscosity is a function of
composition. This means that the dispersion process indirectly depends
on the evaporation process.
The oil viscosity is also altered by mousse formation, but it is not
yet possible to quantify this compositional dependency. The spreading
of the slick results in ever-increasing areas for mass transfer.
The composition of the oil is described in terms of pseudo- components
that are obtained by fractioning the oil in a true-boiling-point
distallation column. This procedure yields cuts of the oil which are
characterized by boiling point and density. This information is then
used to calculate many parameters about the cut.

Sinkers [23]

Spillage containing solid or heavy liquid particles will be deposited
on the sea-bed. The problem of representing the transfer of material to
and from the top layer of bottom sediments is not straightforward. The
transport of suspended sediment in the water column to and from the
sea-bed is a complex problem depending on partical size and tidal
velocities.
It is also possible that pollutants are adsorbed by the sediment in the
interaction of the sea with marine sediments. Sediment- associated
substances can act like long term sources of slowly solving toxic
materials.
Jettisoned materials can float or sink to the bottom. The movement of
objects on the seabed is influenced by factors like: weight, dimension,
depth, turbulence, the slope of the bottom, the cohesive forces, and
current speeds.

The description of the movement of sinking objects or dispersion of relative heavy material is not yet fully developed.

The dispersion of the material in the environmental compartments (long-term) [19]

The exchange of materials between the atmosphere and the marine environment plays an important role in the distribution of pollutants in the compartments.
The prediction of the direction and magnitude of the mass transfer between air and water supplies information about its influence on the variation of the concentration of the pollution in the water phase and its availablity to chemical degradation, sorption and bio-accumulation.
The mass transport between air and water can be described by two processes: evaporation into the atmosphere, and advective transport from air onto water by wet and dry deposition.
The basic assumption is that the conditions in both compartments are "steady state", i.e. concentrations are constant throughout the compartments.
Methods have been developed, based on the two-resistence-model of W.G.Whitman. The principal idea is that the boundary surface between air and atmosphere consists of two stagnant layers: a water-film and a air-film. There are assumed to be no concentration gradients due to the turbulence in the compartments; the transport within the compartments is not taken into account.
The magnitude and direction of the net mass transport can be estimated on base of the mass transfer coefficients estimated by generally applicable empirical relations, the measured or estimated sollubility and vapour pressure.

The consequences to the environment

Different consequence categories can be distinguished with respect to the magnitude of possible size of the area and the the duration of the physical effects caused by the incidental release.
Local material and personnel consequences can be caused by explosions and fire or intoxication aboard a ship due to the release of toxic/flammable vapour clouds. The affected area could be within a radius of maximum a few kilometers. This threat is of interest for the ships crew and for salvors of wrecks.
Consequences of toxic gases for people and animals living in the neighbourhood of estuaries and in coastal areas, are considered to be possible.
The assessment of environmental consequences are regarded as of critical importance in the evaluation of risk associated with the transshipment of dangerous products [9].

In general the complete quantification of environmental consequences of the impact of a release of a chemical will be considerably complicated; the contamination of the environment is determined by many factors. After release a chemical will initially be transported within the environmental compartment in which the emission occurs.
Depending on its basic properties a released chemical generally will be transformed to a certain extent as a result of physical, chemical and/or biological processes. Moreover, in most cases the chemical will enter other environmental compartments, resulting in subsequent transport and transformation.
Regarding the environmental compartments a distinction can be made between abiotic compartments such as air, groundwater and soil and biotic compartments, eg human beings, flora and fauna.
The pattern of exposure of an organism to a chemical largely depends on the fate of the chemical in the environment. Examples of potential exposure routes include inhalation, ingestion via food or drinking water, dermal contact etc. It has been thoroughly recognized by now that exposure to chemicals may have detrimental consequences for human beings, animals and vegetation.
Environmental hazards are predominantly brought on by continuous emissions, but it should be realized that there are examples of incidental releases leading to environmental disasters, e.g. Sevezo, Chernobyl and at sea: AMOCO CADIZ (1978): 200.000 ton of crude oil and IXTOC-platform blowout (1979-80): 500.000 crude oil.
However, a major distinction between a safety and an environmental hazard is that the latter takes acute as well as chronic consequences into account. From the above it will be clear that environmental hazards involve considerably longer time periods than a safety hazards.

Consequently, transformation, transformation of a chemical, as well as transport to other environmental compartments can play a role of importance in an environmental risk analysis.
In the event of a calamity the geographic dimensions within which the impact of a released chemical on the environment manifests itself may extend in orders of magnitude beyond the immediately affected area.

Figure 1: Schematic survey of the estimation of environmental
consequences [17].

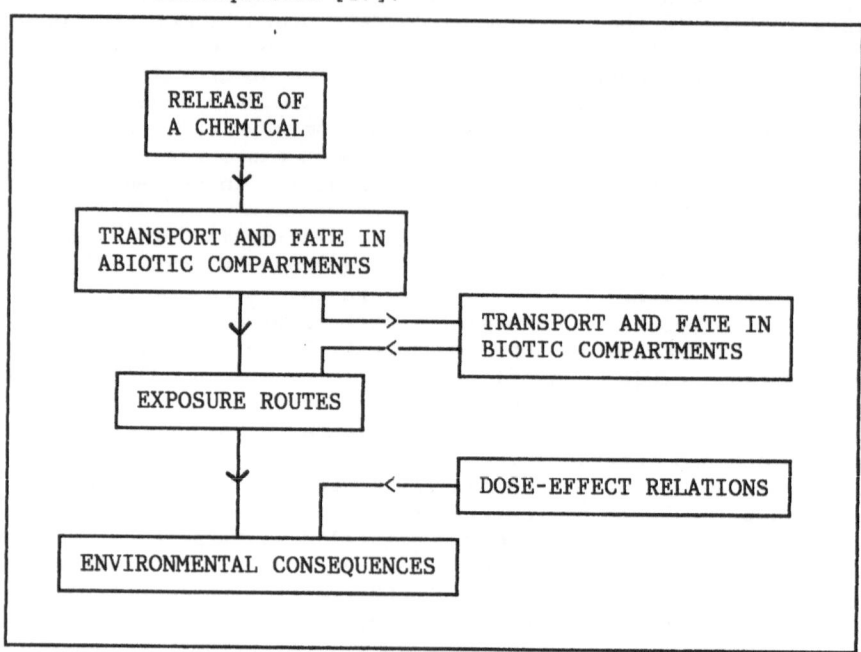

Biological consequences of oil in the marine environment [20]

Petroleum may affect communities of marine organisms in several ways.
Some components are highly toxic and may cause immediately death of
marine organisms. At lower concentrations, they may cause detrimental
biological responses at the populations. Petroleum components can be
accumulated in the tissues of marine organisms, causing tainting of
fishery species and potential consequences to the marine foodchains.
The tarry residues of crude oil, may physically foul the integument or
feeding apparatus of marine reptiles, birds and mamals, causing death
or decreased fitness.
The impact of oil due to toxicity and other biological consequences of
petroleum to marine organisms, are caused by oil spills but also by
chronic inputs to the marine environment.

It is problematical to asses the impact of oil in the marine
environment because crude oils and most refined oil products are
complex mixtures of organic compounds. Besides the hydrocarbons oil
consist of a variety of sulfer-, oxygen- and nitrogen- containing
components.

Once the concentration of hazardous material in the atmosphere, marine
environment or bottom is estimated as function of time, it is possible
to estimate the consequences to the eco-systems when large spills of
toxic chemicals have been accidentely released into the environment.

Table 2: Acute toxicities of aromatic hydrocarbons
to species of marine animals [20].

compound	range of toxicity (mg/l) meas. as 96-h LC50
Benzene	5.8 - 108
Toluene	4.3 - 28.0
Xylene	1.3 - 12.0
Ethylbenzene	0.5 - 13.0
Naphtalene	0.9 - 3.8
Fluorene	0.5

Table 3: Acute toxicities of petroleum to different species
and life stages of marine animals [20].

organism	Fresh crude oil LC50 (mg/l)	Fuel oil/Kerosine LC50 (mg/l)
Finfisch	88 - 18,000	90 - 550
Larvae & eggs	0.1 - 100	0.1 - 4.0
Pelagic Crustacea	100 - 40,000	5 - 50
Benthic Crustacea	56	5 - 50
Bivalve Molluscs	1,000 - 100,000	30,000 - 40,000
Other Benthic Invertebrates	100 - 6,000	5 - 50

With respect to development, growth, and reproduction, sublethal
responses due to exposure to toxic chemicals are caused at
concentrations substantially lower than those causing acute mortality,
as a general rule.
Marine species and sub-ecosystems will show differential susceptibility
to exposure to any specific hazardous substance. In a similar way
restoration times may also vary from one species or one sub-ecosystem
to another. Agents of low specific toxicity could nontheless have long
term and ultimately lethal consequences through physical fouling of
respiratory organs. The likelihood of bioaccumulation of dangerous
chemicals within the food chain has also been recognised [9].

Because of their hydrophobic nature, most hydro-carbons tend to become
adsorbed to particular matter in seawater and precipitate with it to
the bottom. Petroleum hydrocarbons associated with sediments are much

more resistent than hydrocarbons in the water column. Sediment-
associated hydro-carbons may continue to cause consequences to marine
ecosystems for years after a incidental release.

CONCLUSIONS

The assessment of environmental consequences is of great importance in
the evaluation of risk associated with the transshipment of dangerous
products. In estimating the consequence of accidents, models are
necessary to describe the behaviour of the alien hazardous material in
the marine environment.
The existing mathematical models predicting physical effects on the
marine environment, are complex dedicated models mostly available on
mainframe computers only. In accident consequence assessment for
transport of hazardous materials at sea, it is a necessity to
incorporate the available relevant mathematical models into one
computer system.
In emergency situations global algorithms are needed, giving a quick
analysis of the size of the spill in an early stage.
This means that the performance of large computer time consuming models
should be adapted to get the predictions at the right level of
accuracy, within an acceptable period of time.

Literature references.

[1] System identification, parameter and state estimation;
     P.E. Eijkhof; J.Wiley&Sons (1974).
[2] System Identification: Advances and Case studies, R.K. Mehra,
     D.G. Lainiotis; Academic Press (1976).
[3] Revisions and experimental verification of the hazard
     assessment computer system models; F.T. Dodge, J.T. Park,
     J.C. Buckingham, R.J. Magott; Report CG-D-35-83 U.S. Coast
     Guard.
[4] CHRIS, Hazard Assessment Handbook, Report CG-446-3,
     Department of transportation,U.S. Coast guard.
[5] Workshop on evaluation the risk of accidental pollution
     related to the maritime transport of dangerous substances;
     OTAN/CDSM Brest ,march 25-27th, 1987.
[6] HACS/UIM, Users' Operation Manual, R.G. Potts,
     P.D. Little INC, Report U.S. Coast Guard (sept. 1981)
[7] "The Yellow Book".
     Methods for the calculation of the physical effects of the
     escape of dangerous material; published by the Dutch Ministry
     of Social Affairs.
[8] Nato Science & Society Newsletter, issue 17,
     1-st Quarter 1988.
[9] Marine pollution movement simulation models limitations and
     prospects; P. Prudhomme; NATO-CCMS, Ministry of the
     Environment, Report CPT 1381 (1987).

71

[10] Analytical and experimental study to improve computer models
     for mixing and dilution of soluble hazardous chemicals;
     F.T. Dodge, J.C. Buckingham, T.B. Morrow;
     Report CG-D-02-82, U.S. Coast Guard (august 1982).
[11] Transspill, program simulating the fate of oil at sea;
     Rijkswaterstaat, North Sea Directorate, Rijswijk,
     The Netherlands.
[12] The mathematical and physical aspects of SIMSYS2D;
     J.J. Leendertse; U.S. Dep. of Interior/geological Survey,
     report WD-3361-USGS (march 1987).
[13] Open Ocean Oil Weatering in Sub Arctic Marine Environments,
     NOAA/NOS/Ocean Assessments Division, Anchorage, Alaska.
[14] An atmospheric Diffusion Study on a local scale at a coastal
     site; P. Cagnetti, F. Desiato, P. Gaglione, A. Pellegrini;
     Atmospheric Environment Vol. 22, No.6 (1988).
[15] Validation of the TNO heavy Gas dispersion Model.
     Th.L.a. Verhagen and C.J.P. van Buytenen.
[16] Interregional modelling (Draft, 1984),
     K.D. van den Hout, H. van Dop.
[17] Methods for assessing the risk of environmental contamination
     Ir.C.L. van Deelen, paper Symposium Environmental Technology,
     22-26 June, Amsterdam.
[18] Hazards of the transport of oil and toxic liquids;
     P.M.M. Stortelder, Report DDMI-84.07, Rijkswaterstaat,
     Rijswijk, The Netherlands (april 85); [Dutch].
[19] How do estimation models work?; E. de Greef;
     Chemisch Magazine (maart 1988); [Dutch].
[20] Biological effects of oil in the marine environment;
     J.E. Neff; Chemical Engineering Progress,
     pag. 27-33 (nov.1987); [Dutch].
[21] Handbook oil combatting at sea, coastal areas and
     inland waterways (1986); Rijkswaterstaat,
     North Sea Directorate, Rijswijk, The Netherlands; [Dutch].
[23] The behaviour of packed goods which are sinking or have
     been deposited on the seabottom; L. Gielbert;
     Rijkswaterstaat, North Sea Directorate (april 1985); [Dutch].

THE JUXTAPOSITION OF LONG-TERM AND REAL-TIME OIL SPILL MODELLING

PETER C.P. CHANDLER
Martec Limited
5670 Spring Garden Road
Suite 805
Halifax, Nova Scotia
B3J 1H6

ABSTRACT.   Generalized long-term and real-time oil spill models are compared.   The influence of grid element size in the long-term model domain, the size of the time step in each of the models, and the ways in which the data are presented is assessed.   Additionally, the sensitivity analysis of a long-term model, which typically includes only the over and under estimation of environmental forces, is extended to examine factors that are found to vary in real-time spill scenarios. This includes parameters such as the fuel composition and flow rate, and the spatial and temporal resolution of the natural forces which affect the oil behaviour.   The extent to which the feedback mechanism of behaviour verification, available in real-time modelling, affects the perceived usefulness of the hindcast information is described.
    The condensate blowout from the Uniacke G-72 wellsite in 1984, off Canada's east coast, is presented as a case study.   The results from the long-term oil spill model used in the Contingency planning for the well-site are compared to the results generated by the real time modelling of the spill.   The matrix of results is expanded by rerunning the long-term model using the conditions known to exist during the spill incident.   The extent to which the long-term model is sensitive to these real-time conditions is shown to be instructive as an aid to the interpretation of the long-term modelling information contained in the Contingency Plan.

1. Introduction

One of the early considerations in an oil spill response strategy is an assessment of the severity of the situation.   Within the spill contingency plan documentation the results of an oil spill trajectory analysis is designed to aid the response team in evaluating such issues as:
  - shoreline impacts;
  - impact on biologically sensitive areas;
  - sites for control centres for clean up crews; and,
  - the logistical requirements to implement spill countermeasures.

P. Bockholts and I. Heidebrink (eds.), Chemical Spills and Emergency Management at Sea, 73—81.
© 1988 by Kluwer Academic Publishers.

The spill trajectory information typically available in the contingency plan includes maps of areal extent of the slick associated with various probabilities of occurrence of oil. In addition, the average and maximum expected slick speed is provided, as well as some sort of risk analysis of shoreline contact. These results are often presented on a monthly basis. A seasonal 'worst case' scenario may also be described based on a combination of environmental circumstance that present the greatest potential hazard.

Very quickly, however, the response team requires further oil spill behaviour information. The focus soon shifts to the slick of the particular incident, and answers regarding its predicted position, areal coverage, thickness and potential threat to specific location are needed.

This paper addresses the generalized computer methodology used in oil spill behaviour modelling. The emphasis is placed on the basic differences between the long-term models, used to prepare contingency plans, and the real-time models applied to predicting the behaviour during an oil spill incident. The intent of the paper is to identify key features in each of the two models that will allow the end-users to rationalize the differences that are sometimes apparent in the models' results.

2. Overview

In order to compare the two modelling approaches a description of each model is required. Following this, the data typically used to run the models, and the information provided by each model, are reviewed. Finally a comparison is made of the modelling results based on an actual spill; the Uniacke G-72 blowout on the Scotian Shelf in February 1984. This case demonstrates that although the results of the models differed, they could both be used to assess the threat to Sable Island, 20 km to the south.

3. Long-term Spill Models

Reviews of the many existing long-term oil spill models (Davidson and Lawrence, 1982; Spaulding, 1988) underscore the various means by which it is possible to produce computer derived slick behaviour information for contingency plan use. The differences, however, relate primarily to the nature of the environmental input data, and the algorithms used to represent advection and weathering of the oil. The structure of the model, within which these various mechanisms operate, are very similar.

Essentially the model study area, that is the geographical region which encompasses the area likely to be affected by a spill, is digitized into a spatial grid array. The area represented by each grid element (which is coded as being land, water, ice, etc.) is defined by the modeller and dictates the spatial resolution of a model. Figure 1, as an example, shows the domain of two models, one digitized at 1/4° latitude by 1/4° longitude and the other at 1/10° by 1/10°.

The environment forces driving the trajectory, such as wind and current, may be spatially variable at the same resolution of the model (i.e. on a grid element by element basis), or may be applied over several grid elements, or even the entire model domain.

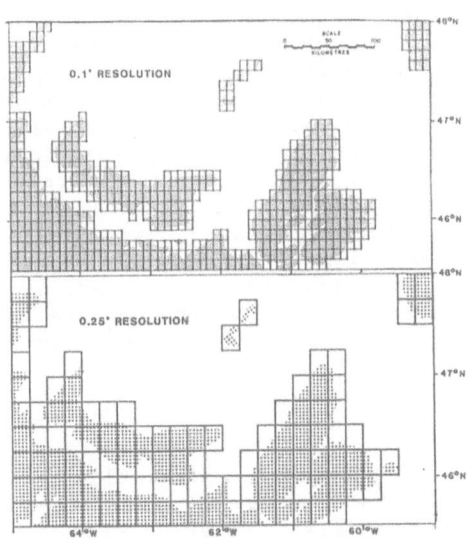

Figure 1.   The Gulf of St. Lawrence long-term oil spill model domain at two spatial resolutions.   a) Grid element size 1/10° latitude by 1/10° longitude.   b) Grid element size 1/4° latitude by 1/4° longitude.

The position of the centroid of a surface slick is evaluated by a 'transport' algorithm.   This simulates the advection of the surface flow, which carries the oil, by vectorially summing the relevant environmental factors.   The time step of the model reflects how often the slick position is updated.   Typical contingency plan models adopt a time step of 1 day, over a simulation period of say 20 years.   The time step is defined by the modeller to complement the environmental data available.

At each time step the characteristics of the slick are updated by using a 'transport' algorithm to determine the slick's new position, and a 'weathering' algorithm to simulate the change in the slick due to evaporation, emmulsification, dispersion, etc.   This computed information (i.e. position, volume remaining on the surface, percent oil evaporated, etc.) is labelled by the model for subsequent processing in terms of a grid element (spatial) and number of time steps elapsed since the start of the simulation (temporal).   After the model has completed the generation and archival of this data, a post-processor is used to produce the information included in the contingency plan.

4. Real-time Spill Models

The purpose of real-time spill modelling is to make useful forecasts of slick behaviour available to the oil response team during an actual spill event. Because results are often needed quickly, speed of execution is a requirement of these models. This speed is achievable in large part to the lesser volume of input and output data processed rather than to a faster method of simulating the spill. In fact, the 'transport' and 'weathering' algorithms used in real-time models are not very different from those used in the long-term models.

The real-time model uses site specific environmental data to predict, on the order of hours or days, the fate of a particular type of oil released at a known location. The spatial and temporal scales are therefore much smaller than those found in long-term models. This change in scale introduces factors in real-time modelling that are not significant in long-term modelling. For instance, by reducing the time step from one day to one hour advective processes, such as tidal currents, and weathering processes, such as the evaporation of the 'high' end volatile components of the oil spilled, need to be considered.

5. Using the Models

In order to achieve an acceptable statistical validity to the information presented in the contingency plan the long-term model simulates many hundreds of spills. Relevant data from each of these simulated spills are combined and analysed to determine an ensemble statistic. On the other hand real-time modelling, which is interested in only one single event, involves little statistical averaging.

Within contingency plans the model produced 'probability maps' are often used by the oil spill response team when it initially assesses the severity of a spill. These maps are intended to reveal a contour around the spill location representing the area within which one could expect oil from the spill to be found with an associated probability. The model produces these maps by assigning a probability of occurrence to each grid element based on the number of times oil entered that grid during the run and then drawing an envelope encompassing all the elements having the required probability based on the total number of slicks. As such they represent where a certain percentage of the slicks may be expected to be found. It is worthwhile to note that, as shown in Figure 2, the spatial resolution of the model can significantly alter the perceived severity of a spill.

Slick trajectories are developed by taking a sequence of driving forces and consecutively applying them to the slick position, starting with the release point of the oil. Figure 3 illustrates this process. Threats to selected areas within the model domain, such as shoreline, are determined by the number of trajectories that encounter, or approach, the region of interest. Trajectories can be generated by real-time models but oil slick position plots may also be provided. Figure 3 also shows the spill position plot. It can be seen that under certain circumstances the difference between these presentations can be significant in assessing risk to selected areas.

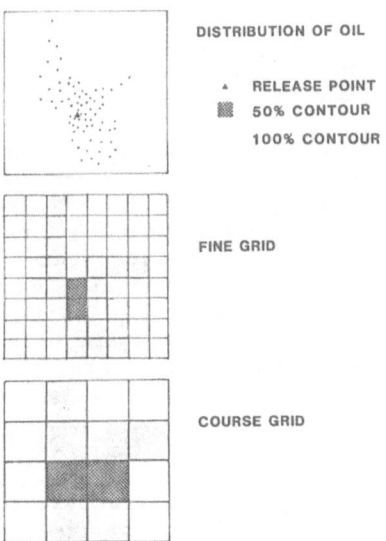

DISTRIBUTION OF OIL

▲ RELEASE POINT
▓ 50% CONTOUR
   100% CONTOUR

FINE GRID

COURSE GRID

Figure 2. A comparison of the contours of the probability of occurrence of oil as a function of grid element size.

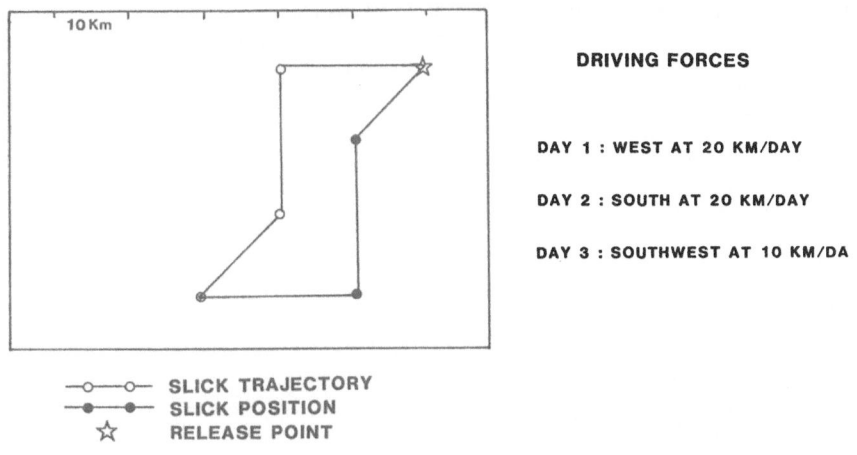

DRIVING FORCES

DAY 1 : WEST AT 20 KM/DAY

DAY 2 : SOUTH AT 20 KM/DAY

DAY 3 : SOUTHWEST AT 10 KM/DA

—o——o— SLICK TRAJECTORY
—●——●— SLICK POSITION
☆ RELEASE POINT

Figure 3. A comparison of the presentation methods for the location of
a slick using a three day event.

Sensitivity analysis, where the best estimate of the input data used in the initial run is forced to vary to allow for the uncertainties expected in the data-set, are undertaken with both long-term and real-time modelling. The former models' usually vary the wind speed and direction, or current speed and direction, and reflect our limited understanding of marine climatology. The model is then rerun, usually by varying only one parameter at a time, and the results included in the contingency plan.

Real-time modelling incorporates a manner of sensitivity analysis based on a feedback mechanism. This is due to the real-time monitoring of the slick that goes on during a spill event. If, for example, the slick is reported to be travelling consistently faster than the model predicts, then the model can be adjusted and rerun in order to match the observations. The real-time models can also vary parameters, such as flow rate or oil composition, with time. Within a few runs, the real-time model may be simulating an oil spill using parameters quite distinct from those expected, and used to govern, the long-term model.

6. The Uniacke G-72 Spill

Shell Canada Resources Limited <u>et al</u>., while testing the Uniacke G-72 well using the semi-submersible drilling unit SDS <u>Vinland</u>, experienced an uncontrolled release of gas at 2230 h Atlantic Standard Time (AST) on February 22, 1984. The flow of gas and condensate through the rotary table on the drilling floor of the <u>Vinland</u> continued for 10 days until the well was brought under control on March 3, 1984. The initial flows from the well were estimated to be between 1.11 and 1.83 x 10⁶ m³/day (40-60 million ft³/day) of gas and upwards of 48 m³/day (300 barrels) of condensate (Martec, 1984). The flow rate was observed to diminish throughout the course of the blowout.

The Uniacke wellsite is located on the Scotian Shelf at 44°11'29.07"N, 59°41'09.4"W, 20 km north of the eastern tip of Sable Island in 158 m of water (Figure 4).

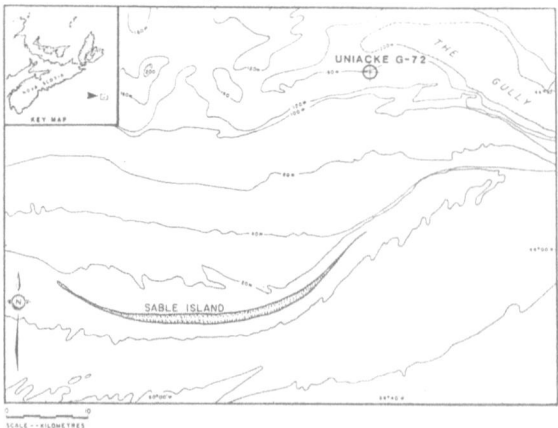

Figure 4. The location of the Uniacke G-72 wellsite.

Immediately after the blowout commenced, Shell's Spill Contingency Plan was implemented. This plan included provisions for containment, clean-up, and an environmental monitoring program. Computer modelled slick information was available from both long-term models (in the Contingency Plan documentation) and by real-time spill modelling.

Table I lists input data to the long-term model as well as output that was of immediate relevance to the spill response team. Due to the end of month start time of the spill, both the results of the February and March analyses were considered relevant.

Table I. Summary of Relevant Long-Term Model Data for Uniacke Spill

Time Step:	1 day
Grid Size:	1/4° latitude by 1/4° longitude
Oil Type:	Scotian Shelf Crude - API gravity 35°
Spill Size:	1000 barrels/day
Wind Data:	20 years of speed and direction recorded on Sable Island on 5° x 5° grid
Current Data:	Synthesis of known residual flow on 1/2° x 1/2° grid
Sensitivity Analysis:	Wind speed ±10%, wind direction ±10° Current speed ±50%, current direction ±45°

Average Slick Speed: 20 km/day

Maximum Slick Speed: 60 km/day

Probability of Mainland Contact: Negligeable

Probablilty of Sable Island Contact*: 30%

Travel Time to Sable Island*: 1-11 days

Percent of Oil on Surface
at Contact with Sable Island*: 3-18%

*based on slick trajectories

The initial conditions for the real-time model, given in Table II, are significantly different from the long-term model parameters. As monitoring information of the slick's fate was received by the response team further adjustments were made which included:
a) flow rate reduction
b) wind vector rotated 20°, then 40°, clockwise
c) slick thickness of 40 nanometres to represent loss of slick visibility
d) slick considered incoherent and not modelled after 5 days elapsed since release.

Table II.   Summary of Initial Real-Time Model Data for Uniacke Spill

Time Step: 1 hour

Spill Size: 300 barrels/day

Wind Data: Synthesis of reports from weather station on Sable Island, monitoring vessels and three drill rigs operating on Shelf. Forecasted wind data from both public and private agencies.

Current Data: Tidal current: M_2 barotropic tide on 4' x 4' grid; residual flow on 1/2° x 1/2° grid as in long-term model.

Predicted Contact With Sable Island: (based on slick position)

Day 6 of Spill: Within 12 hours of release - up to 50% of oil remaining

Day 7 of Spill: 36 to 48 hours after release - 8 to 12% of oil remaining

Day 8 of Spill: 60 to 72 hours after release - 4 to 8% of oil remaining

Day 9 of Spill: 60 to 72 hours after release - 6 to 10% of oil remaining

In comparing the two model results it can be seen that the threat to Sable Island, presented in the contingency plan for a long-term model, is reasonably consistent with the real-time model prediction. In fact, there was no physical evidence of the condensate contacting Sable Island as a coherent slick and, if the spill had not been brought under control as quickly as it was, the real-time model would have been tuned further to reflect this. This does not represent a contradiction to the long-term model results as its calculated probability of contact with the Island was 30 percent.

The long-term model was rerun using conditions observed during the spill event. The results support the contention that the condensate experienced a high evaporative flux as it was jetted onto the sea surface. Having settled as a surface slick evaporation proceeded at a slower rate as many of the volatile components had already flashed. The probability of Sable Island contact was actually increased, but with a reduced volume of condensate to below significantly detectable levels.

7. Conclusion

As a consequence of the points described above, the oil spill response team should expect the results produced by long-term and real-time models to differ. The implication being that with an ability to rationalize these differences both sources of information can be used to full advantage. This can be accomplished via spill exercises and/or seminars that involve the modellers and other members of the response team. Not only does this foster an ongoing assessment of the modelling process, but gives the spill response team a confidence in the models' results which allows them to do their job more effectively.

8. Acknowledgements

Recognition is due the Canadian Oil and Gas Lands Administration, and to Shell Canada Resources Limited, for their efforts in furthering the advance of oil spill modelling technology.

9. References

Davidson, L.W.. and Lawrence, D.J. 1982. 'Review of existing oil spill scenario trajectory models.' Can. Tech. Rep. Hydrogr. Ocean Sci. 9: iv + 60 p.

Martec Limited, 1984. 'Report on the environmental program associated with the blowout at Shell et al. Uniacke G-72.' Prepared for Shell Canada Resources Ltd., Dartmouth, N.S.: 416 p.

Spaulding, M.L. 1988. 'A state-of-the-art review of oil spill trajectory and fate modelling.' Oil and Chemical Pollution, 4: pp 39-55.

EMERGENCY RESPONSE TO CHEMICAL SPILLS

EMERGENCY RESPONSE TO THE RELEASE OF CHEMICAL CARGOES AT SEA

D Cormack
Warren Spring Laboratory
Gunnels Wood Road
Stevenage, Herts UK

ABSTRACT.   This paper proposes an approach to the categorisation of
chemicals both bulk and packaged, which may be released to the sea, as
a basis for decision making at incidents.   Three Search and Recovery
Categories are proposed for lost packages based on allocation of
chemicals according to GESAMP ratings for the marine pollution aspect
and according to the IMDG code for hazard to human life.   With regard
to the release of volatile bulk chemicals the approach depends on
calculations of pool spreading and evaporation rates, followed by
calculation of atmospheric concentrations at varying downwind distance.
Then by reference to fire and explosion concentration limits, and the
various measures of hazard to human life, it is proposed to categorise
such chemicals for treatment as for crude oil spillage, Category 1, or
as requiring additional precautions by operators on scene and the issue
of warnings to others at distance from the scene, Category 2.   A
similar approach to soluble chemicals is proposed and indications given
as to how the UK proposes to proceed from here in the elaboration of
the categories and the allocation of chemicals to them.

Introduction

The range of chemical and toxicological properties of chemicals carried
in bulk and in packaged form is very wide and at first sight appears to
present an almost insuperable problem for those wishing to develop a
rationale for reponse to spillage and to the loss of packages in the
marine environment.
   The purpose of this paper is to attempt to outline an approach to
both problems which is based on the concept of categorisation of the
chemicals in terms of a ranking of risk to the marine environment and
to human life.
   As far as bulk chemicals are concerned the first stage is to
classify them on the basis of the physical properties of density
solubility and volatility into four groups as follows: floaters,
sinkers dissolvers and evaporators.(Cormack 1982)   The second stage is

85

P. Bockholts and I. Heidebrink (eds.), Chemical Spills and Emergency Management at Sea, 85—94.
© *1988 by Kluwer Academic Publishers.*

to note that most floaters will evaporate when they spread, as they will, to produce thin layers on the sea surface. Thirdly it is to be noted that most of the chemicals under consideration whether floaters or sinkers will dissolve to some extent in sea water.

In what follows we will be concerned with evaporation and solution, with the rates of these processes and with the resulting concentrations in the air and in the sea. Finally on the basis of physical properties and of the resulting concentrations in relation to toxicity the chemicals will be categorised into those which may be treated by operational personel as for crude oil spillage Category 1 (Bulk) and those for which warnings will need to be issued for the protection of the general public at distance from the spill source. Category 2 (Bulk).

As far as packaged chemicals are concerned it is proposed that lost intact packages be categorised in terms of those which must be sought and recovered, Category 1 (packaged), those which may have to be, depending on other factors such as proximity to fishing grounds, water exchange rates etc Category 2 (packaged) and those which need not be sought and recovered Category 3 (packaged).

This categorisation approach will define the response problem, provide a ranking of priority for response action and ensure that concern for safety is properly attributed in a cost effective manner.

Packaged Chemicals

The proposed categorisation scheme has been worked out more fully elsewhere 2 (Cormack 1988) and is now under consideration in IMO. The proposal is outlined in summary form here to show the advantages of such an approach, to show how it can work out in practice for packaged chemicals and to act as an introduction to the idea of developing a categorisation procedure for bulk chemical releases.

The Categories are as follows:
Category 1 would apply to substances posing a very severe pollution hazard to the marine environment or a very severe human health hazard if accidentally recovered by fishermen. The recovery of Category 1 substances would be recommended in all cases where such recovery is technologically and economically feasible. If recovery is not feasible consideration should be given to monitoring for possible environmental effects.

Category 2 would apply to substances posing a pollution hazard to the marine environment or a human health hazard if accidentally recovered by fishermen. The recovery of Category 2 substances would be recommended when technologically and economically feasible depending on such factors as quantity likely to escape to the sea, release rate, proximity to fishing grounds or to sensitive areas, water depths, tidal streams, dilution rates etc.

Category 3 would apply to all substances in international trade not included in Categories 1 and 2. The recovery of category 3 substances

would not be considered necessary, nor would environmental monitoring
be required.

The next stage is to consider the method of allocating chemicals to
the Categories. The method which comes most readily to hand is to base
assessment of marine pollution hazard on the work of the Group of
Experts on Scientific Aspects of Marine Pollution, i.e. on the
so-called GESAMP profiles (GESAMP,1982), and assessment of human hazard
on the International Maritime Dangerous Goods Code (IMDG Code). This
approach also has the advantage of harmonising the proposed Search and
Recovery Categories with established work in these areas which has
already been used for a variety of purposes within IMO

It is proposed with reference to Table 1 that Search and Recovery
Category 1 would comprise

- Substances which are bioaccumulated to a significant extent and
  are known to produce a hazard to aquatic life or human health and
  that are highly toxic.(In GESAMP terms substances with a + rating
  in Column A and a 4 rating in Column B.)

- Substances with particularly high aquatic toxicity.(In GESAMP
  terms substances with 96 h TLV of less than 0.01 mg litre^{-1}.)

- Substances identified by GESAMP as being particularly severe
  tainters.

Category 2 substances would comprise the remaining substances which
have GESAMP ratings as follows:

- Substances which are bioaccumulated to a significant extend and
  are known to produce a hazard to aquatic life or to human health.
  (In GESAMP terms substances with a + rating in Column A.)
  OR:
- Substances which are bioaccumulated with attendant risk to aquatic
  organisms or human health, but which have short retention times of
  the order of one week or less. (In GESAMP terms substances with a
  Z rating in Column A.) OR:

- Substances which are liable to produce tainting of sea food. (In
  GESAMP terms substances with a T rating in Column A.) OR:

- Substances which are highly toxic. (In GESAMP terms substances
  with a 4 rating in Column B.) OR:

- Substances which are moderately toxic. (In GESAMP terms
  substances with a 3 rating in Column B.)

Search and Recovery Category 3 would comprise the remaining substances
not included in the other two Categories.

Turning now to questions of human safety, it is convenient to refer
to the IMO Dangerous Goods Gode which applies to safety on board ship

and as such would appear relevant to the situation of fishermen accidentally recovering packages from the sea bed.

The IMO Code divides dangerous goods into classes 1-9 and Annex 1 contains recommendations on packaging of dangerous goods and on the conservation and testing of packagings. For dangerous goods of other than Classes 1,2 and 7 three packaging groups 1,2 and 3 are specified in decreasing order of danger.

On the basis of the IMDG Code it is proposed to allocate substances to the Search and Recover Categories as follows:

Category 1 would comprise:
-IMDG Class 6.1 packaging group 1

Category 2 would comprise:
- Substances in IMDG Class 6.1 packaging group II
- Substances in IMDG Class 8 packaging groups I and II
- Substances in other IMDG classes which have subsidiary poison or corrosive hazards in similar degree

Category 3 would comprise:
- Substances not included in Categories 1 and 2

By means of the above scheme the allocation of substances to the categories is placed on a common base with other systems to which GESAMP and IMDG Code already apply.

The next step is to look at the results obtained when substances are so classified. By way of example a small number of substances listed at the beginning od IMDG Classes 2,3,4,5,6 and 8 have been allocated to Search and Recovery Categories 1,2 and 3 according to the above scheme. Radioactive substances of IMDG Class 7 and explosives of Class 1 are not included because no criteria have yet been produced for these materials. The IMO Working Group has taken the view that it would be appropriate to seek advice from outside the Organisation with regard to radioactive materials, but that experience should first be gained with the Categorisation of the other substances and that similar considerations may apply to explosives.

This approach has been applied to the first 40 substances listed in the IMDG Classes under consideration. The results show that there are in fact very few substances in the proposed Search and Recovery Category 1; that Category 2 is most prevalent in IMDG Class 8 and Categories 1 and 2 predominate in IMDG 6.1 as expected. The distribution within the Categories for the first 40 substances of the IMDG Classes under discussion are shown in Table 1. Table 1 also shows the percentage of these 40 substances in each IMDG Class which falls into Category 3 for which search and recovery would not be required.

TABLE 1

Distribution of the first forty substances in each IMDG class across the search and recovery Categories

IMDG class	Number of substances			Substances in Category 3 (%)
	Category 1	Category 2	Category 3	
2	–	27	13	48.15
3	–	17	23	57.5
4.1	2	21	17	42.5
4.2	2	6	32	80.0
4.3	–	10	30	75.0
5.1	2	22	16	40.0
6.1	7	26	7	17.5
8	–	33	7	17.5

Bulk Chemicals

CONTRAST WITH OIL SPILLS

Crude oil is a mixture of chemical compounds showing a wide range of physical properties. It is well known that some of these compounds are volatile and can give rise to fire and explosion hazards. Procedures have been developed however for the safe handling of crude oil and operational knowledge of what to expect at marine casualty and major accidental releases involving crude oil permits emergency cargo transfer and response to spillage to be conducted safely.

Indeed the fact that 25-30% by volume of crude oil spillages may quickly evaporate into the atmosphere is not seen as a problem but rather as a benefit in reducing the extent of required response. The real problems of crude oil spillage response centre on the persistence on the sea surface and subsequent arrival on shore of the remaining components of the original crude oil which have not evaporated but which go on to form water-in-oil emulsions which are even more persistent. Furthermore with product oils deriving from crudes it is generally agreed that gasoline does not present problems because it evaporates, that diesel oil does not present problems because it disperses into the water column and that only heavy fuel oils do present problems, again because of their sea surface persistence and consequent ability to cover beaches.

POOL SPREADING AND EVAPORATION.

Most chemicals carried in bulk are of low viscosity compared to oils, are likely to spread more quickly and to evaporate quickly (oil components of boiling point $<$ 250°C evaporate quickly enough) and they are not likely to be persistent. The fate of such chemicals is likely to be dispersion into the atmosphere. It is important to see if in this respect they will behave like the volatile components of crude oil and to what extent hazards may be in excess of those related to crude oil.

We will now look at the means of calculating the concentrations in the atmosphere resulting from the evaporation of spilt volatile liquids and compare the results in terms of the spatial extent of explosion and fire hazards with what is common knowledge with regard to the volatile components of crude oil on the one hand, and on the other, consider the spatial extent of toxicity hazards for chemicals which are more toxic than those present in crude oil, with a view to formulating a procedure for warning the general public where required.

The first step is to model the spreading of spilled liquids on the surface of the sea. The approach adopted in the UK is to use the Fay Model (Fay 1971) originally developed to predict the spreading of oil on water surfaces. There are some differences between oils and the chemicals now being considered. Thus viscosity is lower and this will allow chemicals to spread more rapidly. On the other hand when the spreading liquid is soluble the water-liquid interface will become more diffuse, which will tend to reduce the spreading rate. Again the Fay model assumes circular pools and it is well known that oil slicks tend to elongate in the direction of the wind. As a first approximation however it is believed that use of the Fay model will provide useful initial guidance on the nature of our problem.

Evaporation is assumed to be dependent on pool size (diameter,d) wind speed (U) and the physical properties of the chemical in question. Again this model due to Mackay, (Mackay 1980) simplifies the problem. Thus the wind over the evaporating pool is not in a normal atmospheric shear layer, but in a vapour cloud, and such clouds show a stability which reduces normal levels of atmospheric turbulence. The expectation therefore is that evaporation rate would be reduced and show a greater dependence on wind speed and pool size than indicated by Mackay.

Although the errors in the above approach are difficult to quantify the results are thought unlikely to be out by more than a factor of two.

Calculations have been carried out for toluene acrylonitrile and diethyl ether by way of examples. In general, results show that the pool evaporates at a steadily increasing rate while spreading to a maximum diameter. Pool diameters and evaporation times change in the expected fashion with wind speed, spill size and volatility of the spilled material. Pool diameters for spillages up to 1000 m^3 are typically a few hundred meters across and evaporation times vary from about 5 minutes to 2 hours depending on chemical and conditions. Exposure to the wind driven gas cloud would typically be of the same order as the evaporation time which formed the cloud in the first place. Vapour clouds will be at their greatest size with greatest value of the quotient, Evaporation Rate/Windspeed, i.e. for the largest spills in the slowest wind speeds.

VAPOUR CLOUD DISPERSION

The model used to calculate atmospheric concentrations was CRUNCH, produced by the UK Safety and Reliability Directorate (Jagger.1983) for dense cloud dispersion, since all vapours under consideration are distinctly heavier than air. In the later stages the cloud can be treated as a conventional plume. Fig 1 shows the results for toluene and diethyl ether representing the range of volatility likely to be encountered.

The toluene cloud shows a fairly rapid decrease in concentration beyond the edge of the liquid pool and since the saturation vapour presure is only about 1.6% and the Lower Flammable Liquid (LFL) is about 1.3%, the LFL is virtually coincident with the liquid pool boundary, which is about 200 m in diameter. There is only a small heavy gas cloud of about 200 m in extent. The Immediately Dangerous to Life and Health Limit (IDLH) however is about 150 m beyond the pool boundary, while the Threshold Limit Value (TLV) is at 700 m and the exposure threshold for the general public (GP) is at 1.5 km. Incidentally the odour of toluene could be detected at 6 km distance.

For diethyl ether the pool diameter is again about 200 m with the LFL and the IDLH (both nearly the same in this case) a few hundred meters beyond the pool. Boundaries for the TLV and acceptable exposure levels for the general populace (also similar) occur at distances of a few kilometers. Odour would be detected 10 km away.

Thus it would appear that relatively low toxicity hydrocarbons and similar chemicals although highly volatile and inflammable will not produce fire and explosion hazards much beyond the diameter of the evaporating pool and the duration of this hazard in the vicinity of the casualty will be short. Again the toxicity of such chemicals will be low enough to ensure that human health is not at risk at downwind distance from the vicinity of the release. Such chemicals may be placed in Category 1 (Bulk), Category 1 (Bulk) chemicals would therefore be dealt with by salvors, and others associated with the casualty and with spillage response, as for crude oil.

Now let us look at acrylonitrile as a possible candidate for Category 2 (Bulk). In fact acrylonitrile has an LFL boundary close to that of the pool boundary except at the lowest wind speeds when it can extend significantly beyond the pool. Also the IDLH boundary, since it is at a fairly high concentration, is still typically only a few hundred meters beyond the pool. On the other hand because of the low quoted values (about 2 ppm) for the TLV and GP boundaries these occur at considerable distances of the order of 100 Km with an odour threshold at about 10km.

Clearly much depends on the values put on these various measures of hazard and there will be much debate on such questions as whether they really apply to exposures of a few minutes to a few hours duration once in a life-time. Nevertheless, it would appear that if the TLV and GP values for acrylonitrite are substantiated in this application then a case could be made for acrylonitrile and other similarly toxically rated chemicals to be placed in Category 2 (Bulk). This Category being that for which salvors and others would need to take extra care (at least while vapour was present on scene) beyond that appropriate for crude oil incidents and for which warnings ought to be issued to the general public and others at distance from the source. Such warnings might take the form of a caution to remain indoors or below deck for the duration of the exposure time which again would typically be of the order of a few minutes to a few hours.

For odorous compounds in Category 1 however consideration ought to be given to the issuing of statements to the effect that although an odour may be detected by the general public the substance in question is not hazardous at the concentrations present.

SOLUBLE BULK CHEMICALS

Here again the approach will be to calculate concentrations in the water column and duration of such concentrations in relation to toxicity in determining whether the soluable material should be dealt with as for crude oil or as products, Category 1 (Bulk), or should additionally be associated with the need to warn at distance from the source, Category 2 (Bulk).

Work is currently in hand to develop analagous methods of calculation based initially on pool spreading and solution rates to predict concentrations in the sea. Preliminary results suggest again that the two Categories suggested above would be appropriate for soluble chemicals.

Conclusions

The original idea of categorising packaged chemicals as a means of providing response guidance has been extended to bulk chemical releases which give rise to atmospheric and sea water chemical concentrations.

For chemicals which evaporate, pool spreading and evaporation models have been briefly outlined as has the model which permits atmospheric concentrations to be predicted at distance from the spill source. Although these models could be refined, and although the errors are difficult to quantify, the general results obtained do comply with experience of the behaviour of volatile components of crude oil. This suggests that, where toxicity is not a problem, chemicals may be dealt with under the normal safety procedures adopted for crude oil incidents. Such chemicals could be classified as Category 1(Bulk). Although there is still some debate as to how to utilise current measures of toxicity in dealing with once in a life-time exposures of short duration,it is clear that a second category can be distinquished. Category 2 (Bulk), will require greater caution from a toxicity point of view, though no more from a fire and explosion point of view, by operators on scene and will require consideration of the need to issue warnings to those likely to be exposed at distance from the source.

It should also be pointed out that the examples discussed are worst cases in that the entire tank is released and free to evaporate from an extensive pool. In practice rather less may be released or the release rate may be slower, leading to less evaporation and to smaller distances being affected.

We intend to continue this work to classify the chemicals carried in bulk at sea according to the scheme outlined. It is however recognised that further consideration may suggest a greater subdivision of Cateory 2 (Bulk), when the available means of toxicity measurement are evaluated in relation to this particular application.

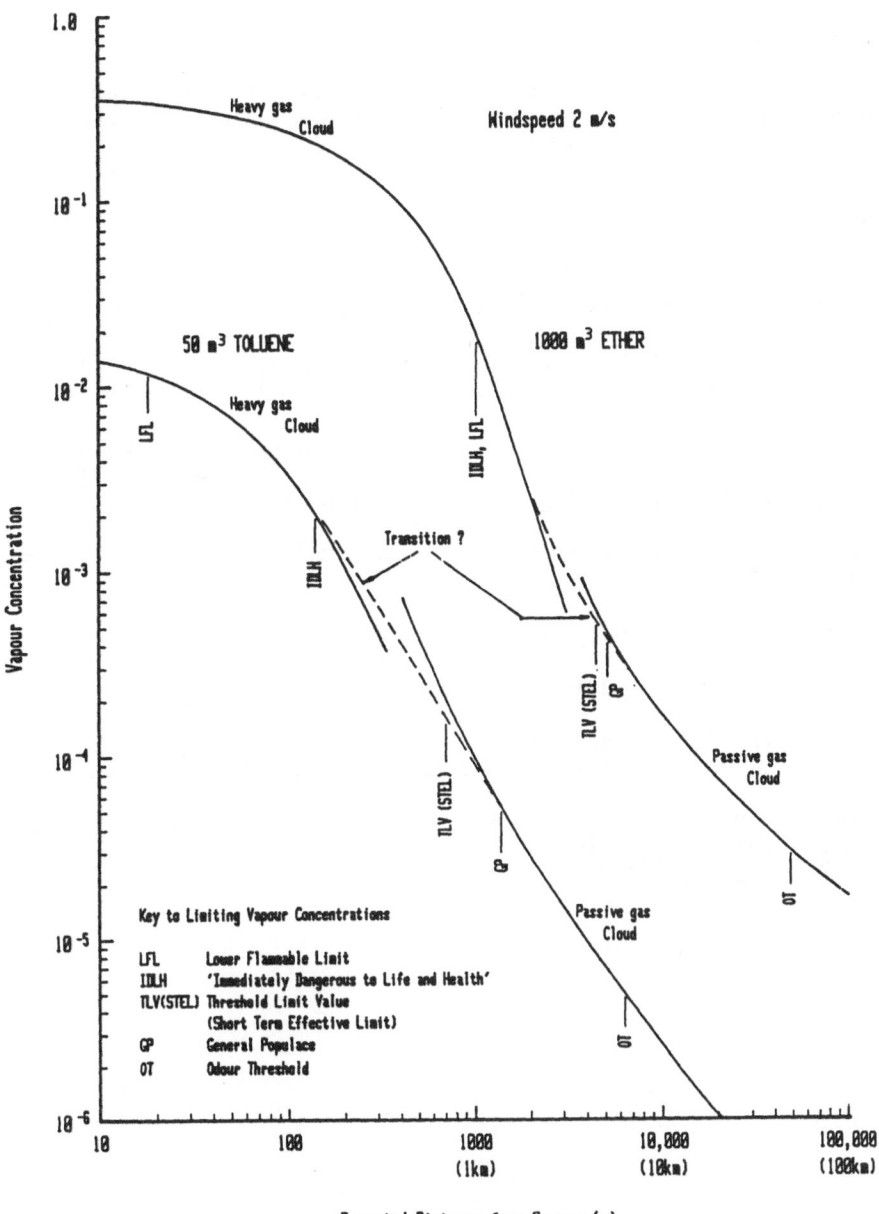

Fig 1. Sample Calculations from CRUNCH.
Variation of Peak Vapour Concentration with Downwind Distance.

94

Acknowledgement

The work described was funded by the Marine Pollution Control Unit, Department of Transport and is published with the permission of the Director.

References

1. Cormack, D,1982 Response Measures for Chemical Spillages In Singleton, B (Ed) Morichem **82** Morichem Secretariat, Rickmansworth UK.

2. Cormack, D, 1988 Response to Hazardous Materials Lost to the Sea: The Current Position Oil and Chemical Pollution **4** 21-38.

3. Joint Group of Experts on the Scientific Aspects of Marine Pollution (GESAMP 1982) Evaluation of the Hazards of Harmful Substances Carried by Ships, Reports and Studies No.**17** IMO London.

4. International Maritime Dangeous Goods (IMDG) Code Code 200-86-10E IMO, London.

5. Fay, J.A,(1971) Physical Processes in the Spread of Oil on a Water Surface in Proc. Joint Conf. on Prevention and Control of Oil Spills. American Petroleum Institute, Washington DC **463-8.**

6. Mackay, D, Paterson,S and Nodeau,S Proc. of 1980 National Conference on Control of Hazardous materials Spills USA 1980.

7. Jagger, S.F, 1983 Development of CRUNCH: A dispersion Model For Continuous Releases of Denser than Air Vapour into the Atmosphere SRD Report No. **R229** Jan. 1983.

REGIONAL POLICY ON MARINE EMERGENCY MANAGEMENT

CAPT. NAMIR A. AL-NAKIB
"ROPME"
P.O. Box 26388
13124, SAFAT, KUWAIT.

ABSTRACT. "ROPME" is a regional organization, established in 1979 with its Headquarters in Kuwait, to protect and develop the marine environment and the coastal areas. The countries of the Region - Bahrain, Iran, Iraq, Kuwait, Oman, Qatar, Saudi Arabia and the United Arab Emirates form the members of this organization and surround the "Sea Area" which they have unanimously agreed to protect and develop. The Kuwait Regional Convention for Cooperation on the Protection of the Marine Environment from Pollution, was adopted in April 1978 by the above-mentioned Contracting States of the organization. The members also adopted a protocol to set-up the Marine Emergency Mutual Aid Centre (MEMAC), which deals mainly with marine emergencies occurring in the "Sea Area". In order that appropriate regional management of resources is achieved a number of issues have to be considered, but when dealing with a marine emergency the policy is to set up three major periods in which the organization becomes involved with management of such an emergency.

1. AN OVERVIEW OF ROPME

1.1 The Regional Organization for the Protection of the Marine Environment "ROPME" adopted three major objectives; the first relates to an action plan to protect and develop the marine environment; the second being the implementation of an instrument to cooperate in the protection of the marine environment, and the third a regional cooperation programme in combating pollution by oil and other harmful substances in cases of emergency.

1.2 The action plan, in relation to the marine environment, is intended to meet the environmental needs and enhance the environmental capabilities of ROPME Member States. The plan is divided into four components namely; environmental assessment, environmental management, legal component and institutional and financial arrangements.

P. Bockholts and I. Heidebrink (eds.), Chemical Spills and Emergency Management at Sea, 95—102.

1.3 The legal component emphasizes the importance and the need for ratification of the Kuwait Regional Convention for Cooperation on the Protection of the Marine Environment from Pollution, as well as provides for the adoption of additional protocols to protect and enhance the quality of the marine environment.

1.4 The Protocol Concerning Regional Cooperation in Combating Pollution by Oil and Other Harmful Substances in Cases of Emergency, sets out an initial step taken by "ROPME" to adopt a regional policy on marine emergency management. For this purpose the Marine Emergency Mutual Aid Centre "MEMAC" was established in 1982 in the State of Bahrain.

2. FUNCTION OF "MEMAC"

2.1 MEMAC is an emergency centre of ROPME established to render assistance in case of marine emergencies.

2.2 MEMAC also disseminates to the Contracting States information that covers the following:

a) Laws, regulations and information concerning appropriate authorities of the Contracting States, contingency plans and other relevant information.

b) Information concerning methods, techniques and research relating to marine emergency response.

c) Lists of experts, equipment and materials available for marine emergency response in the area.

2.3 MEMAC, when requested, assists Contracting States:

a) To prepare laws and regulations concerning matters to prevent or deal with marine emergencies and in the establishment of appropriate authorities.

b) To prepare marine emergency contingency plans.

c) To establish procedures for transfer of personnel, equipment and materials in an emergency situation, exchange reports and promote, develop and coordinate training programmes, as well as prepare anti-pollution manuals.

2.4 MEMAC maintains communication and liaison with competent authorities in the Contracting States and with other regional and international organizations. MEMAC also prepares periodic incidents' report, on marine emergencies for distribution to ROPME Member States and performs any other function that the ROPME Council may assign to it.

2.5 MEMAC may be provided with a future role to initiate operations to combat pollution by oil and other harmful substances at regional level, if such an authority is given by the ROPME Council.

3. NATURE OF MARINE EMERGENCIES IN THE REGION .

3.1 As the "Sea Area" retains, either from on-shore or off-shore the largest reserves of oil in the world a marine pollution emergency has been recognized normally to develop through oil-associated incidents. Only recently, where development of oil-related hazardous materials and other hazardous chemicals or export of hazardous liquid bulk cargoes from the area, have started new schools of thought to develop new policy matters on marine emergency management connected with hazardous materials, other than oil. Two ROPME Member States are developing contingency plans in this regard, whilst almost all Member States have finalized and implemented contingency plans related to oil pollution marine emergencies.

3.2 Leaving aside war-related incidents, marine emergencies which have taken place in the "Sea Area" normally would involve the following:

a) Movement of the hazardous material by ships or through pipes and by other means of transfer from one place to another.

b) Storage of oily or hazardous materials.

c) Exploration or exploitation of the sea bed and through natural seepage at the sea bed.

d) Discharges from on-shore industrial sites to the marine environment.

3.3 A marine emergency can be caused through accidental, intentional or negligent action. Each of these causes may be the end product of a number of issues; or a cause may be a combination of the three. Because of the nature of the "Sea Area" a marine emergency may cause untold damages.

ROPME has conducted baseline studies of the environmental factors in the "Sea Area", upon which such damages may now be established in future incidents.

4. PREPARATORY WORK PRIOR TO THE EMERGENCY

4.1 As a major step to enhance the response to an emergency it is necessary to identify a number of components, with which either ROPME Secretariat in Kuwait, or MEMAC can act towards the incident. The most important component is the part of the Member State, that deals with any contingency matters. This may not be the "Focal Point" or the "Appropriate Authority", but perhaps the organization that is in charge of the response teams or coordinating response. Another component is to identify the institutions that would assist in environmental studies and assessments, in order that impact considerations may be taken in hand. A third component is to list the resources available in the area, which would include manpower, equipment and materials, and identify particular expertise required for particular purposes. The final component is to gather as much information as possible on international entities, that the Region may call upon to assist in a contingency. MEMAC has prepared a directory to serve as an aid to the national authorities and response agencies.

4.2 Determination of high sensitive areas and zones is a task, which is encouraged by ROPME to be conducted by national authorities. Information of this nature is retained by ROPME Secretariat or MEMAC, when provided by the Contracting States. Where necessary ROPME provides expertise to assist Contracting States in such studies. So far a number of ROPME Member States have conducted sensitivity studies in respect of oil pollution.

4.3 When an emergency is confronted on a regional level, then preparations must be taken in hand to facilitate the transfer of manpower, equipment and materials amongst Member States. This policy item is an essential but complicated sector of regional response. Although the Contracting States may have a common will to assist each other, yet their laws and regulations may differ in one way or another towards sending or receiving resources needed for a contingency. MEMAC is preparing a detailed working agreement on the loan of personnel and equipment and their movement from one State to another, including the transfer of needed materials.

4.4 Bahrain, Iran, Kuwait, Oman, Saudi Arabia and Qatar, have finalized and issued their contingency plans to deal with oil pollution of the marine environment; Iraq and the United Arab Emirates is almost finalizing their plans. Kuwait and Saudi Arabia are studying the introduction of contingency plans to deal with hazardous substances polluting the marine environment. MEMAC disseminated national contingency plans for oil-related incidents amongst member states for their benefit and has provided assistance with national plans when requested. MEMAC is planning a project for a regional contingency plan, with the intention to conduct a regional exercise, however the war situation in the area has placed some limitation on such a project. MEMAC, therefore participates in exercises of national contingency plans responding in accordance with its role.

4.5 The most valuable work that ROPME (Secretariat/MEMAC) provides to the Member States is in the hosting of workshops, seminars and technical meetings. More recently such events have moved from the oil-related subjects to those of the hazardous substances. A number of workshops were conducted in the operation, maintenance and storage of pollution combating equipment. Seminars were convened to spell out some issues relating to regional response and the use of dispersants. The events in most cases included practical exercises and on-scene training. Through such activities the Member States were assisted to set their own policy matters in respect of their own training programmes, acquisition of equipment and materials, instructions to their national authorities and the use of certain materials(e.g. oil dispersents).

4.6 Exchange of information among the Member States through ROPME has now become an important aid, whereby incidents are reported to provide warnings of an imminent danger. An annual report is issued highlighting all the incidents that have taken place. The report also contains war-related incidents. Other form of information transfer, which is being persued by ROPME is in the development of existing trajectory models for predictions of oil slick movement. ROPME intends to take-up an operational national model in one of the Contracting States and enhance its capabilities to operate over the whole "Sea Area".

5 ACTION DURING AN EMERGENCY

5.1 Pollution reporting is of course the essential issue when an emergency takes place. The dissemination of such a report becomes the second essential issue. ROPME through MEMAC has designed and circulated a reporting format for ships, oil installations and others. The report constitute the important elements required for this situation and when received is passed on to all ROPME Member States through their focal points. MEMAC then responds to all questions raised in respect of this report.

5.2 Since it has been decided that for the time being MEMAC provides a coordinating role rather then a response role, an emergency will be dealt with by national resources alone. Where necessary information pertaining to such actions will be transmitted to other interested parties that may be affected at a later stage.

5.3 MEMAC if called upon, will endeavour to assist with the logistics of an emergency; particularly if funds are provided for the contingency by Member States. This may include transportation of personnel, equipment and materials.

5.4 National, regional and international coordination, especially in cases where it is needed to provide experts, information to public media and specialized equipment, as well as the provision of contacts with international organizations, may be conducted through MEMAC under the supervision of ROPME Secretariat.

5.5. A final function of the coordinating role is to accumulate information on the surveillance and monitoring; for example when an oil related incident takes place then surveillance conducted by national authorities is relayed to other ROPME members. On the other hand available mathematical models' information is disseminated amongst the Member States exposed to the danger of pollution. ROPME also undertakes to provide experts, to assist national authorities in assessing impact of the emergency upon the marine environment; particularly during its occurrence or whilst response to the incident is conducted.

6. THE AFTERMATH OF AN EMERGENCY

6.1 Collection of data through investigation as to how an incident has occurred and dissemination of such data in order to avoid future occurrence, is a policy function of MEMAC. In fact investigative reports are requested from national authorities for this purpose. Other information on use of equipment and materials, with new clean-up and disposal techniques are documented for use in future training workshops and seminars.

6.2 Assisting Member States to return equipment and materials to country of destpatch is a coordinating role and may become a role that would relieve the sending and receving countries of a number of problems.

6.3 Assisting Member States in settling claims either amongst themselves or towards third parties in international forums. ROPME is in the process of forming a Judicial Commission consisting of nationals from Contracting States, who are highly qualified and experienced in juridical matters, to deal with disputes arising from pollution of the marine environment of the region.

6.4 The most important policy function is the final assessment of the marine emergency capabilities. Where the problems exist, where the mistakes were made, have the contingency plans responded well or is it time to review them and intoduce new concepts.

7 CONCLUSIONS

7.1 A regional effort in setting up policy matters on marine emergency management is a necessary requirement that should be well detailed.

7.2 The extent of responsibilities and tasks given to an implementing body of such policy, will provide equal if not less response to the emergency, in accordance with what has been designated for such implementing body. Therefore, there is no use pretending that an emergency will motivate the implementing body to provide more than the levels designed for its functions.

7.3 It may prove useful and advantageous to offer a level of flexibility particularly in functions, freedom of control, financing and other venues, by which a policy implementing body can operate. At the same time such flexibility may be clearly monitored to avoid excessive and unnecessary actions.

7.4 In a geographical area where its circumstances have caused a situations hampering the implementation of the regional marine emergency management policy, alternatives will need to be devised to partially maintain the management process.

References :

- Final Act of the Kuwait Regional Conference of Plenipotentiaries on the Protection and Development of the Marine Environment and the Coastal Areas, Kuwait April 1978.

- IMO - UNEP Meeting on Regional Arrangements for Cooperation in Combating Major Incidents of Marine Pollution.

- MEMAC Report to the Sixth Ordinary Meeting of the ROPME Council - Kuwait April 1988.

- MEMAC Incident Report to the Sixth Ordinary Meeting of the ROPME Council.

- Legal Aspects in the Formation and Practice of Regional Oil Spill Contingency Plans. Dr. Badria Al-Awadhi

- Reports of the Task Teams on Marine Monitoring and Research Programmes.

POLICY IN THE NETHERLANDS WITH RESPECT TO RESPONSE OF CHEMICAL SPILLS

W. KOOPS
North Sea Directorate - Rijkswaterstaat
Ministry of Transport and Public Works
P.O. Box 5807, 2280 HV Rijswijk
The Netherlands

1. Introduction

This paper describes the response preparation in the Netherlands regarding response to hazardous spillages at sea , the present state of knowledge in the field of chemical pollution countermeasures and the white spots at this moment.
In order to minimize the conseqences of an incident, a disaster control organization has been set up to enable prompt and effective action . In view of the extention of the Bonn Agreement with chemical substances it is necessary to know the precise scope of risks involved with the transport of chemical substances at sea and the hazard potential of the carried chemicals when they are released. Possible counteracting measures are therefore an essential part of the study concerning the preparation activities of the North Sea Directorate.
The transport of chemicals by sea has increased in recent years and will increase in the near future. The possibility of a calamity, where chemicals detrimental to the environment may be released has increased accordingly.
Before adapting and optimizing the existing oilresponse organization with the view to deal with chemical spillages aswell the following questions had to be regarded:
1. Which chemicals could be spilled into the sea ?
2. What is the change of a particular chemical spillage in the marine environment?
3. What is the fate of a chemical spillage ?
4. What are the hazards ?
5. Is it necessary to combat the spilled chemicals ?
6. Is it possible to combat the spilled chemicals ?
7. Are there other landbased organizations dealing with chemical spills ?
In the following these questions will be discussed.

P. Bockholts and I. Heidebrink (eds.), Chemical Spills and Emergency Management at Sea, 103—114.
© 1988 by Kluwer Academic Publishers.

2. Policy analysis

In the Netherlands a policy analysis has been carried out to determine the policy objectives of the North Sea Directorate, which have lead to recommendations to protect the marine environment against damages by harmful substances.
The policy analysis, of which a part is a risk-analysis has been carried out in three fases;
The first fase of the policy analysis addresses the question how much response potential is required to eliminate the effects of a spill consisting of hazardous materials (given the risk on such a spill and the expected damage in case no response will take place)
In the second fase of the policy analysis various recovery/response methods have been evaluated in terms of costs, efficiency, usability and added value to the response potential.
The third fase of the policy analysis gives the number and specification of the equipment to meet the objectives of fases one and two.

2.1 FIRST FASE OF THE POLICY ANALYSIS

In the first fase of the policy analysis the question of the required response potential for the Dutch part of the North Sea area has been evaluated, taking into account the different scenarios under which spills are likely to occur, and the effects if no response takes place.

To get an idea about the effects of the different kind of spills the following criteria have been used:
The chance on a particular scenario (frequency);
The effects on the marine environment;
The effects on the tourist industry;
The costs involved with cleaning the coast;
The cost involved with response at sea;
The chance that response at sea is possible.

The following frame has been used to give support to the decision about the required amount of combatting material

What is going on ?   **What are the effects ?**

-amount, number, chance   -effects on environment
 and frequency of pollution   -effects on tourism
-location of the pollution   -costs of response material
-windforce and direction   -costs of cleaning shores
-type of pollution

    RISK-ANALYSIS   **EFFECT-ANALYSIS**

It was impossible to find hard figures to give concrete
form to the environmental effects. By the use of a step by
step approach it is tried to quantify the environmental
effects of a pollution. The result of these approach is
that different scenarios might be compared to each other.
The final results are relative from which it can be derived
that an ultimate pollution of the estuaria is much worse
than an open sea spill. It also shows that the effects are
not proportional with the amount released.
The North Sea directorate has made an inventory of the
quantities and types of chemicals handled in the port of
Rotterdam to get an idea about the chemicals which could be
spilled into the sea. This inventory gives an idea of the
transportation pattern of chemicals in bulk over the North
Sea, the number of times a particular chemical product is
transported annually by sea and the actual volume per
year.
On the basis of this information an attempt has been made
to indicate which substances constitute the highest risk to
human and to marine environment.
Because of the relatively flat, sandy coastline and the
comparatively high density of traffic on this part of the
North Sea, incidents are most likely a result of
collision.
An estimate of the relative risk on the basis of quantities
carried, is given in table 1 and 2

| | MAIN HAZARD | | |
SUBSTANCE	HUMAN HEALTH	MARINE POLLUTANT	RATIO CHEMICAL/OIL
Sulphur (fluid)		X	1:275
Benzene	X		1:293
Xylene	X		1:350
Methanol		X	1:353
Toluene	X		1:357
Caustic soda		X	1:441
Styrene	X		1:487
Evaporators (all)	X		1:236
Floaters (all)	X		1:48
Dissolvers (all)		X	1:93
Sinkers (all)		X	1:159
Unknown (trade names)			1:276
Total (chemicals)			1:22

Table 1 Comparison of quantities of crude oil with the
other hazardous substances.

SUBSTANCE	QUANTITY (tonnes)	MAIN-GROUP	NUMBER OF TRANSPORTS
1. Sulphur (fluid)	684.000	sinker	62
2. Benzene	648.000	floater	189
3. Xylene (o, m, p)	537.000	floater	66
4. Methanol	532.000	dissolver	182
5. Toluene	527.000	floater	107
6. Caustic soda	426.000	dissolver	<10
7. Styrene	386.000	floater	54
8. Phosphate acid	234.000	dissolver	35
9. Cyclohexane	208.000	floater	66
10. Acrylonitril	207.000	floater	54
11. Vinyl acetate	151.000	floater	29
12. n-Butanol	121.000	floater	24
13. Sulphuric acid	117.000	dissolver	22
14. Butadiene	113.000	evaporator	<10
15. Butane	101.000	evaporator	71
16. Ethanol	97.000	dissolver	15
17. Phenol	96.000	sinker	49
18. Vinyl chloride	95.000	evaporator	<10
19. Isopropyl alcohol	95.000	dissolver	29
20. Propane	92.000	evaporator	35

Table 2 Chemicals with the largest risk based on
transported quantities (to and from Rotterdam, 1979)

It can be concluded from these tables that "Floaters" are
in the majority and are the most likely to be released.
Only chemicals in bulk have been dealt with in this risk
analysis. About 10% of the chemicals are shipped in
packaged form. These are mostly the more dangerous
substances.

Table 3 shows the major accidents reported in the period
1976-1988. This table shows that 9 out of the 21 cases did
not involve chemicals transported in bulk.

POLLUTANT	AMOUNT	NAME VESSEL	YEAR
1. Crude oil	2,000 m3	Pacific Colocotronis	1975
2. Crude oil	10,000 m3	Olympic Alliance	1975
3. Sodiummethylate	200 drums	Coaster	1976
4. Sulphurdichloride	54 drums in container	Franciszek Zubrzycki	1977
5. Ortho-Cresol	1 tankwagon	Gealic Ferry	1977
Ethylene Oxide	1 tankwagon	Gealic Ferry	1977
Paints	drums	Gealic Ferry	1977
6. 50 different	bulk/package	Attican Unity	1977
7. Heavy fuel oil	1000 m3	Eleni V	1978
8. Nemagon	99 drums	Bodrum	1978
9. Styrene	1 cargo tank	Stolt Busan	1979
10 Chlorine	50 cylinders	Sindbad	1979
11 Calcium Carbide	drums/367ton	Stanislaw Dubois	1981
Caustic Soda	sacks/534ton	Stanislaw Dubois	1981
Methylethylketon	162 drums	Stanislaw Dubois	1981
12 Heavy fuel	1650 m3	Katina	1982
13 Dinoseb	80 drums	Dana Optima	1984
14 Uranium-hexafluoride	30 cylinders	Mont Louis	1984
15 Crude oil	1200-2100ton	Orleans	1986
16 Flyash with heavy metals	3555 ton	Olaf	1986
17 Gasoline	2400 m^3	Olympic Dream	1987
Fuel oil	300 m^3	Olympic Dream	1987
18 Fuel oil	155 ton	Hell/Skyron	1987
19 Waste material	370 ton	Junior	1987
20 Fuel oil	120 ton	Borsea	1988
21 Acrylonitrile	550 ton	Anna Broere	1988
Dodecylbenzene	500 ton	Anna Broere	1988

Table 3 Accidents reported in the period 1976 -1988

Part of the policy analysis carried out in 1987 was the assessment of the risks of a particular spill. The method used in quantifying the risk of accidents is based on historical figures of accidents in the past in the Dutch area of the North Sea and on a study of Beyer and Painter. An analysis of tanker accidents have been made by them over the period 1966-1973 (ref. "Estimating the potential for future spills from tankers, offshore developments and onshore pipelines, Oil Spill Conference 1977). They established from this study figures on which it is possible to estimate the percent probability that a spill will be smaller than a certain size. It is clear that from 1977 a number of measures have been implemented which reduces the values. Therefore the values are worst case estimations in view of the safety improvements since 1977. On arriving

probability figures the following assumption has been made:
"annually two accidents in which an oil pollution is
involved occur". The chances (indicative) for the
Netherlands, of a given quantity of oil being released as a
result of a tanker accident in the soutern part of the
North Sea are:

Quantity released	chance annually	probability
15000 tonnes	once in 32 year	0.03
30000	70	0.01
60000	130	0.008

2.2 SECOND FASE OF THE POLICY ANALYSIS

In the second fase of the policy analysis various
recovery/response methods have been evaluated in terms of
costs, efficiency, usability and added value to the
response potential.

When a pollutant is spilled at sea it poses a threat to
marine life in general, fishery resources, and to tourist
industry. As long as the pollutant remains on the surface
it poses a threat to sea birds. If the pollutant is
persistent in relation to its drift time, it may reach the
shore or other vulnerable areas. In general the effects of
a pollutant are likely to be greater in the coastal zone
and estuaria than at open sea with regard to marine life
and public amenity.
The aims and objectives of response on pollutants is to
safeguard human health, to prevent and or to reduce the
effects of the pollution by protecting the coastal and
estuarial areas, the marine environment and in particular
the bird populations and amenity resources.
A number of response methods have been compared with each
other based on a number of general criteria.
The applicability has been evaluated for the different
categories of pollutant concerned, i.e. "Gases",
"Floaters", "Dissolvers", "Sinkers", and "Packaged goods"

In the following the limitations on the application of the
different methods will be described per category of
pollutant. The suitability of a response method concerns
the restrictions of such a method encompassing e.g.
meteorological, geographical circumstances or restrictions
regarding the pollutant (too viscous, too thin layer,
fire/explosion hazard etc.).

Response to "Gases"
Taking into consideration the number of transports of
volatile hazardous materials the risk that a gascloud
escape can roughly be estimated to be once in 18 years.
The danger of a gascloud is related to the explosion and
toxic hazard. 5 response methods are possible to reduce the
effects of a gascloud.
The applicability for these methods is given in table 4

	Applicability
The doing nothing option	50%
Burning the pollutant	*
Evacuation/ blocking off areas	80%
Use of foams	*
Diluting /inserting inert gas	*

Table 4 The applicability of each response method

* The usefulness of these methods is very limited as these
methods require time for logistic support which is very
often not available.

Response to "Floaters"
Taking into account the number of transports and the
carried amounts these "Floaters" involve the greatest risk
for the marine environment and human health. The risk of an
accident with floating hazardous substances (except oils)
is once in 5 years time. The risk of an accident with an
oil spill is 22 times greater than the risk of an accident
with "chemicals". The danger aspects of the group
"floaters" is on one hand the explosion danger due to
evaporation and on the other hand due to the solubility in
the watercolumn affecting the aquatic environment.
Moreover floaters may drift by wind and current and can
reach sensitive areas along the coast or wetlands. Floaters
can be divided in four subclasses:
-Pure floaters   14%
-Floaters/evaporators 85%
-Floaters/dissolvers   1%
12 response methods are possible to reduce the effects of a
floater.
The applicability for some of these methods is given in
table 5

	Applicability
The doing nothing option	43%
Burning the pollutant	2%
Evacuation/ blocking off areas	80%
Use of foams	*
Diluting /inserting inert gas	*

Table 5 The applicability of each response method

Response to "dissolvers"
Accident risk for dissolvers spillages is estimated to be once in 13 year in accordance with the transported amounts. Two subgroups of these substances can be distinghuised e.g. pure dissolvers (87%) and dissolvers/evaporators (13%). The greatest danger of this group dissolvers lies in the high concentration of the hazardous substance in the seawater during the pouring out phase and the possible forming of gases over the watersurface. 7 methods are available to deal with dissolvers.
The applicability for these methods is given in table 6

	Applicability
The doing nothing option	85%
Mechanical dispersion	30%
Other methods	*

Table 6 The applicability of each response method

* The usefulness of these methods is very limitad

Response to "Sinkers"
The risk-analysis provided a risk factor for sinkers released accidentally into the marine environment as once in 19 years, taking into account the number of transports and the amounts carried. Within the group of sinkers, two subgroups can be distinghuised; pure sinkers and sinker/dissolvers, respectively 94% and 6%. The substances in this group are in general hazardous to the marine environment, where as direct danger to human species is minimal. 7 response methods are possible to reduce the effects of a dissolver
The applicability for these methods is given in table 7

	Applicability
The doing nothing option	6%
Removal from shore	1%
Removal from seabed	50%
Burial	5%
Restrictions in spreading	5%
Biological degradation	5%
Di-hazardousation	5%

Table 7 The applicability of each response method

Response to "packaged goods"

The protection of objects and personnel has the highest
priority when packaged hazardous substances are lost at
sea. In case the packaged substances are not dangerous or
hazardous to the environment it is advisable to leave them
alone unless they are an obstacle for the navigation.
9 response methods are possible to reduce the effects of a
packaged goods containing hazardous substances.
The applicability for these methods is given in table 8

	Applicability
The doing nothing option	10%
Removal from the water surface	36%
Removal from shore	10%
Removal out of the water column	10%
Removal from seabed	30%
Di-hazardousation	4%
Sinking	-
Burial	-

Table 8 The applicability of each response method

3. General approch to hazardous material spills.

When responding to a hazardous material spill incident,
there are several critical decisions to be made which will
influence the total spill response and may well mean the
difference between a disaster, saving human lives or a
succesful clean-up action. A general approach to chemical
spillages is required by which it is possible to handle a
broad spectrum of incidents. (see table 9)

The stages of such a systematic structured approach are:

Notification

Situation analysis (define the problem)

Action plan (consider different solutions)

Decision making (choose for the "best" solution)

Execution of the plan (implement the solution)

Evaluation at the spot-------Feed back

Table 9 Standard approach to an incident

3.1 NOTIFICATION

A national Coastguard Centre has been set up. It is housed in the Scheveningen Radio building in IJmuiden. It functions as a report and information centre. The centre is equipped with the information-processing (decision simulation support system) and communications equipment required. For each kind of hazardous material spill checklists have been prepared to enhance the gathering of necessary information. The centre will be manned round the clock. Additional staff can be made available at short notice if circumstances do require (e.g. in the event of a major accident)

3.2 SITUATION ANALYSIS (DEFINE THE PROBLEM)

When a chemical spillage occurs it is important to be familiar with the physical and chemical properties of the substance together with the danger aspects in case the chemicals are released into the sea. Besides handbooks which are available in the contingency centre, the North Sea Directorate is preparing a data base system in which these information is stored and available on short notice. Physical and chemical properties were extracted from literature lists. Chemicals are classified in groups according their physical and chemical properties in order to develop specific procedures and counter-pollution measures. The North Sea directorate developed a decision simulation support system and simulation models to predict the behaviour of the chemicals released. Based on computer outputs of mentioned models an impression is formed about the hazards of chemical spillages and the necessity and possibility to combat them.

The situation analysis for chemical substances is much more critical than for an oil spill. Product knowledge, physical-chemical properties, hazards to people, property and environment are difficult to find in handbooks, databases etc.
The developed decision simulation support system "Seabel" can be used as a helpful tool, particular in cases where time-pressure exists. Experience with this system is not yet available.
At this moment one problem still exists and that is the lack of information about the behaviour of some categories of chemical spillages. There is still a need for various mathematical process descriptions that enables the prediction of the behaviour.

3.3 ACTIONPLAN (CONSIDER DIFFERENT SOLUTIONS)

At the moment there is a lot of experience available regarding oil recovery response. But for chemical spillages this is much more complicated. Those methods will be evaluated in the "SEABEL" system. A number of criteria such as IMO class, windforce etc. are used to evaluate applicable actions:
By dividing the chemicals in different classes, it is possible to select easier appropriate response methods and to get a better understanding of the danger aspects in the marine environment. In the "SEABEL" system standard response procedures are developed to handle a broad spectrum of incidents by grouping the substances with simular behaviour and danger aspects. The process of the problem-solving will be assisted by an "expert".

Also an inventory of means and expertise has been carried out. The gathered information is stored in a database, of the Seabel system. The theoretical list of response methods has been interconnected with the database of the inventory of means and expertise. In spite of the available knowledge about chemicals it is not in all cases applicable at the moment for appropriate response. One of the problems is that there are not at any time enough "specialists" available, on 24 hour basis, to deal with all the necessary information needed for quick response.
On the other hand there are also hazardous substances for which appropriate response is possible, in particular packaged goods, where time-pressure to make decisions is not present. In such cases situation analysis and response can be done with outdoor "specialists" in due time. The "Junior" and the "Olaf" are good examples of such cases.

4. Recommandations for adapting and optimizing the existing "oil" response organization are :

On-duty officers, on-scene commanders will be trained as "specialists" in controlling and combating spillages of hazardous substances other than oil.
Special attention will be given to situation analysis and decision analysis. The use of "Seabel" will be a useful tool in this training.

Striketeam crews and others involved will be trained in measuring techniques, safety aspects and hazards.

Hazard identification and response decision system implemented on a fully manned, round the clock, computer has been set up to assist the on-duty officer, on-scene commanders and decision-making team in effective and quick response.

An inventory of means and expertise has been carried out in the Netherlands. This socalled "Community Information System for the control and reduction of pollution caused by spillages of hydrocarbons and other harmfull substances at sea" has been implemented in the "Seabel system"

The On-shore teams specialized in countermeasures (as a result of the inventory) will be made available in case marine casualty (e.q. fire brigades, chemical industry etc.).

In all cases where decision analysis and execution of the plan can be postponed for a few days "outdoor specialists listed in a database" will be consulted.

Emergency procedures and response routines will be set up where a distinction will be made in immediately response and normal response.

Decision and action procedures should be clearly stated to facilitate rapid response in case of an emergency

A BALTIC SEA MODEL OF THE RESPONSE TO CHEMICALS

Cpt. K. Schroh
Head of the Federal Unit for
Marine Pollution Control

0. Abstract

In the Baltic Sea the protection of the marine environment is implemented on the basis of the Convention
on the Protection of the Marine Environment of the
Baltic Sea Area (Helsinki Convention, ratified by all
the Baltic Sea States by 1980).

For the purposes of this convention the Baltic Marine
Environment Protection Commission was established.

Three permanent committees and ad hoc subsidiary
bodies advise the Commission on scientific aspects, on
pollution damage, on the prevention of pollution from
ships and on reasonable and adequate countermeasures
against oil pollution and spillages of other harmful
substances. A special sub-working group is now dealing
with the elaboration of guidelines to combat chemicals
(CC Chem).

These guidelines will be finalized in 1989 as a Baltic
Sea Area Supplement to the IMO Manual on Chemical
Pollution. It will contain

- an inventory of chemicals carried on the Baltic

- transport patterns for chemicals carried by
  tankers and gas carriers

- a risk analysis for the transport of chemicals

- a classification system of chemicals transported

- data sheets with all relevant chemical and

P. Bockholts and I. Heidebrink (eds.), Chemical Spills and Emergency Management at Sea, 115—126.
© *1988 by Kluwer Academic Publishers.*

physical properties and all criteria relevant to response measures containing also response and combatting measures.

It is to facilitate and speed up - as part of the Baltic Sea Combatting Manual - decisions on adequate countermeasures to be taken by operational control authorities (OCA) and on-scene-coordinators (OSC).

1. Introduction

The "Helsinki Convention" was the first international convention on the protection of the marine environment embracing pollution from <u>all</u> sources: landbased (comprising all discharges form the land reaching the sea waterborne, airborne or directly from the coast, including all outflows from pipelines); ship-borne; and pollution from exploration and exploitation of sea-bed resources.

The cooperation in combatting marine pollution is laid down in the convention; it obliges the contracting parties to take measures and cooperate in order to eliminate or minimize pollution of the Baltic Sea Area by oil or other harmful substances.

For the purposes of this convention the Baltic Marine Environment Protection Commission was established. In three committees recommendations and proposals are elaborated which require endorsement by the commission; one of these committees consists of experts in combatting matters. They prepare for the commission matters related to the combatting of marine pollution caused especially by spillages of oil and other harmful substances at sea. This committee decided at its annual meeting in the autumn of 1985 to establish an ad hoc working group (CC Chem) for the elaboration of guidelines to combat spillages of other harmful substances. The exceptional geographic, hydrographic and ecological characteristics of the Baltic Sea and the sensitivity of its living resources to changes in the marine environment gave the initial input for this special task.

2. Content of the Baltic Sea Area Supplement to the IMO Manual on Chemical Pollution

The ad hoc WG "CC Chem" had to investigate chemical and gas transport in the Baltic Sea and cargo handling

in all sea ports in order to get a realistic and reliable picture of shipments, numbers and amounts of the
chemicals carried. This information is the basis for
the risk assessment in various regions of the Baltic
Sea.

The States parties to the Baltic Sea Convention need
an overview of those areas with a higher likelihood of
an accidental outflow in order to implement the Baltic
Sea Recommendation Concerning the Development of
National Ability to Deal with Spillages of Harmful
Substances Other than Oil.

The governments of the contracting parties should when
developing this capability - amongst other obligations
- take into consideration:

    a) the trade pattern of harmful substances other
       than oil within their response regions;
    b) the kind and amount of substances transported;
       the relevant IMO and Helsinki Commission;
    c) guidelines on the combatting of such substances
       when finalized.

2.1 Inventory of Chemicals Carried

The inventory comprises all noxious liquid substances
carried in bulk as listed in Appendix II of Annex II
to the Marpol Convention. It also includes other
liquid substances carried in bulk and listed in
Appendix III of the Convention and likewise liquefied
and/or pressurized gases.
The decision to select the b u l k chemicals was
based on the fact that
    a) the Marpol Anti-Pollution Manual dealing with
       bulk chemicals has meanwhile been introduced;
    b) the immediate damage to living resources, the
       hazards to aquatic life and/or human health
       resulting from the accidental release of bulk
       chemicals are in principle much higher than those
       caused by the loss of packaged goods. An accident
       involving a tanker carrying the above -mentioned
       chemicals requires in general urgent counter-
       measures in order to avoid severe damage to the
       marine environment whereas packaged goods washed
       over board require at first search and salvage
       activities.

The total volume of chemicals carried per year on
routes within the area and the number of shipments

broken down into Marpol categories is illustrated in
Table 1 of Annex 1. Westbound quantities leaving the
Baltic Sea generally account for just over half the
total quantities carried. Additionally, about 130,000
tons of chemicals were carried, which were not categorized in the Marpol system at the time of the investigation.

The total number of substances carried in 1987 in the
area is about 163, comprising

```
liquid substances listed in App.  II - Cat. A - ....12
   "        "         "     "   "    "  - Cat. B - ....31
   "        "         "     "   "    "  - Cat. C - ....35
   "        "         "     "   "    "  - Cat. D - ....41
   "        "         "        " App. III       .......16
   "        "           not categorized at present .......20
Gases                                            ....... 8
```

2.2 Chemical Trade Patterns in the Baltic Sea

In order to analyse the accident risk of a chemical
outflow and its consequences to the ecologically sensitive areas of the Baltic Sea, it is important to
know the main shipping routes with the relevant a-
mounts of the chemicals carried. Kinds of chemicals,
number of shipments and amounts carried in each individual case must be referred to a reporting period of
one year, to get a reliable picture of the chemicals
carried. For this purpose the main shipping routes in
the Baltic were divided into several segments to
illustrate the transport of chemicals in different
geographical areas. The ends of individual route segments are determined by branching off subroutes to national loading or unloading ports.

The main stream of shipping enters the Baltic Sea via
the Kiel Canal and diverts in north-easterly direction
into one main path towards the Gulf of Finland and one
along the Swedish east coast. A second main stream of
lesser magnitude enters the area from the North Sea
via the Sound and the Danish Belts. This Section
contains a general overview of the transport pattern,
based on information from the reporting of all shipments of chemicals in Baltic Sea ports throughout
the year 1987 conducted by the Baltic Sea States.

The general transport network for chemicals is shown
in Annex 1, comprising a total of 24 route segments.

Each shipment has been accounted for on each route segment between the port of departure (or entrance into the Baltic Sea) and the port of destination (or exit from the Baltic Sea).

The transport patterns are shown in Annex 2, illustrating the flow to and from the various coastal zones for substances of Marpol Category D.

2.3 Risk Analysis for the Transport of Chemicals

Extensive studies have been carried out by the Parties to the Helsinki Convention regarding the hazards of transport in the Baltic Sea. The risk model that has been developed, based on statistics and risk calculations, is briefly described in this section.

The accidents resulting in cargo outflow in the Baltic Sea are mainliy groundings and collisions. Groundings are relatively more frequent in the Baltic Sea than in world-wide shipping patterns.

Therefore the regional risk assessment in the Baltic Sea will be made on the following basis:

- determination of an accident risk factor on a per voyage basis with adjusted statistical risks for groundings and collisions involving outflow in the area, taking also into account the total loss of a chemical tanker,

- type of tanker and cargo properties in water (lower or higher density than water) are taken into account by the application of an outflow factor,

- quantities of chemicals transported on route segments have to be multiplied by accident and outflow factors.

The total expected accident rate in the Baltic Sea area involving tankers carrying chemicals of Categories A-D is about 0.6 per year with an expected annual outflow of 120 cu.m. comprising per Category A:1 cu.m., B:12 cu.m., C:67 cu.m. and D:40 cu.m. The highest expected accident rate is in the south-west part of the area, including the approaches to the Kiel Canal, accounting for almost half of the expected outflow quantity. The chart in Figure 2 illustrates the expected statistical annual outflow in cu.m. per main route segment for substances of Categories A, B, C and D taken together.

The risk of accidents with gas tankers resulting in outflow of cargo is smaller and has been calculated to be one in 100,000 voyages or one in about 160 years.

The calculated expected outflow per Category and per route segment have been added up, taking into account the different environmental hazards posed by substances of different Categories. Applying the relative hazard factor of 1000 for Category A substances, 100 for B, 10 for C and 1 for D substances gives the hazard distribution of the environmental hazards as shown in Annex III.

2.4 Classification System of Chemicals Carried

A classification system of chemicals is necessary to provide guidance to OCA regarding the response strategies.

A "European Classification System" has been elaborated in order to classify chemicals according to their physical behaviour in water when spilled into the sea.

The main principle of the system is a characterization of spilled loose chemicals, such as evaporators, floaters, dissolvers and sinkers. From this basic characterization and from other details regarding physical properties, the chemicals are classified according to the following 12 "Property Groups":

G	= gas,		GD	= gas/dissolver
E	= evaporator,		ED	= evaporator/dissolver
FE	= floater/evaporator		FED	= floater/evaporator/ dissolver
F	= floater		FD	= floater/dissolver
DE	= dissolver/evaporator		D	= dissolver
SD	= sinker/dissolver		S	= sinker

The Classification System is thus based on the physical behaviour which chemicals show when they are spilled into the sea. The purpose of the system is to arrange chemicals in Property Groups, where chemicals in the same group show the same physical behaviour in water and could be responsed to in a similar way.

The Property Groups are defined exactly by certain limits of vapour pressure (VP), density (D) and solubility (S) measured for the chemicals at a certain temperature. The temperature chosen for the Baltic Sea is 10° C.

The groups DE and D comprise those chemicals whose solubility exceeds 5% (D = dissolver). Chemicals with a solubility of 5% or lower belong either to E, ED (E = evaporator), FE, FED, F, FD (F = floater) or SD, S (S = sinker), depending on their densities.

The method of classifying chemicals by physical property limits is shown in the flow diagram of Annex IV.

2.5 Data Sheet Format and Content

A complete set of data sheets will be added to Part I containing all relevant data necessary for the identification in international literature, namely:

- hazards relevant to human health and to combatting and response strike teams

- GESAMP hazard profiles for the marine environment,

- health hazards like human carcinogen, etc.

- all physical and chemical properties which could be important for the decision on adequate countermeasures.

Part II contains combatting and response measures with the following details:

- description of the appearance of the chemicals,

- health hazards to strike teams, hazards to the environment,

- advice on the risks of fire or explosion,

- emergency actions concerning fire fighting and spill response,

- applicable recovery methods in case of major spills.

These data sheets are intended to enable OCAs and OSCs to react without delay and in the most proper way to eliminate the hazards of a major outflow or to minimize the damage to the marine environment as far as possible.

3. Final Goal of the CC Chem

The goal of the WG is to include the "Baltic Sea Area
Supplement to the IMO Manual on Chemical Pollution" as
a separate volume in the already existing "Combatting
Manual" of the Baltic Sea States. All annexes to this
presentation must be regarded as provisional, since
some minor modifications might be made until the final
meeting of the WG in June 1989. A system of updating
and adjustment must be introduced in order to provide
States parties to the Helsinki Convention regularly
with new findings on scientific and technical developments or progress in the field of response to chemicals accidentally spilled at sea.

With the outcome of the WG, reference is made to
article 11 of the Helsinki Convention, namely the
obligation of the Contracting Parties to take measures
and co-operate in combatting marine pollution in order
to eliminate or minimize pollution of the Baltic Sea
by oil or other harmful substances.

Annex I

Category	Quantity	ts	No of shipments
A	170,000		100
B	780,000	"	450
C	1,820,000	"	700
D	1,340,000	"	770
Appendix III	900,000	"	420
gases	1,900,000	"	600

Table 1. Summary of quantities of chemicals carried.

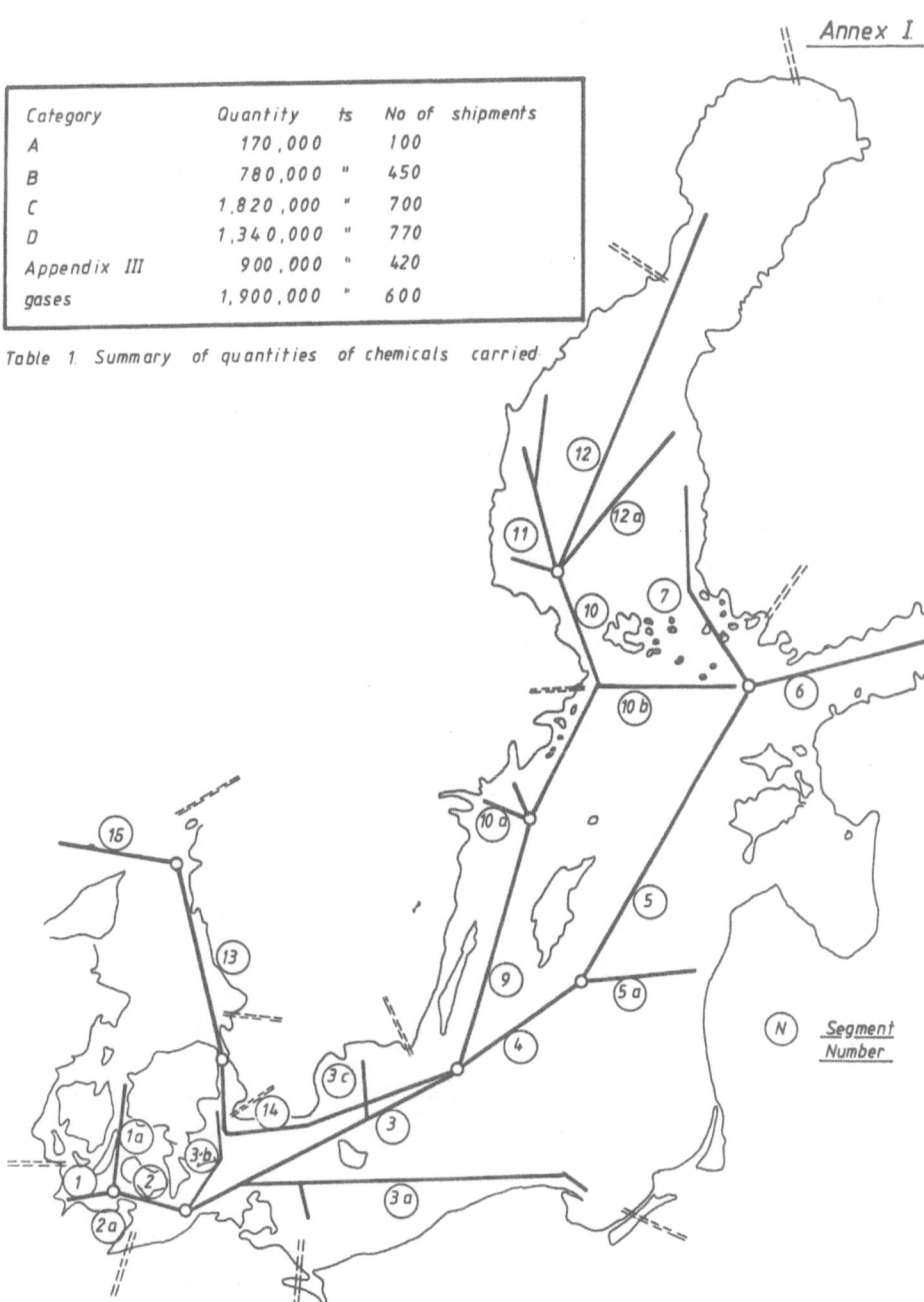

Figure 1. Transportation route system.

=====: domestic / coastal traffic

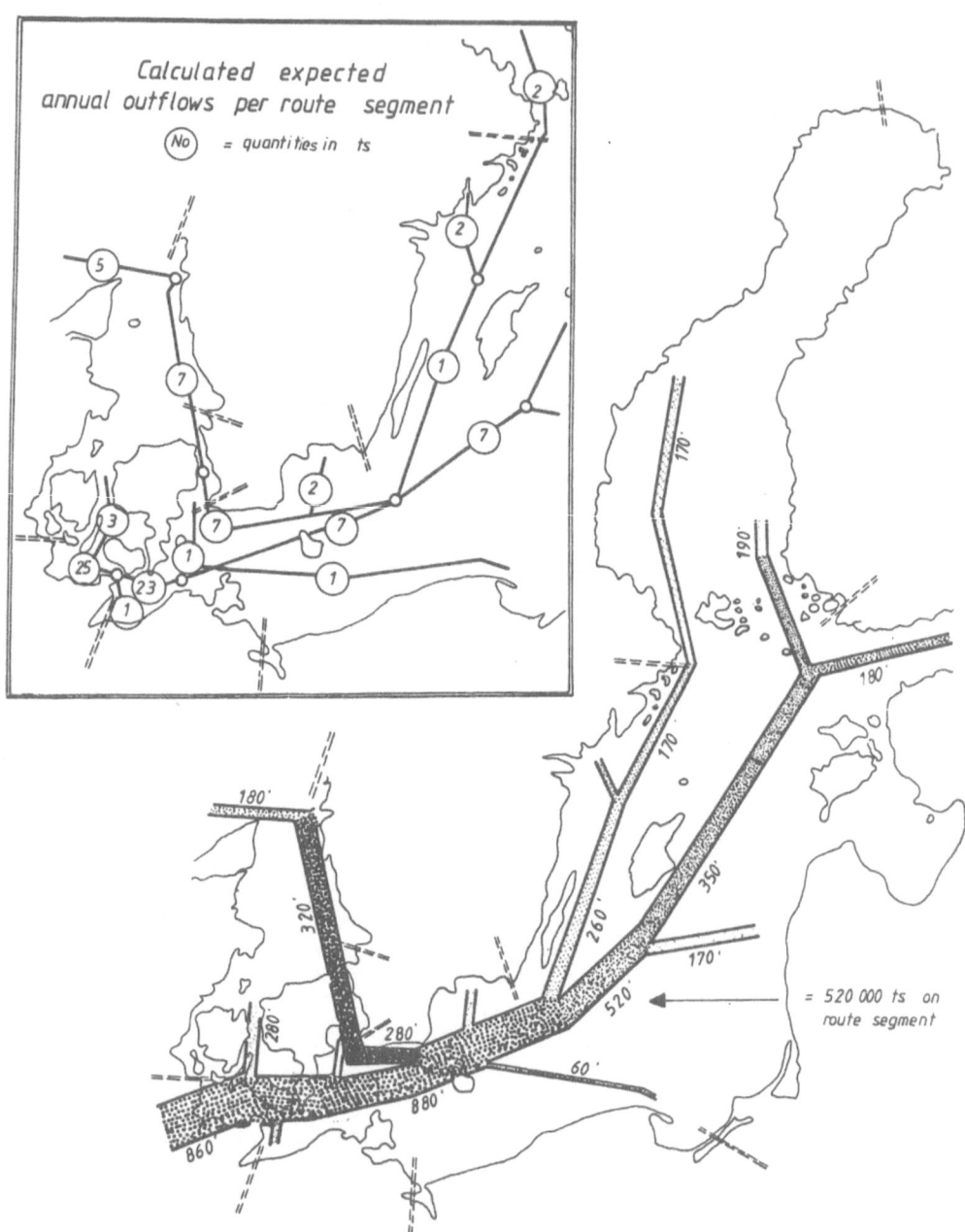

Figure 2. Transportation plattern for Category D substances.

Indicating transported quantities in 1000 ts per route segment

Risks to the Marine , Environment of the Baltic Sea

depending on Categories and Quantities of Cargo transported

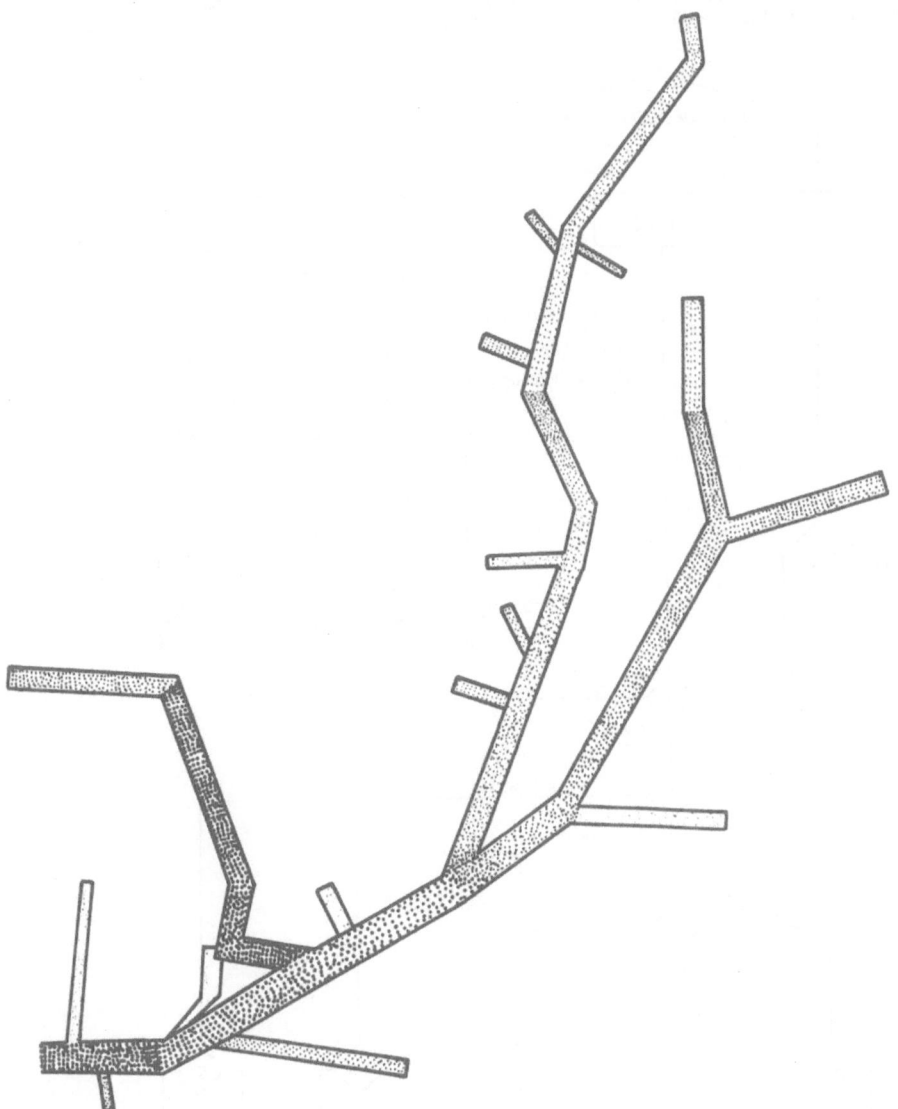

OVERVIEW OF TRANSPORTATION HAZARDS , CATEGORY A-D
SUBSTANCES.
The dull segments are representing the highest hazards to the marine environment.

EUROPEAN CLASSIFICATION SYSTEM - FLOW DIAGRAM

This flow diagram shows the principles of the European Classification System for chemicals which are spilled in the sea. By means of the system and starting from physical properties that describe behaviours in water, chemicals can be classified into 12 groups (G, GD, E, ED, FE, FED, F, FD, DE, D, SD and S). The purpose of this classification is to relate the groups of chemicals to different response strategies and thus simplify contingency planning and preparedness for chamical accidents.

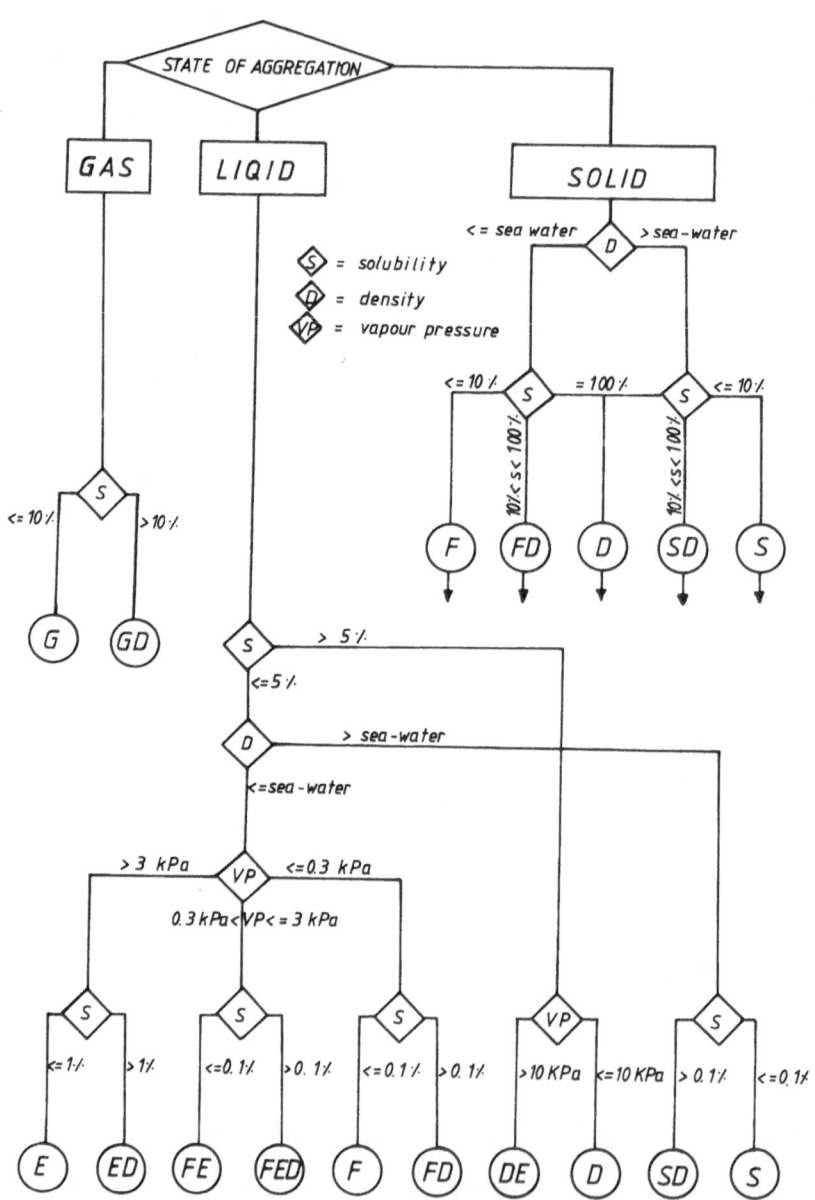

OFFSHORE INDUSTRY RESPONSE TO EMERGENCIES

H.J. Helder
Contingency Planning Coordinator – NAM
Purple Sector Coordinator – Nogepa
P.O. Box 28000
9400 HH Assen
The Netherlands

SUMMARY. All industrial activities entail a certain risk of
incidents/accidents and emergencies. Offshore oil and gas operations
are not excluded, although every effort is made to prevent calamities.
    It would be irresponsible of the industry if no plans and
arrangements were made to cope with contingencies.
    The offshore industry has prepared their plans and made their
arrangements both within the individual companies, and in the wider
context of mutual assistance and governmental liaison.

OFFSHORE INDUSTRY RESPONSE TO EMERGENCIES

No elucidation and explanation is needed when the statement is made
that there is only one good answer on the subject of emergency and that
is that all precautions shall be taken to prevent an emergency.
It has recently been shown in a dramatic way that all staff must remain
alert on safe constructions and pay appropriate attention to safety
regulations and safety awareness.

I refer to the tragic event with the Piper Alpha platform.

It must be emphasised that all efforts shall be made to prevent calamities by:
    - welltrained staff (own and contractor personnel);
    - the best practical industrial techniques;
    - only the best practical materials and equipment;
    - high standards and clear and rational regulations;
    - good supervision.
Continuous attention must be paid to these factors.

P. Bockholts and I. Heidebrink (eds.), Chemical Spills and Emergency Management at Sea, 127–136.
© 1988 by Kluwer Academic Publishers.

However it would be unrealistic to assume that <u>all calamities</u> can be prevented:

Equipment can still fail, and human errors are often the origins of emergencies and accidents.

Companies must therefore be alert and must be prepared to cope with incidents and emergencies at any time.

Emergencies can occur in many types and sizes, but undoubtly, one of the most serious calamities that can happen in the offshore industry is a <u>Blow-Out</u>.

The following pictures (fortunately not of our area) might give an impression of what a Blow-Out can look like.

<u>Picture 1</u>   <u>Onshore Blow-Out</u>

Picture 2   Offshore Blow-Out

   In such cases, a rapid response is required and every effort shall
be made to regain control as soon as possible.
   A list in detail of all the equipment or materials required if
companies are confronted with such calamities isn't included. These
details can be found in many places and in many books.
   With respect to the combatting of calamities, offshore operators
have a number of resources at their disposal and there is an inter
relationship between them.

<u>Resource nr. 1</u> is the <u>Company contingency plan</u>.

1. COMPANIES OWN CONTINGENCY PLAN

Lets give some thoughts and ideas about a Contingency Plan:

* Contingency plans should be more than a list with only the telephone numbers of managers so that managers can call each other to tell them that an emergency is going on.

* On the other hand, contingency plans should not go into too much detail. It should not be a "Cook book".
  It is not the intention that an operator goes around with a big volume in his hand and his nose in the plan looking for the next step to be taken.

* A contingency plan is not an operating manual and should only contain that information relevant to combatting the emergency. The shorter the better.

* The contents of a contingency plan can consist of:
  - information schemes and communication lines – who will inform who and when?
  - organisation schemes – with very short task description.
  - if possible a list of emergencies that can be foreseen;
  - equipment in stock or available at short notice;
  - maps with sufficient details;

* As the contingency plan is normally sent to third parties such as the Coastguard, some explanation is required concerning the company and drilling/production techniques. For the same reason key words etc. should be explained.

* An abstract of the original contingency plans can be a valuable document for operators.

* A very important condition for a proper contingency plan is that it must be updated on a regular basis. This is an exceedingly important aspect and needs sufficient time and attention. It can be a fulltime job for large companies.

No matter how well documented by means of impressive contingency plans/scheme's etc., people who are trained in using the contingency plans and the equipment provided to cope with an emergency are the cornerstone in effective handling of emergencies.
    Exercises on larger scale involving <u>onshore</u> management, supply base operation and offshore installation are held at least once a year, to test the implementation of the contingency plans on a wider basis.

Training

Standard requirements for offshore safety training have been developed
by the Netherlands Oil Industry Association (NOGEPA).
Everyone working offshore for more then 15 days a year should have
training in Sea Survival and abandon platform techniques. Apart from
this everyone, including day-visitors, receive a safety introduction
upon arrival on the offshore installation.

Exercises

Within the Offshore industry weekly drills are held for fire fighting,
man overbaord and abandon platform situations, and on drilling rigs
blow-out prevention equipment is tested weekly.
        Exercises on larger scale involving onshore management, supply
base operation and offshore installation are held at least once a year,
to test the implementation of the contingency plans on a wider basis.

Compared with onshore operations, the situation offshore normally
requires more sophisticated equipment for the combating of emergencies,
for example fire fighting vessels, diving support vessels etc.
This equipment needs a large investment.
To share these costly investments arrangements have been made between
operators on the Continental Shelf.

This brings us to resource nr. 2 - OCES.

2. OFFSHORE CLEAN SEAS & EMERGENCY SERVICES - OCES

It might be useful to briefly explain role of the
National Associations, as probably not everybody is familiar with
these.
        In the Netherlands -The Netherlands Oil and Gas Exploration and
Production Association- (NOGEPA) was founded (in 1974 in its present
form) to take care of the common interests of Companies operating
onshore and offshore.
This association provides a means of communication with governmental
and other relevant agencies.
The role of Nogepa covers many aspects and areas of the oil and gas
industry and the work within each particular subject is carried out by
committees and workgroups. These are assisted and directed by the permanent secretariat and report ultimately to an executive committee.
The workgroups on Safety and Offshore Emergency Services are particularly relevant to the subject of Emergency Response.
        Some 10 years ago Nogepa established the Code of Practice (COP),
in which it was stated that:
- each operator is fully responsible for his own operations;
- each installation holds its own fire fighting and rescue equipment
  for first line defence.

- each operator will provide mutual assistance but only:
    . if requested;
    . if own operations will not be jeopardised.

Similar associations have been established in:

United Kingdom - UKOOA
Norway   - NIFO
Denmark   - NSOC-D
Ireland   - I OOG
Germany   - WEG

They all have approximately the same principles and work on a same basis.
    These associations have made arrangements for mutual assistance in times of serious emergencies, for example blow-out, collision, explosion, oil spills etc.
    In 1971 an organisation was established with the name North Sea Operations Clean Seas Committee (NSO CSC). This organisation was formed to coordinate and enable mutual assistance in the cases of major oil spills, originated by oil exploration and production activities.
Members were the national associations of the countries around the North Sea and adjacent waters UKOOA, NIFO, NSOC-D etc.
NOGEPA was also a member although no oil was produced at that time on the Dutch Continental Shelf.
    Another organisation, founded somewhat later, was the Offshore Sector Club Organisation (OSCO). This organisation was formed to provide mutual assitance during calamities in general for example blowouts, explosions etc. and had more or less the same membership of the national organisations.
The basis for this organisation was a joint declaration code, published in 1978, which contained the principles as outlined previously under the NOGEPA's Code Of Practice.
    It was hard to justify the existence of these two separate organisations (OSCO and NSOCSC) since in many cases e.g. an oil well blowout both organisations could be/should be involved.
    For this and other reasons in 1987 both organisations, the NSO CSC and the Offshore Sector Club Organisation (OSCO), were merged into one new organisation the OCES which stands for

OFFSHORE CLEAN SEAS & EMERGENCY SERVICES

    For good registration/administration the North Sea and adjacent waters was divided into sectors, each designated by a colour.
    Initially there were five sectors, now there are ten.
    The Dutch Continental Shelf and German Continental Shelf together form the Purple Sector.

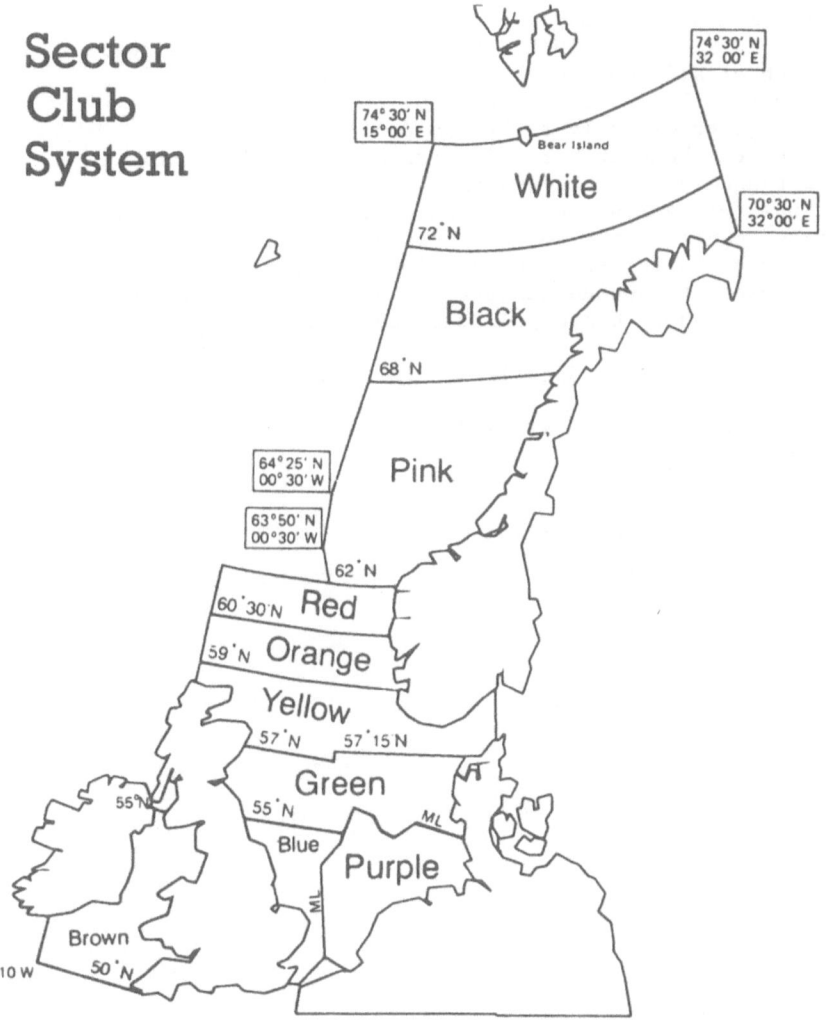

Sector Club System

74° 30' N
32 00' E

74° 30' N
15°00' E

Bear Island

White

72 N

70°30' N
32°00' E

Black

68 N

64°25' N
00°30' W

Pink

63°50' N
00°30' W

62 N

60 30'N Red

59 N Orange

Yellow

57 N   57 15 N

Green

55 N   55 N   ML

Blue

Purple

Brown

10 W   50 N

<u>Picture 3</u>   <u>Sector Club System</u>

Each sector has appointed a coordinator who is responsible for the
administration and registration of all resources in his sector. He
maintains good contact with operating companies in his sector and other
coördinators (colleagues) in other sectors.
The information, relating capabilities, locations, movements and
availabilities of the resources was initially collected and distributed
by telexes but this entailed a lot of paperwork being generated on a
weekly basis.

Since 1984 all this information is stored in the British Prestel
Computer System via LLP (Lloyds of London Press).

The ORGANISATION OF MUTUAL AID information is as follows:
- Registration of all resources in LLP-system.
- All coordinators linked to LLP-system.
- Assistance either directly or by coordinator.
- Address and telephone number of coordinator in LLP-system.
- Release arrangements in LLP-system.
- Instructions for central switchboard or telephone-operator.

The first panel of the system shows the overall set up:
1-10 sector reports for the ten individual sectors
11-17 update page, telephone numbers, vessel specs., call out
      procedures, shore based pollution control equipment.

A sector report gives details according a fixed pattern

OFFSHORE EMERGENCY VESSEL LOCATION SYSTEM
<u>Purple Sector Positions</u>

<u>Key</u>:
01   Fire fighting/diving-semi-subs
02   Fire fighting/diving-monohulls
03   Fire fighting-monohulls

<u>Picture 4</u>   FF monohull

04   Diving-monohulls
05   Specialist pollution control vessels
06   Derrick barges
07   Accomodation vessels
08   Drilling rigs

For the Purple Sector items 03, 07 and 08 are particularly applicable.

In fact, any equipment that can be required during emergencies, can be made available through this system registered or not.

Shore based oil recovery equipment, is now also listed on a special page in the computer system, including call out adresses (page 17 of the system).

The system now has a high degree of perfection, bearing in mind the objective of OCES, the variety of national legislations and companies, and the complexity of emergencies.

OCES is a good example of how industry can cooperate to protect the environment and to organise mutual assistance if need arises.

The E&P forum should be noted for completeness sake. This is a world wide organisation of the Oil Industry. Membership is open to National Associations and individual oil companies. It is a forum for policy matters and technical and operational problems. The E&P forum is not involved in regional or national emergencies.

Every one will agree that in the case of an emergency the first steps must be taken to safeguard and rescue personnel. That is the first objective and that involves resource nr. 3 – Governmental Agencies.

3. GOVERNMENTAL AGENCIES

The crew of a production platform or drilling rig have a number of options for evacuation.

If evacuation can be foreseen, transport helicopters which are under contract, can be directed to the platform or rig.

However, for emergency evacuation the timely presence of contractor helicopters cannot be guaranteed. For real emergencies the only escape routes are by boat, whose presence cannot be guaranteed either, or by emergency equipment provided on the platform or rig, such as marine survival capsules and dinghies. The only agency that can then rescue the people, under often severe weather conditions, from the survival equipment is the SAR – (Search And Rescue).

SAR is a section of the Dutch Coastguard. Their helicopters can be on site within 1 hour after notification. They have very well trained crews to rescue people from dinghies and capsules, and bring them to neighbouring safe platforms or onshore. Large scale exercises have been held on a regular basis to test the readiness and effectiveness of the SAR response.

Picture 5   Large scale exercises

It should be recognised that assistance can be offered in both
directions. Platforms can serve as a safe place to drop victims of an
aircraft crash or shipwreck during rescue operations, thus saving
flight time and fuel. Moreover most platforms can offer refueling
facilities for the rescue helicopters of SAR.
It is quite clear that without the close cooperation between
operating companies and governmental agencies, a very important link is
missing in the chain of combatting emergencies and the rescue of
people.

In conclusion:

Operational risks, in whatever industry, no matter how low those risk
may be, cannot be denied.
Emergencies and calamities cannot ever be totally excluded.
Therefore once faced with such events the offshore industry is prepared
to cope with then in a proper way, either:
by their own organisation and utilising their own resources
or by close cooperation with other companies and governmental agencies.

INTEGRAL NORTH SEA EMERGENCY PLAN
THE COORDINATED APPROACH BY THE NETHERLANDS GOVERNMENT

ir. I.F.H.C.C. van den Enden
Ministry of Transport and Public Works
Directorate-General Shipping and Maritime Affairs
Director IJmond region D.G.S.M.
Seinpostweg 36, P.O. Box 121, 1970 AC IJmuiden.

ABSTRACT
To respond to emergencies on the North Sea the Netherlands Government
has implemented an integral emergency plan.
The concept for this plan is based on the proven division between policy and
operation in decision and management.
For the execution of the plan access to the available government resources
is possible through the communication infrastructure and the coordination
capabilities of the Netherlands Coast Guard.
The resulting response is in better proportion to the magnitude and nature of
the emergency and provides a more effective use of resources.

INTRODUCTION
The Netherlands have been living and coping with water for many centuries.
The governmental involvement with the North Sea has intensified over the
past decades. Several user functions became competitors for the same space
and as the use increased it became a threat to the environment. Not only
from the use itself but also from the increasing risks involved.
Recent accidents and near accidents have shown that these risks are real.

More or less independently several government branches developed policies
for the North Sea within their sphere of interest.
Considerable loss of efficiency was the obvious result. The solution -specific
for the Netherlands - was not found in creating a specialised ministry for
the North Sea but in a process of harmonization of policies, coordination
and cooperation between operational services.

The Minister for Transport and Public Works was appointed as coordinating
minister for North Sea Affairs.
After presenting the harmonization statement on North Sea Policies to the
parliament in the early 80's, an action programme has so far shown many
results in combining efforts to safeguard the North Sea, its use and its users.

P. Bockholts and I. Heidebrink (eds.), Chemical Spills and Emergency Management at Sea, 137—141.
© *1988 by Kluwer Academic Publishers.*

HARMONIZATION

The results of the harmonization process cannot be measured in terms of power, command structure or hierarchy. The process however is not aimed at results in that sense. The aim is a better governmental product for the North Sea in terms of policy, legislation, services and law enforcements.

The use of the North Sea may generally be categorized as follows:

- transport;
- production of fish;
- mining natural gas, oil, sand pebbles;
- spillage and waste disposal from the first three functions, but also specifically through burning, dumping, from the river effluents and deposition from the atmosphere;
- recreational use;
- military exercises;
- scientific marine research.

In general these functions are in conflict when competing for the same space. The notion of a wide open sea with no restrictions has changed into a crowded sea, but the basic international laws for the freedom of the sea still apply.

The fact that the North Sea has a national as well as an European border complicates matters.

To support the coordinating minister for North Sea Affairs in her task of harmonizing these policies a special committee chaired by a former Prime Minister was established.

THE INTERDEPARTMENTAL COMMITTEE FOR NORTH SEA AFFAIRS

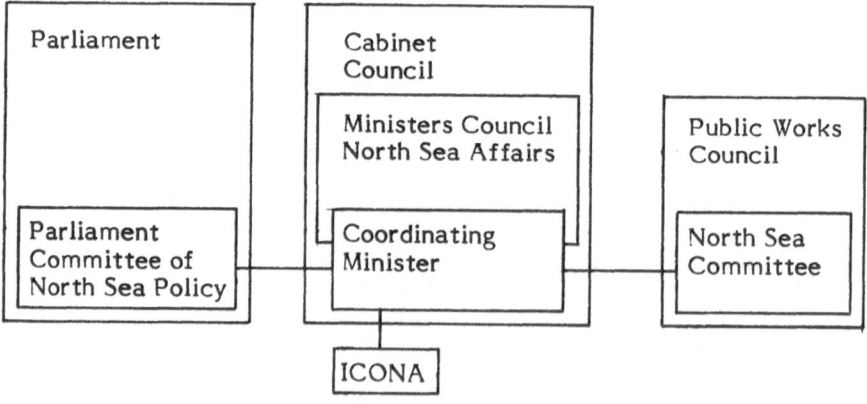

In this committee all ministries are represented by managers on the level of director-general. This committee has been most succesful in supporting the coordinating minister in combining the governmental efforts with regard to the North Sea.
Some of these results are:
- a steady coordinating process in the budget proposals to parliament on North Sea subjects;
- the Netherlands Hydrographic Institute, encompassing all governmental hydrographic activities on the North Sea;
- the Netherlands Coast Guard;
- an integrated emergency plan for the North Sea.

The Netherlands Coast Guard provides the (operational) nautical means and the communication infrastructure.
The emergency plan provides the decision structure, the management and land side means to react in response to critical, emergency and disastrous situations on the North Sea.

RESPONSE TO EMERGENCY SITUATIONS

The Netherlands Coast Guard
Apart from the basic governmental tasks of formulating policies on the use of land or sea and legislation, the Netherlands Government provides services and law enforcement on the North Sea. In 1987 the Netherlands Coast Guard was established especially for these operational tasks. Its organization reflects the policy of harmonization, coordination and cooperation. The Netherlands Coast Guard is a combination of existing government services and/or branches, each maintaining its own responsibilities, but bound by an agreement signed by the six ministers involved. These are: Minister of Transport and Public Works, Minister of Justice, Minister of Finance, Minister of Defense, Minister of Agriculture and Fishery, Minister of Interior Affairs.
Between these ministers it was agreed that the planning and scheduling of operations was to be carried out by a coordination centre.
This centre is also the national "Search and Rescue" (SAR) coordination centre and is equipped with extensive communication facilities.
It is the only place in the Netherlands where the communication systems of the participating services are linked. By co-locating the Coast Guard Centre with the public coastal radio station access to all civil communication facilities is provided.
The Coast Guard Centre has an operational overview on all government activities on the North Sea. The services involved are:
- North Sea Directorate;
- Directorate-General of Shipping and Maritime Affairs of the Ministry of Transport and Public Works;
- State Police (naval branche);
- Customs;
- Royal Netherlands Navy (limited to SAR tasks);
- Private Lifeboat Associations.

Of these services the North Sea Directorate is the specialized agency for combatting oil and chemical spills.

The Coast Guard Centre monitors the position and status of all ships involved. Combined with the operational plans of the services involved it has access to all the available government resources, with sea going capabilities.

The Centre works in close cooperation with the private lifeboat associations and provides alerting and coordinating services to these organizations.

Integral Emergency Plan

Although already part of the tasks to be undertaken by the services participating in the Coast Guard Centre, the disaster with the "Herald of Free Enterprise" provided an extra stimulans to develop an integral plan for emergencies at sea.

This plan includes separate plans of the various services already in effect before the aforementioned plan came to be in existence.

The integral emergency plan not only includes the resources of the Coast Guard but also provides access to resources for emergencies on land.

Experience has taught that a major emergency at sea requires extensive support on shore for the activities at sea. It also requires capacity for the handling of casualties ashore and other shore based facilities such as a public information centre.

To this effect the Coast Guard Centre has communication links to the operational centres for emergencies on land, thus providing the proper exchange of information, since disasters are not bound by the shore line and effects from emergencies at sea may be felt on land and vice versa.

Relevant information is gathered at the Coast Guard Centre in case of an emergency. The operational procedures of the Coast Guard Units and affiliated organizations are such that this kind of information is immediately transferred to the Coast Guard Centre.

The basic task of the Coast Guard Centre is to assess the situations and initiate the proper measures.

First priority of course is saving lives. Second priority is protection of the environment by trying to reduce the escalation of emergencies as far as possible.

Meanwhile, the gathered information is checked for consistency and - if possible - confirmation by independent sources is sought.

Procedures of the plan include the dissemination of the gathered information to the relevant specialists and organizations at the earliest possible moment. This service, provided by the Coast Guard Centre, is the most vital part in the governmental response to emergencies on the North Sea.

If a situation cannot be handled adequately by the available operational means of the Coast Guard, a team of specialists and the chairman of the policy team are notified. The head of the Coast Guard and the relevant specialists together form the operational team. This team then convenes at short notice. The operational team will assess the situation and will develop strategies for handling the emergency.

Depending on the severity and complexity of the situation the chairman of the policy team will call a meeting of the policy team.

The policy team consists of appointed representatives of virtually all ministries; members act on behalf of their resp. minister. If the occasion arises all team members will be notified, but the actual attendance will depend on the nature of the emergency. In this way the policy team has access to far more resources than the operational team or the Coast Guard Centre. It can also provide the operational team with extensive command facilities.

The combination of specialists and Coast Guard Centre forming said operational team provide all the necessary functions for managing the response of the government, once the strategy is chosen and the resources are allocated by the policy team.

This procedure is not a new way of organizing the response to emergencies, it is a well proven concept. However, it is the new emergency plan that provides the joining of the efforts and the possibility to mobilize far more resources at short notice than ever possible before.

This is a result of better communication, better assessment of the situation, not because there are more resources. This results in a prompt convening of the authorities responsible for making the decisions, thus providing better chances for the right decisions on strategies and the allocation of resources.

Evaluations

The Coast Guard Centre has the operational task to evaluate procedures and the response to incidents it can handle with the available operational means. If an emergency results in meetings of an operational and the policy team, this response of the government is evaluated too. Special emphasis is placed on the decision process and management of the situation, thus striving to improve the procedures for a better response.

CONCLUSION

The North Sea is of vital importance to the Netherlands. To ensure a lasting use of the North Sea with its many functions the Netherlands Government has taken far reaching measures.

Measures not only pertaining to the prevention of incidents and emergencies, but also measures to enable the authorities making the decisions to use all available resources if and when the situation arises.

The organization concept used to achieve a harmonized policy for the North Sea, to reach coherent decisions and to combine operational forces, is specific for the Netherlands. It is used as a growth concept that may lead to possible other organizational structures in the future.

For the time being it is regarded as the optimum solution for the most effective response to emergencies.

CLASSIFICATION OF CHEMICALS FOR SITUATION ANALYSIS

W. KOOPS
North Sea Directorate - Rijkswaterstaat
Ministry of Transport and Public Works
P.O. Box 5807 2280 HV, Rijswijk
The Netherlands

1. Introduction

In case of an emergency the first step should be to collect the
necessary information of the chemicals involved. The selection
of response methods is highly dependent on the nature and the
extent of the spill, the local circumstances and the properties
of the pollutant. Troughout the time span of an incident, from
first notification to final response, it is necessary to obtain
information for evaluating the impact created by the incident.
Because of the enormous diversity of chemicals, a variety of
classification systems have been developed, each with a
different purpose. The systems are based either upon the
physical hazards represented by the chemicals, their reactivity
or their physical properties in the aquatic environment.
When a chemical spillage occurs it is important to be familiar
with the physical and chemical properties of the substance
together with the dangers involved in case of release of the
chemicals into the sea. Various handbooks and chemical cards are
available for this purpose.
Where pollution occurs, a knowledge of the physical and chemical
properties of the substance concerned and the dangers inherent
in its escape is important. For situation analysis the
following kind of information about a spilled substance must be
available:

- identification
- reaction
- hazards to human
- hazards for the marine environment
- physical behaviour
- response
- other relevant physical and chemical properties

P. Bockholts and I. Heidebrink (eds.), Chemical Spills and Emergency Management at Sea, 143—156.

144

Because of the enormous diversity of chemicals, a variety of classification systems have been developed. Since a number of these classification systems simplify the situation analysis and are therefore important for response purposes they will be examined briefly

The classifications applicable in the field of response to hazardous material spills in the aquatic environment are:

-Classification by transport
-Classification by physical hazard
-Classification as marine pollutant
-Classification by behaviour
-Classification by reactions

2. Classification

Roughly, several hundreds of chemicals are shipped in bulk while over several thousands of different chemicals are shipped in "packaged" form, in containerised or non-containerised shipments. However, at present over ten thousand different substances are on the market, and the number of different wastes, products, complex-mixtures is also increasing. Each chemical or mixture can be characterised by its own discrete physical/chemical properties. In order to provide an uniform approch for understanding the nature, properties and potential risks of hazardous substances, and for developing contingency plans, categorization of these discrete, relevant substances is required. By dividing the chemicals in different classes as described in this paper it is possible to develop more appropriate response methods and get a better understanding of their possible harm to the aquatic environment.
Table 1 summarizes the classifications useful for situation analysis.

Transport	Physical hazard	Marine pollutant	Behaviour	Reaction
gases	explosives	category A	Gases	not reacting with water
bulk fluids	gases	category B	Fast fluid evaporators	reacting with water
bulk solids	flammable liquids	category C	floater/ evaporator	self-reacting
oils	flammable solids	category D	floater	reacting with other chemicals
packaged goods	oxidizing substances	non-marine pullutants	floater/ evaporator/ dissolver	
	toxic substances		floater/ dissolver	
	radioactive substances		dissolver	
	corrosives		dissolver/ evaporator	
	miscellaneous dangerous substances		sinker/dissolver	
			sinker	

Table 1 Response classifications summarised.

There are several critical decisions to be made in response to a hazardous material spill incident. They will influence the total spill response and may well mean the difference between a disaster, saving human lives or a succesfull response. A general and simplyfied approach to such spillages is required by which it is possible to handle a broad spectrum of incidents. During the initial phases of an incident approach, the problem must be analysed to define if there is any imminent or potential hazard to human health and or threat to the aquatic environment. It must be determined whether immediate actions are needed to prevent or reduce the hazards and or to protect safety of response personnel. When a discharged hazardous substance has been identified, the hazards presented can be evaluated, using the classifications already mentioned. After this classification one can determine the hazards involved, define the extent of the problem and take appropriate response actions. The classification is the first step, and a simplified approch, in

order to be able to choose the "best" solution for a particular
spill, at short notice.

The aim of those classifications is also to create a system
which give immediate information about the course of action to
be taken in the initial phase of an intervention in a spill of
hazardous materials. The classifications give information
about:
  -Health hazard;
  -General precautions;
  -Marine pollutant hazard;
  -Behaviour;
  -Fire/explosion hazard;
  -Physical form;
  -Transport regulations;
  -Response.

3. How to use the classifications ?

In order to evaluate a particular substance in the aquatic
environment the following step-wise approch is applied:

1. Find out if the released substance reacts, if so than the
original substance as well as the reaction product(s) have to be
evaluated.

2. Find out the way of transport, such as packaged substances,
gases, bulk chemicals or oils.

3. Determine the hazards involved for safety and decision making
purposes.
First the hazards to human and secondly the hazards to the
marine environment

4. And finally for response and detection the behaviour class

These steps will be discussed below.

3.1 REACTION.

Classification by reactions is very useful for this purpose:
There are a number of chemical reactions that can occur and that
can change the composition and quality of a transported product,
such as:
Polymerisation:
A process whereby a number of molecules of the same material
react together to produce a much larger molecule.
Decomposition:
A process whereby a single molecule breaks up into a number of

smaller molecules
Oxidation:
A process involving a reaction of products with oxygen (usually from the air).
To prevent unwanted changes occurring and to maintain the quality of the product, certain additional chemicals are used. Under certain conditions of temperature, exposure to air or contact with a catalyst, certain cargoes tend to undergo polymerisation, decomposition, oxidation or other chemical reaction. Care should be taken if such cargoes are involved in an accident
Transport of hazardous substances may involve dangers for the ship and her crew and for the surrounding environment if a substance escapes during transport, e.g. in case of accidents. Hazardous substances which can otherwise be inert, can react to form new substances which can be more dangerous than those involved in the initial spill. Reaction products may pose additional physical and toxic risks to emergency response teams. Therefor it is important to recognise and consider the aspects of reactivity and compatability in the response to hazardous material spills in the aquatic environment.
Solid and liquid hazardous substances can, for this purpose, be subdivided as follows:

> 1.-Reactive with water;
> 2.-Self-reacting (polymerization, etc.);
> 3.-Reactive with other substances.
> 4.-Non-reactive with water;

3.2 WAY OF TRANSPORT

Classification by transport:
There are several regulations for the transport of dangerous cargo.
-Regulations for the control of pollution by noxious liquid substances in bulk.
-Regulations for transport of solids in bulk.
-Regulations for prevention of pollution by oil.
-Regulations for transportation of gases.
-Regulations for the prevention of pollution by harmful substances carried by sea in packaged forms, in freight containers, portable tanks and road and rail tank wagons.

The conventions dealing with these regulations have the objective to achieve the complete elimination of intentional pollution of the marine environment and the minimization of accidental discharges, not only by oil, but also by the other harmful substances.

148

In the regulations for transport of hazardous materials the
following "classes" can be distinguished:

-Liquid bulk chemicals (IBC-Code);
                                   Type 1 ships
                                   Type 2 ships
                                   Type 3 ships

-Solid bulk chemicals

-Oils   (MARPOL annex 1);

-Gases in bulk   (IGS-Code);
                                   Type 1G ships
                                   Type 2G ships
                                   Type 2PG ships
                                   Type 3G ships

-Packaged goods   (IMDG-Code).
                                   Stored on deck
                                   Stored in the ship

3.3 HAZARDS

3.3.1 Classification by human hazard:
The classification based on the physical hazards is the most
used one. However the main objective is the safety during
transport and the handling. This classification can be used for
identification and situation analysis in responding to hazardous
material spillages. The international convention for the safety
of life at sea (Solas 1974), sets out the various classes of
dangerous goods. These classes are:

Class 1: explosives
Class 2: gases, compressed, liquefied or dissolved under
         pressure
Class 3: inflammable liquids
Class 4: inflammable solids or substancese
Class 5: oxidizing substances (agents) and organic peroxides
Class 6: poisonous (toxic) and infectious substances
Class 7: radioactive substances
Class 8: corrosives
Class 9: miscellaneous dangerous substances
Above mentioned classification applies to dangerous goods which
are carried in packaged form or in solid form in bulk. Packages
containing dangerous goods shall be durably marked with the
correct technical name and placards.

3.3.2 Classification as marine pollutant:
The International Conference on Marine Pollution in 1973 adopted
the International Convention for the Prevention of Pollution
from ships, 1973 (MARPOL 73). The convention in its annex II
contains detailed requirements for the discharge criteria and
measures for control of pollution by noxious liquid substances
carried in bulk. For this purpose noxious liquid substances are
divided into four categories depending upon their hazard to the
marine resources, human health, amenities and other legitimate
uses of the sea as evaluated by the Ad Hoc Panel of IMO. The Ad
Hoc Panel evaluated substances according to the hazards they
might pose when released into the sea for the following
considerations:

    A- Bioaccumulation
    B- Damage to living resources;
    C- Oral intake/hazard to human health;
    D- Skin contact and inhalation/hazard to human health;
    E- Reduction of amenities.

Based on the procedures proposed by GESAMP, each substance is
given a hazard profile, an example of which is shown in the
following table 2

	profile	rating
Bioaccumulation	A	0
Damage to living resouces	B	4
Oral intake	C	not available
Skin contact and inhalation	D	II
Reduction of amenities	E	**

Table 2 Hazard profile of Chlorine

Inwhich:
A - Bioaccumulation rating:
    (0) - Not known to be significantly bioaccumulated, or to
    support one of the above ratings

B - Damage to living resources rating:
    (4) - Highly toxic   TLm (96h)< 1   mg/l

D - Skin contact and inhalation rating:
    (II)- Hazardous -carginogen

E - Reduction of amenities rating:
    (**) Moderately objectionable because of the above
    characteristics, but short term effects leading only to
    temporary interference with use of beaches: also used when
    there is circumstantial evidence that the substance is an
    animal carcinogen but where there is no clear evidence that

the substance has caused cancer in human beings.
The column D rating by GESAMP has been disregarded, since this
categorization for that rating relates to hazard to human health
(skin contact and inhalation) and not to aquatic pollution.

Based on above mentioned hazard profiles GESAMP has categorized
the substances falling into A, B, C, D pollutants. To establish
whether a substance is category A, B, C, D it is necessary to
relate the GESAMP hazard profile with the table 3.

HAZARD PROFILE

A	B	C	E	
+	-	-	-	
-	4	-	-	
T	3	-	-	CATEGORY A
Z	3	-	***	
T	-	-	-	
Z	-	-	-	
-	3	-	-	CATEGORY B
-	2	-	***	
-	2	-	-	
-	1	4	**	CATEGORY C
-	1	3	**	
-	1	-	-	
-	-	4	-	
-	-	3	*	CATEGORY D
-	-	-	***	
-	-	-	**	
-	D/BOD	-	-	

Table 3. pollution categorization table.

To qualify a substance in any category it is necessary to comply
with all the requirements in any one row at one and the same
time.
To establish the individual substance categorization table 3
should be read horizontally so that, for example, a rating "T"
in column A and a "3" in column B results in categorization "A";
whereas a single rating of "T" or "3" results in categorization
"B".

Category A substances are bioaccumulated and liable to produce a
hazard to aquatic life or human health, or are highly toxic to
aquatic life (defined by TLm less than 1 ppm). Additionally
certain substances which are moderate toxic to the aquatic life
(defined by: TLm <10 ppm) when particular weight is given to
additional factors in the hazard profile or to special
characteristics of the substance.

Category B substances are bioaccumulated with a short retention
time in the order of one week or less, or they are liable to
produce tainting of the sea food, or they are moderate toxic
(defined by 1 ppm< TLm <10 ppm). Additionally certain substances
which are slightly toxic to aquatic life (defined by: 10 ppm <
TLm < 100 ppm) when particular weight is given to additional
factors in the hazard profile or to special characteristics of
the substance

Category C substances are slightly toxic to aquatic life
(defined by: 10 ppm < TLm < 100 ppm). Additionally certain
substances which are practically non-toxic to aquatic life
defined by: 100 ppm < TLm <1000 ppm when particular weight is
given to additional factors in the hazard profile or to special
characteristics of the substance.

Category D substances are practically non-toxic to the aquatic
life defined by: 100 ppm < TLm < 1000 ppm), or causing deposits
blanketing the sea floor with a high biochemical oxygen demand
(BOD), or highly hazardous to human health. With a LD(50) value
of less than 5 mg/kg, or produce moderate reduction of the
amenities because of persistance, smell or poisonous or irritant
characteristics. Possibly interfering with use of beaches, or
moderately hazardous to human health, with a LD(50) of 10 mg/kg
or more, but less than 50 mg/kg and produce slightly reduction
of amenities.

Above mentioned guidelines for the categorization of noxious .
liquid substances are mainly used for the discharge requirements
of noxious liquid substances. The hazard rationale was developed
for the particular purpose of the International Convention for
the Prevention of Pollution from ships, MARPOL 73/78. As a
consequence the hazard profile are intended to be used solely
for that purpose. For response purposes one shoud be aware of
the limitations and restrictions.

3.4 BEHAVIOUR

This classification into groups is based upon the affinity of a particular substance to air, watersurface, water, bottom or combinations of these to determine which compartment is threatened most. To establish the criteria on which hazardous bulk materials will be categorized according to the classification mentioned they are divided into four main categories.

- Substances which tend to evaporate quickly
- Substances which float on the watersurface
- Substances which dissolve or disperse into the watercolumn
- Substances which tend to sink to the bottom

The aim of this classification is to divide chemicals into categories according to their physical-chemical behaviour in the event of spillages in the aquatic environment. It is intended to use various behaviour classes based on response methods to be employed to reduce the effects once a spill occurs. Because chemicals can also fit into (several) other categories a division is made into subcategories (for example substances which float and also evaporate, or substances which sink and also dissolve). The European classification is the result of close collaboration within the Bonn Agreement between the authorities. Figure 4 shows the agreed European classification criteria for the division of chemicals into subgroups.

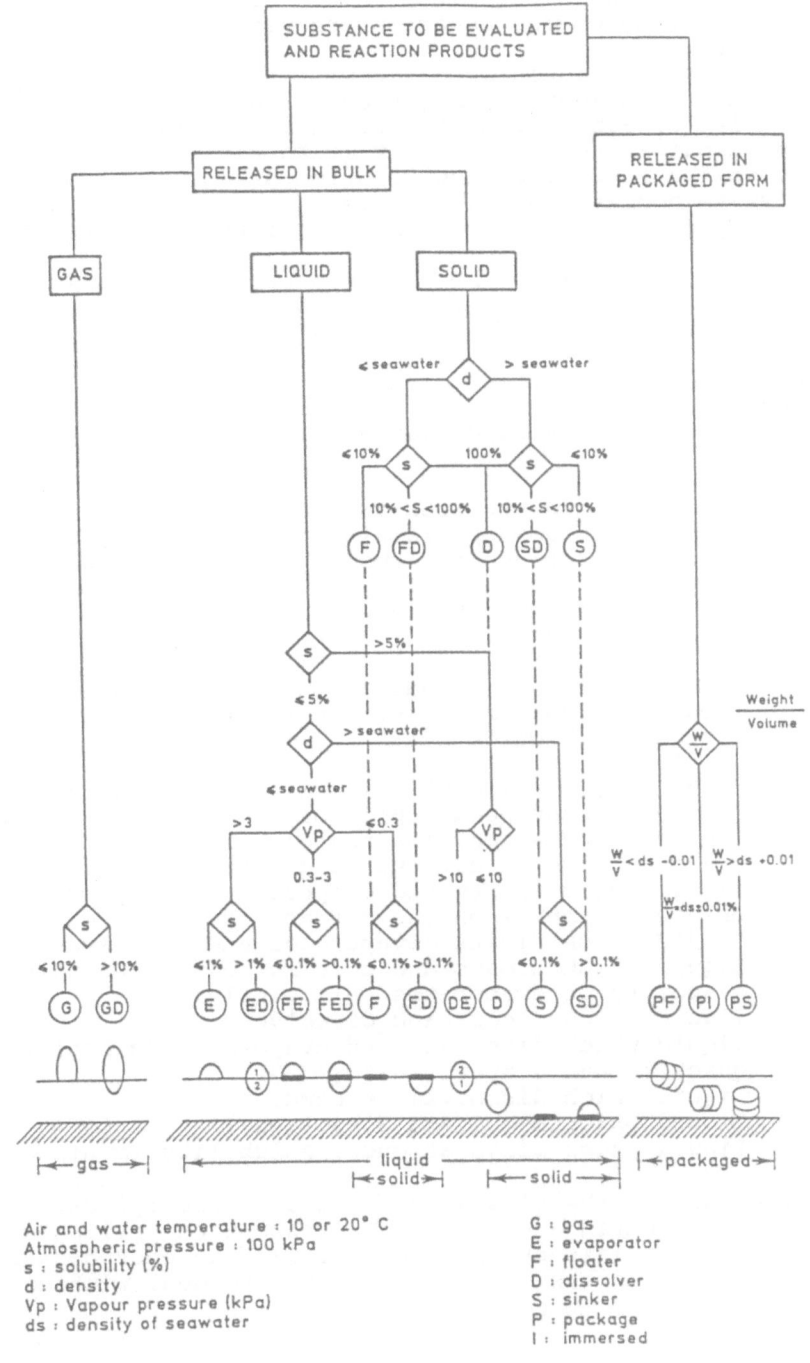

Figure 5 The agreed European Behaviour Classification

The determined characteristic values of substances, which have a special bearing on the behaviour of hazardous substances in water are:
- Phase: solid, liquid or gaseous;
- The main behaviour types (products which Evaporate, Float, Dissolve or Sink and their relevant combinations);
- The reactivity of the product (with air, water or with itself and taking account of the reaction products;
- Relative density in relation to sea water, i.e. the bulk density of the substance compared to the bulk density of water
- Vapour pressure i.e. the pressure of the vapour in equilibrium with the liquid phase at a given temperature;
- Solubility, i.e. the maximum weight of the dissolved substance in 100 g or 100 ml of sea water. When this solubility exceeds 100%, the substance is said to be "miscible in all proportions"
- The inclusion of "intact parcel" in the classification.

Gases
A distinction is made between two subclasses:
1. G: Gases which do not dissolve or only slowly dissolve in water; this class relates to gases both lighter and heavier than air.
2. GD Gases which dissolve in water.

Liquids
The various types of behaviour in the event of a spill of a liquid product in bulk are given in figure 4. The behaviour possibilities for liquids are:

3. E   Liquid which evaporates immediately (in about one hour for an instantaneous spill of several hundred tonnes)
4. ED   Liquid which evaporates and dissolves (evaporation takes place in a few hours and is quicker than dissolution)
5. FE   Liquid which floats and evaporates (the product remains only a few hours on the surface of the water).
6. FED Liquid which floats, evaporates and dissolves.
7. F   Liquid which floats and does not dissolve and/or does not evaporate or evaporates only slowly.
8. FD   Liquid which floats and dissolves
9. DE   Liquid which dissolves and evaporates (dissolution is quicker than evaporation).
10 D   Liquid which dissolves at once.
11 SD   Liquid which sinks and dissolves
12 S   Liquid which sinks at once; solubility is nil or low.

An other use of the classifications is a simplyfication of the response in contingency plans. To respond on spillages with hazardous materials one has to deal with an enormous amount of different substances. It is impossible to deal with all these

substances separately in a contingency plan. By grouping the
substances with simular behaviour and danger aspects (see table
5) in response groups, standard response procedures can be
developed to handle a broad spectrum of incidents.

HAZARD	G	GD	E	ED	FE	FED	F	FD	D	SD	S	PF	PI	PS
Explosion hazard	X	X	X	X	X	X								
Fire hazard	X	X	X	X	X	X	X	X				X		
Health hazard	X	X	X	X	X	X								
Water pollution hazard	X		X		X		X	X						
Air pollution hazard	X	X	X	X	X	X								
Bottom pollution									X	X				
Hinderance amenities	X	X	X	X	X	X	X	X			X	X	X	X
Radioactive hazard											X	X	X	

Table 5 Grouping the substances with simular danger aspects

An other example where use is made of the classifications as
described before is the recovery classes approach:
In this approach recovery classes are subdivided in three
groups. It shows the necessity of removing the substance from
the environment in case of a spill. It therefore is an
indication of the hazard of a substance. The subdivision is as
following:

The GESAMP hazard profile and the IMDG-code are used to assess
the above mentioned subdivision.

Search and recovery class 1:

To apply to substances posing a very severe pollution hazard to
the marine environment or a very severe human health hazard if
accidentally recovered by fishermen. The recovery of Category 1
substances is recommended in all cases where such recovery is
technologically and economically feasible. If recovery is not
feasible there may be a need to monitor environmental effects
Some examples of substances from this class are:

Name	IMO Class	Column A	Column B	Packaging group
Aldrin	6.1	+	4	2,3
Tetramethyl lead	6.1	Z	3	1
Chlorine	2.3	0	4	-

Search and recovery class 2:

To apply to substances posing a pollution hazard to the marine environment or a human health hazard if accidentally recovered by fishermen. The recovery of Category 2 substances is recommended where such recovery is technically and economically feasible. Depending on such factors as the following: quantity, likely to escape, release rates, proximity to fishing grounds or to sensitive areas, water depths and tidal conditions, etc need to monitor environmental effects
Some examples of substances from this class are:

Name	IMO Class	Column A	Column B	Packaging group
Acrylonitrile	3.2	0	3	1
Cresols	6.1	T	3	2
Styrene monomer	3.3	T	2	2

Search and recovery class 3:

To apply to all substances in International trade not included in categories 1 and 2. The recovery of category 3 substances would not be considered necessary.
Some examples of substances from this class are:

Name	IMO Class	Column A	Column B	Packaging group
Ammonia	2.3	0	2	—
Ethanol	3.2	0	0	2,3
Toluene	3.2	0	2	2

CHEMICAL EMERGENCIES AT SEA:

THE SALVOR'S PERSPECTIVE

BY

KLAAS J. REINIGERT

MANAGING DIRECTOR, SMIT TAK B.V.

Captain Klaas Reinigert is Managing Director of Smit Tak, the world's largest ocean towage and salvage company. Smit Tak has been responsible for the salvage of a series of vessels carrying hazardous substances. The company has made a significant contribution to marine environmental protection, by preventing or minimising pollution from casualties such as the Anna Broere, Cason, Olaf, Ariadne and Mont Louis.

----ooo----

P. Bockholts and I. Heidebrink (eds.), Chemical Spills and Emergency Management at Sea, 157—169.
© *1988 by Kluwer Academic Publishers.*

Over the past 20 years, there has been dramatic change in the characteristics of the international shipping industry. This includes the emergence of large crude carriers and continued growth in the carriage of hazardous and toxic substances by highly specialised, complex ship types.

The professional salvage industry has geared itself to the new challenges - ranging from major tanker fires developing temperatures hot enough to melt steel to the loss of chemical tankers carrying mixed cargoes of dangerous substances.

Today, salvors face an almost infinite variety of potential incidents. The hazards of ocean transportation include collision, stranding, fire, explosion and leakage or loss of chemical substances and other pollutants. The vessel types involved include large black oil and product tankers, chemical carriers, gas tankers, bulk carriers, general cargo vessels and ro-ro/passenger ships. At any time, a casualty may occur anywhere in the world, from the busy waters of North West Europe to a remote port in an undeveloped country on the other side of the globe.

In order to respond effectively to a crisis at sea, the salvor must make a large and on-going capital commitment, to maintain huge salvage resources and an extensive logistical infrastructure. Given the nature of the work, he must often be prepared in the initial phase to deploy those resources in the absence of a commercial contract.

Against the salvage world's uncertainties, the salvor must strive to generate sufficient regular income to preserve his international capability. His resources include experienced personnel, skilled in a variety of engineering and technical disciplines; a large fleet of specialised salvage vessels (from fire-fighting craft to heavy lift cranes); and strategic stores of costly salvage equipment and materials. In today's difficult market, there are very few private contractors who can assemble such resources and offer truly worldwide response to marine emergencies.

However, the major contractors have suffered severely in recent years. Work in the Middle East Gulf - the salvage of the Iran-Iraq conflict's shipping casualties - bought some time, but there is now a ceasefire and future prospects look grim for salvage.

Yet, in recent years the major salvors have achieved much in enhancing their ability to deal with chemical and other emergencies at sea. Powerful, air-portable fire pumps, of a new generic type, have been developed. Personnel have received intensive training in fire-fighting, dealing with hazardous and toxic cargoes and pollution prevention. In addition, relationships have been strengthened with external interests - acknowledged experts in fields ranging from the recovery and handling of radioactive substances to chemical monitoring on site. Certainly, in the late 1980s, even the largest salvors can no longer hold in-house all the specialists required to deal with complex chemical emergencies at sea.

People are our most important resource, as experience is everything in the salvage business. Beyond broad contingency planning and ensuring that stores and equipment are maintained at adequate levels, we can make no specific plans in advance of a salvage. Therefore, individual experience and expertise is crucial. Inspired improvision in the early hours of an emergency, for example, can make all the difference between success or failure as an operation unfolds.

The modern salvage industry can be described as an international marine fire brigade, on call round-the-clock for immediate action. The industry's main strength, as indicated above, is its pool of highly trained personnel. They have unique skills, which are of even more value today as a result of reduced manning levels onboard ships. It is ironic that, as the hazards (environmental and otherwise) posed by ships and cargoes have increased, the crews responsible for their safe operation have become ever smaller.

Who benefits from the work of the salvor? There are many beneficiaries: shipowners, cargo interests, underwriters, central and local government, the public at large and, of course, the marine environment. Yet, paradoxically, in recent years the salvor has received little recognition of the value of his services.

The International Salvage Union recently issued its annual statistics, completing a 10-year survey of the industry. It announced that ships and cargoes worth some £5.5 billion has been salved over the past 10 years. The salvor's reward was just six percent of that figure! Indeed, the average level of remuneration for salvage continued to decline last year, to 5.3 percent of salved value (excluding Gulf cases).

Present difficulties stem from three factors: the end of the Gulf conflict, lower remuneration and a rapidly shrinking market. The Gulf conflict benefitted very few ISU members (who have been uncomfortably aware of their dependence on this activity). Total salved value last year amounted to £716 million, of which £273 million resulted from the rescue of Gulf war casualties. The industry's view was expressed by ISU President Mr Rom W. Scheffer, who said that the survival of the world's marine emergency service should not hinge on war between nations. In any event, salvors should not be forced to expose personnel and vessels to the hazards of war in order to subsidise the cost of providing essential salvage cover elsewhere in the world. I am sure that anyone who has witnessed at first hand a hazardous salvage operation at sea will testify that it presents more than enough peril in its own right!

On the second point - falling levels of remuneration - the maintenance of adequate salvage cover is now under threat. Those who benefit from salvage services would do well to join salvors in reviewing the future of the industry, in order to ensure the salvage services continue to be available when needed.

Meanwhile, the market for salvage services around the world continues to decline. Salvage operations under the traditional "no cure-no pay" form of contact fell by one-third (to 106 cases) in 1987. The number of Lloyd's awards last year dropped to 65, as against a figure of 158 in 1980. In part, the declining market reflects improved levels of safety in shipping operations, following the introduction of traffic separation schemes, new rules governing shipboard operational procedures and the wider acceptance of new technological aids to safe navigation.

However, accidents at sea will always occur; it is impossible to legislate against natural perils and human error. This brings me to a central question: Who will salvage the salvage industry before its level of income drops to the point where the world's busy shipping areas and environmentally vulnerable coastlines are no longer protected? The fact that marine accidents are declining in number does nothing to comfort, say, the master of a chemical tanker finding himself in serious trouble in the North Sea. Put another way, if a town suffers fewer fires because of the efficiency of fire prevention officers, it does not necessarily follow that it is a good idea to stop paying the wages of the local fire brigade!

The ISU has 34 member firms, based in over 20 countries. It may surprise you to learn that they share an annual global market for Lloyd's Open Form salvage services worth only $50 million. The wreck removal sector is worth another $75 million annually, but this is a business plagued by cut-throat competition and financial risks of frightening proportions.

The market of $50 million a year should be viewed in relation to high salved values and the billions of dollars that the salvage contractors save the insurers and the oil and chemical industries. In preventing and minimising pollution, they help them avoid huge claims.

If the private salvage industry collapses, state run salvage must fill the void. However, state salvage will be more costly and less effective. State-run salvage companies suffer from an inherent lack of experience, as crews may spend long periods awaiting a casualty. In addition, state salvage resources are not available internationally, which rules out the development of global operational experience. No Government in Western Europe, for example, could risk the use of its salvage fleet overseas. It may not survive the political backlash, should a chemical casualty threaten gross pollution of its coast while state-owned salvage craft are busy helping someone else in another part of the world. In short, state-run salvage is a very poor substitute for a viable private salvage industry.

What can be done to salvage the salvage industry? I could devote the whole paper to answering this point. However, the following measures give some idea of what could be done to ease current difficulties:

* Acceptance and swift implementation of the International Maritime Organisation's draft Convention on Salvage, due to be considered at a diplomatic conference in London next Spring.

* Revision of Lloyd's Open Form, last amended in 1980, to provide a "safety net" for salvors engaged in the prevention of chemical pollution (such a safety net already exists in the case of oil pollution - salvors working to prevent oil pollution know they will receive a guaranteed sum).

* Acceptance that certain types of casualty - including vesseᵢ carrying dangerous cargoes - represent complex and specialised wreck removal tasks. They are more akin to civil works than traditional salvage.

Salvors engaged in special circumstances like the Anna Broere operation should work under a form of contract similar to that employed by land-based construction companies, rather than the "all-or-nothing" gamble of no cure - no pay or a fixed lump sum. The latter are negotiated in a crisis situation, before anyone has had an opportunity to gauge the true extent of the work involved.

We have no desire to abandon Lloyd's Open Form for conventional salvage, as it avoids "haggling" and allows the salvor to press on and deal with the emergency. However, no cure - no pay is quite inappropriate for casualties such as the Herald of Free Enterprise, Anna Broere and Cason. In their weakened state, salvors cannot afford to take such gambles; failure in such circumstances could spell financial collapse. Certainly, Smit Tak will not gamble with its future should another Herald-type incident occur.

In reacting to an emergency, a salvor must assess the task and complete his negotiations within hours or days, rather than the weeks or months that commonly apply ashore. The costs of unforeseen difficulties, from bad weather to inaccurate initial information from the client, are borne by the salvor. This situation is totally unacceptable. Can anyone imagine the likes of Wimpey, Costain, Volker, HBG, Bos Kalis or any other major construction group undertaking contracts on such unreasonable terms?

An answer to the problem is readily at hand. Salvors should negotiate contracts which include variation orders - a device frequently used in the construction industry.

In part, our problems are rooted in the salvage industry's romantic, image. People tend to forget that the salvage company is no different from any other commercial entity: it must be able to earn a living. Contracts including variation orders would do much to help. After all, what relevance has Lloyd's Open Form to a chemical pollution containment operation, demanding a fleet of costly vessels on station for weeks or months at a time?

Recently, Smit Tak was contracted under no cure - no pay to remove the wreck of a collapsed oil platform. The project was said to involve a lift of 2000 tonnes. Subsequent investigations revealed that the true weight was 4000 tonnes. We were then told by the client that this was "our problem". Indeed it was - the project economics were transformed yet we were helpless without a variation order. Today, the salvor works under a regime which can only be described as a buyer's dream!

A few months ago, Smit Tak completed the salvage of the Anna Broere, thus preventing a serious pollution incident in the North Sea. As you are all aware, public sensitivity to North Sea pollution has never been higher. The Dutch Government commissioned us to undertake the work and the total cost of the operation, including the cost of naval deployment, was Fls 15 million. Our reward was a fixed sum of just Fls 4 million.

The difficulties encountered on the Anna Broere project stretched us to the limit. This small chemical carrier, carrying a dangerous liquid cargo, sank on May 27 following a collision. She went down in the mid-North Sea, laden with 700 cu m of acrylonitrile, 1000 cu m of industrial detergent and a variety of other hazardous substances. The loss triggered a full-scale emergency response.

While we signed a contract four days after the sinking, our vessels were at the scene within hours of the collision. A large salvage fleet was mobilised and a 10-mile exclusion zone enforced around the wreck.   Initially, the prospects looked good for a swift and successful conclusion to the salvage. The floating sheerlegs crane Taklift 4 was on hand to lift and manoeuvre the Anna Broere over the submerged main deck of the semi-submersible barge Giant 2, which would then transport the wreck to Rotterdam.

The salvage fleet on site also included a large vessel outfitted as an emergency hospital, with a medical evacuation helicopter stationed at the ready. Other units included Dutch frigates, various traffic control craft, a floating crane acting as accommodation base, and a technical support/chemical monitoring vessel. Salvage workers were equipped with protective suits and breathing apparatus and Taklift 4 was rigged with an automatic gas alarm system and sophisticated facilities for the analysis of water samples. Chemical specialists were appointed to join the salvage team.

We anticipated a 10-day job. In the event, it took 10 weeks due to unseasonal weather. Strong southerly winds and unfavourable sea conditions resulted in countless delays to lifting operations. June and July dragged on with no hope of effecting the salvage - yet we were forced to keep our fleet at the highest state of readiness. At one point in mid-June we came close to success, but wreckage twisted under the hull of the Anna Broere prevented us from placing the wreck on the deck of Giant 2.

The constant weather delays were compounded by another serious setback. On July 17, sampling of seawater over the wreck revealed high concentrations of acrylonitrile. At a distance of 100 m, the concentrations were no higher than 2 ppm, but values of up to 35 ppm were found directly above the Anna Broere. In addition, atmospheric concentrations of up to 15 ppm were recorded. This prevented the placing of cables and the salvage coordinator was forced to withdraw his units for a few days. On return to site, a salvage team began the difficult task of discharging a cargo tank underwater. But bad weather continued to hamper operations on site and the salvage was not completed for another month.

Other chemical carrier casualties handled by Smit have also presented extreme difficulties. The Cason, for example, grounded at a very exposed position off the Spanish coast. She suffered repeated explosions. Containers and drummed cargo were removed from the wreck during a lengthy and dangerous operation.

In 1986, Smit Tak recovered the vessel Olaf and its cargo of flyash. The vessel went down off IJmuiden; its successful recovery avoided any serious contamination of the sea. The previous year, Smit Tak played a leading role in the recovery of uranium hexafluoride from the Mont Louis and the removal of 2000 tonnes of dangerous goods from the Panamanian vessel Ariadne. The latter grounded off the Somali port of Mogadiscio, while carrying a cargo which included tetraethyl lead.

To conclude, I believe that the world (and western Europe in particular) has been fortunate in avoiding a very serious chemical pollution incident. No-one can say how long this luck will hold. When the day comes, as it surely will, salvors will be ready to play their part in averting catastrophe. But this assumes their survival in the industry's present crisis. What will you do if the major private contractors have disappeared?

I should add that part of the solution to the salvage industry's crisis lies in our own hands. Salvage companies should have the courage to refuse to be party to commercial agreements drafted following a "play-off" and containing a range of entirely one-sided conditions. If the customer is unwilling to pay a fair rate for the job, he must seek his solution elsewhere.

REPORTING AND COMMUNICATION/IMMEDIATE RESPONSE

REPORTING AND COMMUNICATION

Inspector John Østergaard
National Agency of Environmental
Protection
29 Strandgade
DK-14ol Copenhagen K
Denmark

ABSTRACT.This paper discuss a standardized format
for reporting oil and chemical pollution at sea
between Contracting Parties to a multilateral or
bilateral agreement. The reporting format - The
Polrep system - has been developed for use within the Bonn Agreement and has later been adopted
for use within the Helsinki Convention and the
Copenhagen Agreement. Furthermore the paper gives
a description on how the command and communication
problems during a joint operation have been solved
in the three Nordwest European Regional Agreements.
Finally the paper descripes the type of
exercises in use within the same Regional Agreements.

BACKGROUND. All countries with coastlines facing
sea areas with various activities such as shipping traffic and offshore industries have more
or less been forced to deal with situations where
oil or chemical spills occur. These spills can be
the results of accidents at sea such as groundings
or collisions, accidents on offshore platforms or
illegal discharges. Much effort has been made to
prevent marine pollution by implementing new legislation which improve techniques construction,
operational standards, aids to navigation etc.,
but no matter how successful this new legialation
has been, pollution at sea will continue to occur.
Most countries have also been aware of the threat
from oil or chemical pollution to the marine environment and have for these reasons established
a special organization which can respond to a
detected or reported spill at sea and ashore. The
magnitude of this national response organization

P. Bockholts and I. Heidebrink (eds.), Chemical Spills and Emergency Management at Sea, 173—191.
© *1988 by Kluwer Academic Publishers.*

is normally based on a kind of risque analysis.
Due to the difference in calculation methods and
the economic and political possibilities the response capability is very different in countries
around the world. As early as in the beginning og
the seventies it became obvious to many countries
that no single nation is able to establish a national response capability that can handle the
biggest pollution judged to be possible in the
area. Cooperation between neighbouring countries
was the only and most natural sollution to that
problem. This understanding has been the main
reason for countries in Northern Europe around
the North Sea and the Baltic Sea when establishing
regional agreements on cooperation in oil spill
prevention and combating. In the later years some
of the regional agreements have been further developed also to cover the combating of chemical
spills. Within the framework of these agreements -
the Bonn Agreement, the Copenhagen Agreement and
the Helsinki Convention - the contracting parties
have found it necessary to create and establish
special systems for mutual information and communication about oil and chemical spills, request for
assistance, command structure in joint operations
and the performance of exercises. It seems also to
be possible to reach consencus about very similar
systems in these topics between the different
agreements and that is of great value due to the
fact that several countries are contracting parties in two of these agrements and Sweden and Denmark in all of them. For that reason it is also
necessary to develop and describe these systems in
a general form which makes it possible to transfer
experiences drawn from the use of the systems within one agreement to another. Experiences gained
during the last years have clearly indicated that
the existence of such systems are absolutely necessary for a successful cooperation between contracting parties creating and performing a joint
operation in a real case situation. Guidelines for
pollution reporting, requesting and providing assistance, command structure in joint operations and
exercise performance are included in the operational manual for each agreement. In these manuals
you also find information regarding national response organization, combating resources etc. As a
supplement to the Bonn Agreement and the Helsinki
Convention the Federal Republic of Germany and
Denmark have agreed on a joint Maritime Contingency
Plan called the DENGER Plan. Most of the systems

in use in the multilateral agrements are also applied to in the DENGER Plan and a similar plan
which at the moment is under negotiation between
German Democratic Republic and Denmark.

REPORTING SYSTEMS. Within the frameworks of all
regional agreements in the North European Region
all contracting parties have found it necessary
to establish a uniform reporting system for reporting pollution at sea between the responsible
authorities and further, by the reporting format
is should also be possible to give additional information on the development of the spill situation and finally it should be possible to raise
a formal request for assistance. A proposal for
such a Pollution Reporting System was develop
within the Copenhagen Agreement in the late seventies. But in the early eighties it was recognized
that a uniform Pollution Reporting System covering all the Regional Agreements should be developed. Since the 1st January, 1985, the Pollution
Reporting System (POLREP) has been in force within all the agreements in the North European Region.

It should be understood that the POLREP is for use
between the responsible authorities of the Contracting Parties to exchange quick and adequate information when pollution at sea has occurred or
when such a threat is present.

The POLREP is divided into 3 parts:
(a) Part 1 or POLWARN POLlution WARning
(figures 1-5) gives first information or warning
of the pollution or the threat.
(b) Part II or POLINF POLlution INFormation
(figures 4o-6o) gives a detailed supplementary report as well as situation reports.
(c) Part III or POLFAC POLlution FACilities
(figures 8o-99) deals with requests for counter
pollution facilities or resources as well as matters of operational character.

Summarized list of figure explanation on POLREP
is given in annex 1.

The division into three parts is only for identification purposes. For this reason consecutive figures are not used. This enables the recipient to
recognize merely by looking at the figures whether he is dealing with Part 1 (1-5), Part II

(40-60) or Part III (80-99).This method of division shall in no way exclude the use of all figures in a full report or the separate use of single
figures from each part or the use of single figures from different parts mixed in one report.

Part I is used as a quick and easily understood
first notice or warning and Part II is the logical
consequence of Part I in which the contracting party concerned can inform the other relevant parties
of its assessment of the nature and extent of the
situation. Part III is used exclusively for the request of assistance and related matters.

The working language when using POLREP is normally
English and time information is given in UTC.
The identification group contains information as
to what agreement is concerned by using the code
words BONN, NORDIC, BALTIC, DENGER. The serial number group gives information about the country
transmitting the POLREP by using the international
abbreviation for that country followed by an adequate number, enabling the receiving country to
check whether all the reports on a particular pollution have been received. By answering a POLREP
the serial number from the transmitting country is
used as reference in the answer.When the POLREP
is used for exercise purpose the text always will
be introduced with the word EXCERCISE and finished
with this word repeated three times and the following reports in the same excercise will start and
end with EXCERCISE as well.

The POLREP is always and only intended to be used
between responsible authorities in each country.
Many of these authorities have to deal with other
national missions than oil combating and due to
experiences drawn from excercises it has been found
that POLWARN normally must be given traffic priority URGENT.

A POLREP message is normally transmitted by telex
communication, but of course other communication
systems can be used. Experiences have shown that
the use of telephone very often has caused a lot
of misunderstandings, for this reason a communication system where the message is transmitted in
writing is recommended. In the later years the use
of telefax and telecopies has shown great advantages as a quick and reliable method of transmission.
Lessons learned from exercises and real operation

have demonstrated that the acknowledging of messages is only necessary in the initial stage. Further it is recommended to establish a good log-
keeping on/ in- and outgoing messages. Finally it
should be mentioned that the POLREP system has
worked well during exercises as well as during
real operations.

Examples on differend POLREP messages are given as
annex 2.

COMMAND STRUCTURE AND COMMUNICATION ORGANIZATION.

A demand for assistance often will result in a
joint operation at sea. Judging from experiences
learned from real case operations the best way to
render assistance in joint operations is to set up
national strike teams consisting of personnel,
ships, equipments etc. under national command. By
doing so, difficulties which could arise due to
differences in equipment, training, operational
procedure, languages etc. can be avoided or limited. Nevertheless, the execution of a joint operation also requires a clear and simplified command
structure agreed upon beforehand. Another very
essential thing in an operation is to have a reliable radio communication between the different
command levels, strike teams and single units.
This radio communication network must also meet
the requirements of having separated frequencies
which could be operated without disturbing each
other. The contracting parties have been well aware of these practical and operational problems and
have tried to reduce the problems by adopting or
recommending a certain Command and Communication
System. The system is based on the following main
points.

- two main co-ordination and command levels - Operational Control ashore and Tactical Command on
scene.
- liason officers from participating countries
should be integrated in the staff of the Operational Control.
- the over-all Tactical Command is executed by a
designated Supreme On-Scene-Commander (SOSC).
- national strike teams should operate under a National On-Scene-Commander (NOSC) who operates under the SOSC.
- the Operational Control should be executed by the
lead country which means the country within whose

response zone the operation takes place.
- an officer from the lead country should act as
SOSC.
- change of Operational Control and Tactical Command might - when practical and agreed - take place if the operation moves from one response zone
to another.
- communication between Operational Control - SOSC,
SOSC - NOSC and NOSC - team units should, if possible, be performed on separate radio-frequencies.
- if possible, wireless teleprinter, telex or telefax should be used between Operational Control -
SOSC.
- international VHF-channels lo, 67 and 73 should
be used between SOSC and NOSC and between NOSC and
team units special domestic frequencies, if possible.
- the working language should be english or the
most appropriate one.

Based on the objectives, a scheme for command
structure and radio communication has been worked
out and is presented in annex 3.

Due to the broad aspects of radio communication
problems the contracting parties have informed the
national telecommunication authority about the
scheme which seems to have been accepted in countries concerned.

For practical reasons only few joint operation
exercises have been performed. In most of these
exercises only the tactical command structure has
been tested but the main impression is however
that it has worked well.

EXERCISE PROGRAM. As mentioned earlier one of the
main points in all the agreements is to cooperate
in real spill situations and - if needed render
assistance to contracting parties threatened by
oil spills or engaged in an oil combating operation. This might lead to a joint operation where
strike teams from different countries participate.
All lessons learned from real cases point to the
fact that if such a joint operation shall be successful, exercises must be executed regarding information, staff functions and equipment. For that
reason in all agreements exercise programmes have
been adopted. The programmes concerning the Helsinki Convention and the Copenhagen Agreement are
extensive and very similar while the Bonn Agreement

programme is more restricted.

In a general form the programmes consist of the
following types of joint exercises.
- Synthetic
- Alarm
- Equipment
- Operational.

The different types of exercises can briefly be
presented and described as follows:

Synthetic Exercise. This exercise type is a "paper exercise", the aim of which is to create a
basis for discussion of matters relating to organization, communication, logistics etc. in joint
combating actions involving two or more countries.

The outline of the exercise is preplanned in such
a way that the players will be presented with a
pollution incident giving such facts of the incident that most probably would be at hand in the
initial phase.

The situation in the initial phase will be followed up by presentations of the situations as it
has developed at certain stages.

The national follow-up actions are then presented
and discussed before a presentation of the next
following chosen stages of the incident situation
will take place.

This exercise type is intended for the use at meetings when all Contracting Parties are represented.
The exercise planning and conduct could be the responsibility of the host nation, an intersessional
working group or the Secretariat in co-operation
with appointed contact persons from all or some
of the Contracting Parties.

To the full extent these types of exercises have
only been carried out within the Copenhagen Agreement. The experiences show that these exercises
are most helpful as a background for preparing the
other types of exercise and for understanding national differences in organization, command structures and tactical attitudes.

Alarm Exercise. The aim of this exercise type is
to test the agreed procedures and lines of commu-

nication for reporting and requesting and providing assistance, and to get a picture of the
current response readiness of the Contracting
Parties to call for assistance.

The exercise further aims at familiarizing the
personnel with the use and national handling of
the adopted POLREP reporting forms.

It is not the intention with this exercise that
combatting equipment and its handling personnel
should be activated. The exercise could be initiated by an appointed Contracting Party or by
the Secretariat.

The exercise can be carried out in turn between
two or more of the Contracting Parties and the
arrangement and initiation of the exercise are
undertaken by representatives of the countries
involved.

The exercise is executed without notice within
a specified period of time, e.g. 3 months, and
the participants in the exercise are not noti- -
fied of the incident before the execution of the
exercise.

When receiveing an Exercise POLFAC, the participating countries should make an realistic evaluation of types and amount of equiment as well as
personnel at their disposal for rendering assistance called for as well as the time for its arrival at the scene of the accident.

After termination of each exercise, the Contracting Parties shall submit a report to the Secretariat for distribution to all Contracting Parties
with the aim to have the national reports presented
and discussed at the first following meeting of the
Contracting Parties.

A yearly exercise schedule for the exercises is
adopted within each agreement and by co-operation
between the Secretariat the schedules are coordinated in order to avoid time collisons.

Within all agreements several exercises have been
performed and many problems have been identified
resulting in improvement both in the reporting
system itself and the national organization.
Starting with simple exercises, the exercises have

been more and more advanced including accelerated time condesing the simulated time period,
cost calculation, solutions for disposing recovered oil, contacts with national authorities for
border and customs clearance etc. The main impression is that this type of exercise has been most
valuable for the operational co-operation and must
be conducted regularly in the future in order to
maintain a high standard of co-operation.

Equipment

The purpose of this exercise is to test the co-
operation between combating units of the Contracting Parties with respect to both communication
and equipment. It is intended to involve staff
functions only to a very limited extent.

The exercise is carried out between two or more
Contracting Parties with bordering Response Regions.

Notice at to time and event is to be given well in
advance of the exercise and Contracting Parties not
taking part in the exercise and the Secretariat shall
be invited to send observers to the exercise.

Reports on the exercise should be sent to the other
Contracting Parties in order to have the reports
presented and discussed at the following meeting
of the Contracting Parties.

The exercise is arranged end executed after direct
consultation between the Contracting Parties involved in accordance with a yearly exercise schedule
worked out.

One of these parties acts as lead country and is
responsible for the preparing of the exercise. To
avoid time collisions between the agreements the
Secretariat has been given the same responsibility
as regarding the alarm exercises.

Due to the cost aspects and for geographical reasons this type of exercise normally is carried out
on a bilateral or trilateral basis. Within the Copenhagen Agreement exercises have been performed between two or three countries. Within the Bonn Agreement minor exercises have been executed on a bilateral basis while in the Baltic Convention until
now no such exercises have been initiated.

Within the Copenhagen Agreement the first exercises created many equipment problems and command
and communication confusions. During the years from
1978 - 1986 great improvements have been achieved
and within the last years the equipment exercises
have slowly changed towards the operational exercise.

Operationel Exercise

The aim of this exercise type is partly to test
the alarm procedure, the response capability and
the response time of the Contracting Parties, partly to test and train the staff functions and the
co-operation between combating units of the Contracting Parties.

This exercise - which is a combination of alarm
and equipment exercises - is the most advanced and
sophisticated one in the exercise programme. It creates special problems for the lead country to build
up for a condensed time period a relistic scenario
which can be accepted and judged logical and consequent for all participants. Special attention has
to be paid to the Operational Control staff function if the personnel in this function shall get the
right feeling in their work.

Since 1984 this type of exercises has been executed once a year within the Copenhagen Agreement.

At the 1986 meeting of the Contracting Parties
to the Copenhagen Agreement it was decided to invi-
te participants from the Helsinki Convention and
the Bonn Agreement.

In the 1987 exercise one oilcombatting Vessel from
USSR participated as an observer and in this years
exercise oilcombatting vessels from several countries within the Helsinki Convention participated
in the exercise which took place outside the swedish port of Stromstea close to the Norwegian border.

CONCLUSION AND THE FUTURE. Performed exercises
within all agreements clearly show the necessity
of having mutual   agreed systems for reporting,
communication and command structure in joint operations. It also clearly advises the Contracting
Parties to go further in testing the systems and
their own capability to act within them.

I am very convinced that the only way to achieve
the objective of the various agreements concerning cooperation in joint operations in the future is to continue and to further develop the

exercise programmes. Well trained and cooperated personel is the best and only garantee for quick mutual response to big oil or chemical pollution incidents which no single nation alone can over-come. Prepared for that, we stand much stronger in our efforts to protect man property and environ-ment against damages from such a pollution, which, unfortunately will come sooner or later.

Summarized list on POLREP

Address   from ...   to.....

Date Time Group

Identification

Serial Number
----

1. Date and time

2. Position

PART I   3. Incident

(POLYWARN)   4. Outflow

5. Acknowledge
----

40. Date and time

41. Position

42. Characteristics of pollution

43. Source and cause of pollution

44. Wind direction and speed

45. Current or tide

46. Sea state and visibility

PART II   47. Drift of pollution

(POLINF)   48. Forecast

49. Identity of observer and ship on scene

50. Action taken

51. Photographs or samples

52. Names of other states informed

53-59. Spare

60. Acknowledge

PART III

(POLFAC)

80. Date and time

81. Request for assistance

82. Cost

83. Pre-arrangements for the delivery

84. Assistance to where and how

85. Other states requested

86. Change of command

87. Exchange of information

88-98. Spare

99. Acknowledge

POLREP
Example No. 1
(full report, (Part I, II and III)

Explanation	Message text
Adress	From: DK
	TO: FRG and NL
Date time group	181100z june
Idenfication	POLREP BONN AGREEMENT
Serial number	DK 1/2 (DK 1/1 for FRG)

1. Date and time	1. 181000z
2. Position	2. 55º 30' N - 07º 00 E
3. Incident	3. Tanker collision
4. Outflow	4. Crude oil, estimated 3,000 tonnes

----

41. Position and/or extent of pollution on/above/in the sea.

41. The oil is forming a slick 0.5 nautical miles to the South-East. Width up to 0.3 nautical miles.

42. Characteristics of pollution

42. Venezuela crude. Viscosity 3,780 Cs at 37.8ºC. Rather viscous.

43. Source and cause of pollution.

43. Danish tanker ESSO BALTICA of Copenhagen, 22,000 GRT, call signXXXX, in collision with Norwegian bulk carrier AGNEDAL of Stavanger, 30,000 GRT, call sign yyy.
Two tanks damaged in ESSO BALTICA. No damage in AGNE-DAL.

44. Wind direction and speed.	44. 270 - 10 m/sec.
45. Current direction and speed and/or tide.	45. 180 - 0.3 knots.
46. Sea state and visibility.	46. Wave hight 2 m. 10 nautical miles.
47. Drift of pollution.	47. 135 - 0.4 knots.
48. Forecast of likely effect of pollution and zones affected.	48. Could reach the island of Sylt, FRG or furter South, NL on the 23rd of this month.
49. Identity of observer/ reporter.	49. AGNEDAL, figure 43 refers.
50. Action taken	50. 2 Danish strike-teams with high mechanical capacity on route to the area.
51. Photographs or samples.	51. Oil samples have been taken. Telex 64 471 SOK DK.
52. Names of other states and organisations informed.	52. FRG.
53. Spare.	53. DENGER PLAN is activated.

----

81. Request for assistance.	81. FRG is requested for 2 strike teams with high mechanical pick-up capacity.

188

82. Cost.	82. FRG is requested for an approx. cost per day of assistance rendered.
83. Pre-arrangements for the delivery of assistance.	83. FRG units will be allowed to enter Danish territorial waters for combatting purposes or Danish harbours for logistics informing SOSC beforehand.
84. To where assistance should be rendered and how.	84. Rendezvous 57° 30' N-07° 00' E. Report on VHF channels 16 and 67. SOSC, Lieutenant Commander Hansen in GUNNAR SEIDENFADEN, call sign OWAJ.
99. ACKNOWLEDGE.	99. ACKNOWLEDGE.

```
            POLREP
        Example No. 2
   Abbreviated report (single figures from part III)
```

Explanation   Message text

Adress   From: FRG

                         To: DK

Date Time Group   182230z june
Identification   POLREP BONN AGREEMENT
Serial number   Your DK 1/2 refers
BT   BT
80. Date and time 80.182020z
82. Cost   82.Total cost per day will be
                         approx.....
84. To where as-   84.ETA FRG units at POLREP BONN
     sistance should   AGREEMENT DK 1/2 will be 182100z.
     be rendered and
     how.

POLREP
Example No. 3
Exercise report

Explanation	Message text
Address	From: DK
	TO: N
Date Time Group	21 0940z june
	URGENT
	EXERCISE
Idenfication	POLREP BONN AGREEMENT
Serial number	DK 1/1
BT	BT
1. Date and time	1. 210830
2. Position	2. 57º 50 ' N – 10º 00' E
3. Incident	3. Tanker collision
4. Outflow	4. Not yet
5. Acknowledge	5. Acknowledge
	EXERCISE EXERCISE EXERCISE

VISIONAL SCHEME OF COMMAND RADIO COMMUNICA-
TIONS IN JOINT COMBATTING OPERATIONS

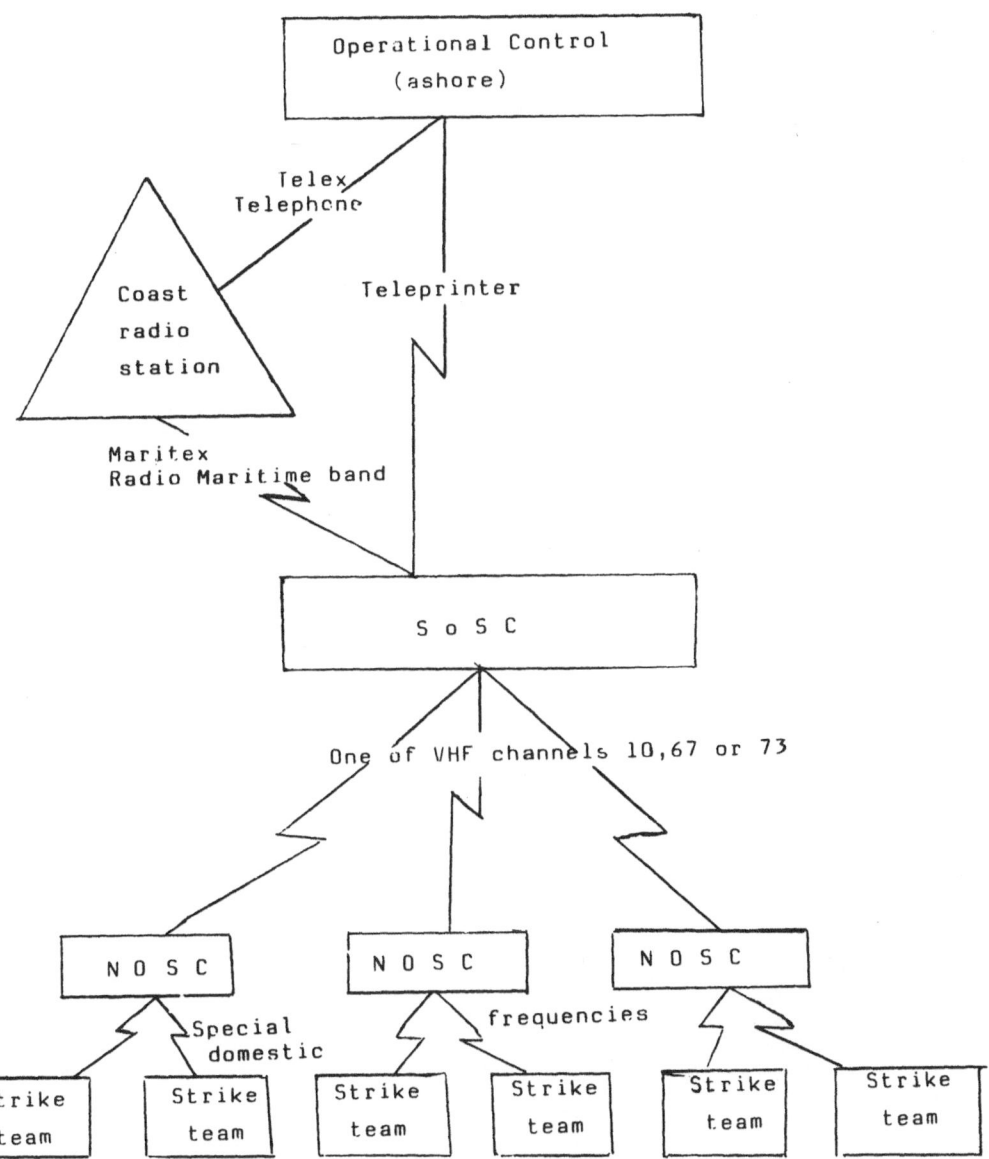

IMMEDIATE RESPONSE IN CHEMICAL ACCIDENTS

Arne Kjær Sørensen
Water Quality Institute
11, Agern Allé
DK-2970 Hoersholm
Denmark

ABSTRACT

The paper describes elements of an emergency response system with particular emphasis on a group of experts, whose members are suggested to
be chemists. Education and training of this group and of other members
of the system are described. The value of realistic exercises is mentioned. Review is made of the available information systems on the basis
of which decisions are made on how to combat the accident. A thorough
description of the dialogue between the officer in charge and the
advisor, involving informations necessary for the chemist to make the
right decisions and informations about chemical characteristics,
suggestions given by the advisor on how to fight the accident, etc.,
constitute the essential part of the paper. Hazards to humans on land
and actions to be taken in the case of formation of gas clouds are
presented.

1.   INTRODUCTION

Transportation, storage and processing of chemical compounds and
products are a very important element in the infra-structure of modern
society. It it to be expected that future transportation of dangerous
goods on land and at sea will increase. During the last few years
several high-scale chemical accidents have occurred worldwide, and for
all countries the basic philosophy as far as chemical accidents are
concerned has to be:

"It is not a question of whether a catastrophe will happen, but when it
will happen."

In order to minimize the consequences of such events, it is necessary to
form emergency response systems and to work out regulations and
guidelines for transportation of dangerous goods. Many countries have
formed such response systems and examples of existing regulations and
guidelines concerning transportation of dangerous goods at sea are:

P. Bockholts and I. Heidebrink (eds.), Chemical Spills and Emergency Management at Sea, 193–198.
© *1988 by Kluwer Academic Publishers.*

- Safety of Life at Sea (SOLAS)
- International Maritime Dangerous Goods Code (IMDS)
- Emergency Procedures for Ships carrying Dangerous Goods (EMS)
- Medical First Aid Guide for Use in Accidents involving Dangerous Goods (MFAG)

In order to be prepared to combat chemical accidents which will be seen in future, some initial steps have been taken, but we are still far from having a satisfactory emergency response system. In the following, some important elements of such a system are described and suggestions are given regarding how we should immediately respond in the case of chemical accidents at sea.

2.   EMERGENCY RESPONSE SYSTEM

An authority such as a National Agency of Environmental Pollution (NAEP) should be in charge of a preparedness against chemical accidents at sea. NAEP has to initiate the formation of an Emergency Response System (ERS) and together with other authorities and organizations taking part in the preparedness, NAEP should ensure that this ERS is currently improved as far as education, training, equipment and information systems are con- cerned.

In general an ERS is based on the existence of:

- Emergency forces involving personnel from the Navy, Civil Defence, Fire Brigades, NAEP, etc. with experience in combating chemical accidents

- An advisory group whose members are specially assigned chemists

- A network of hospitals with facilities for taking care of victims heavily polluted with toxic material.

Chemical accidents at sea differ from accidents on land in many ways. Most important is the fact that the crew members in some cases have to fight the accident themselves. Fire-extinguishing equipment, personnel protection clothing, antidotes, first aid equipment, EMS and MFAG-plans, etc. have to be carried by each ship transporting dangerous goods.

A basic course in chemistry and toxicology for the officer in charge and training of the crew members in combating fire, spills on deck, etc. are recommended. Such courses should be similar to courses arranged for civil defence and fire brigade personnel. The most important elements have to be demonstrations of chemical reactivity, e.g. reactions of acid chlorides with water (and foam with even very small amounts of water) and reactions between chemicals from which poisonous gases are formed. It should also be emphasized that fire is serious, but poisonous and corrosive chemicals are even more dangerous and to act accordingly. Most important, however, is the establishment of contact with the advisory group. In any case occurring, the chemist should be called for and at least placed in a stand-by position.

In Denmark, very few hospitals are prepared to take care of victims heavily polluted with toxic material, and the staff of medical doctors are not sufficiently educated and trained. It is obvious that medical doctors in sufficient number should be trained in operating polluted victims, dressed in protective suits at the hospital as well as in the field. Plans of preparedness should be worked out and decontamination facilities and other equipment ensuring a proper treatment of the victims as well as a safe environment for the staff should be installed.

The advisory group constitutes in my opinion the most important element in an emergency response system. The group is indispensable to the system, but should by no means replace any other element in the ERS, and the members must realize that they are not in charge of the operation. What they have to do is to give advice to the emergency forces, the officer in charge and the medical doctors. Combating chemical accidents can only be successful if these different groups cooperate very closely and respect each others' technical skills.

The members of such advisory groups must be chemists. Even such an educational background is not sufficient and it is suggested that technical high schools, universities, etc. introduce a special education in "accident chemistry".

The advisor should be extensively experienced in chemical reactivity and have a thorough knowledge of toxicology, first-aid, on-scene tactics and regulations and he should currently participate in exercises.

The group of experts should be called for on a 24-hour basis and the members should have the necessary access to data bases, such as OMMTADS, ECDIN etc. and to different types of chemical emergency cards, handbooks in chemistry, toxicology, regulations etc.

In my experience, data bases are of great value in order to predict the down-wind hazard of gas clouds and the spreading of spills at sea, and physical-chemical characteristics, toxicological and ecotoxicological values could also - in some cases - be achieved, but in general an unknown compound could not be identified with sufficient certainty by data bases, based on information such as density, colour, odour, chemical reactivity with water and other material, solubility in water, flammability, etc.

This fact is one of the main reasons why non-chemists do not have the qualifications necessary to predict the possible effects of a chemical accident. For the chemist this means once again that experience and thorough knowledge of chemical reactivity etc. are the only basis on which he can make the right decisions and give the best advice.

For the advisory group as well as for the other members of the emergency response system, exercises have to be carried out from time to time. Such exercises could be of the phone, paper and pencil type, but more realistic exercises should be arranged, say annually. An example may be a collision close to shore between a bulk carrier loaded with a flammable compound and a ship loaded with packed chemicals which are themselves corrosive and toxic, and which form poisonous gases by reaction with water and react violently with the flammable compound.

The arrangement of such exercises is the best way in which to point out what has to be improved and it is of importance to each member of the emergency response system because he is faced with all the problems, and he himself can see (or is afterwards told) that extinguishing a fire instead of helping victims polluted with a corrosive, or working bare-handed with victims polluted with a toxic and skin-penetrating compound are fatal.

As a supplement to training, education and exercises, it is recommended that a special test and training centre be constructed comprising laboratory and exercise facilities manned with, among others, some of the chemists taking part in the advisory group. At such a centre the necessary research and development in the field could be carried out.

3. HAZARD IDENTIFICATION AND IMMEDIATE RESPONSE

Combating a chemical accident could only be successful if observations regarding location are absolutely precise. The following information should be addressed to the group of experts as soon as possible:

- Type of compounds
- Quantity of compounds
- Precise description of the accident
- Exact statement of the position
- Personal injuries
- Weather conditions
- Characteristics of compounds and reaction products (colour, odour, formation of smoke/vapour/heat, solubility in water, reaction with water and other materials, viscosity, flammability)

Within say 15 seconds, the advisor should provide the most urgent advices to the officer in charge of the ship and within 15 minutes supply with thorough information.

The information given should comprise:

- Toxicity, caustic properties (inhalation, skin, eyes, ingestion)
- Symptoms (to be expected/to keep an eye on)
- First aid
- Personal protection
- Chemical reactivity (reaction with water, reaction between compounds involved)

- Fire hazard (flammability, explosion, oxidation and ignition of other materials)
- Formation of combustible/toxic /caustic compounds
- Evacuation of the ship
- Neutralization if possible (type of material, quantity, procedure, hazard of the neutralized mixture)
- Adsorption/absorption (materials, quantities)
- Fire extinguishing (water, water mist, carbon dioxide, dry chemicals, halon agents, sand)
- Chemical and physical properties of compounds and any reaction products (volatility, solubility in water, relative density of liquids and vapour [water = 1, air = 1]).
- Handling of waste (neutralized mixtures, water from fire extinguishing)
- Cleaning of personal protection clothing, equipments, tools, etc.
- Damage to the environment

For a well trained and educated crew, several of the above-mentioned informations will be obvious as soon as the chemicals and any reaction products are identified, and the physical and chemical characteristics are described.

However, it is very important to create a calm dialogue between the chemist and the officer in charge, and he will normally be very pleased and satisfied when the chemist verifies his viewpoints and suggestions.

The information to be given as described above is of course also a checklist for the expert.

Chemical accidents and spillages at sea do not alone affect the ship and the crew. Also humans on land could very easily be affected in the case of chemicals leaking from tanks or simply by evaporation from the water surface after a spillage. Gas clouds could be formed and reach the coastline within a few minutes (wind speed 4 m/s is equivalent to approximately 1 km in 4 minutes).

The consequences of this possibility have to be taken into consideration by the group of experts at a very early stage. Using mathematical models, the downwind hazard should be estimated, the emergency response system activated and any precautions must be taken in order to save lives and reduce damage to property.

Release of e.g. huge quantities of ammonia from a leaking tank close to the coastline of a densely populated area is a very frightening perspective. Immediate response to such an accident is the first and most important condition for combating the accident successfully.

4. CONCLUSION

The formation of emergency response systems is a matter of high priority for all countries. As far as transportation of chemicals at sea is concerned, accidents will occur without regard to borderlines, which means that neighbouring countries are dependent on each other. Cooperation is essential and a certain degree of uniformity in the structure of the national response systems is preferred. It is recommended that, in future, uniform guidelines on how successfully to combat chemical accidents at sea be worked out on an international basis, special attention being paid to the structure of an emergency response system.

A structure consisting of 3-4 modules covering information on accident parameters, evaluation of possible effects, identification of hazards and emergency response, respectively, will possibly be the outcome of such a study. By means of modern computer techniques, actions to take and decisions to make in each particular case could be maximized, taking all aspects into consideration. Even with such a system in operation, it is however of utmost importance to supply the officer in charge with informations needed by direct contact with the chemist within a few seconds.

Communication and evacuation

C. G. Ulf Bjurman
Deputy Assistant Under-Secretary
Ministry of Defence
103 33 STOCKHOLM
Sweden

ABSTRACT

It is quite obvious that the questions concerning communication and
evacuation in case of chemical accidents have not been given much
attention previously and that there is a need for research and development in this field. This is especially so when it is a transport
accident. The problems, that follow from the fact that, as a rule,
there are different responsible authorities for emergency response at
sea and on land and the communications and co-operation beween these,
have to be dealt with. Experience shows that even though chemical
accidents at sea are not frequent, they can be serious with far
reaching consequences. There is a more or less general view in
different countries that the public, which might be affected by an
accident, has a right to get information about the risks which it is
exposed to and the measures that should be taken in case of an accident. The forms for giving such information need to be developed. The
methods and means for giving the alarm and assessing the situation
must be improved. This is vital for transport accidents especially at
sea in order to be able to warn the population and to decide on what
action to take. Due to the time factor, evacuation is not considered
to be a primary option. It is proposed that the public should, when
the warning signal is given, seek protection in the nearest building
and wait for instructions on the radio. All the same, the emergency
services have to have plans for evacuation. In the paper, exercises
are underlined as the most efficient way of getting an adequate emergency response in case of chemical accidents.

INTRODUCTION

The title is meant to indicate that this paper is about the information to and handling of the affected population in connection with
chemical accidents at sea. It is quite clear that this is a field
that has not been given much attention previously. This paper aims at
indicating some of the problems and sollutions and forming a base for
discussions.

P. Bockholts and I. Heidebrink (eds.), Chemical Spills and Emergency Management at Sea. 199—205.
© 1988 by Kluwer Academic Publishers.

Even the emergency response to accidents at installations on land needs to be improved in this respect, which was one of the conclusions in the US Environment Protection Agency (EPA) report this May to the US congress on the findings from a review of emergency systems at facilities that produce, use or store extremely hazardous substances. Proposals for improvements were made by EPA. The problems do not seem to be solved satisfactorily in practice anywhere in Europe, even if the Seveso-directive has provisions about information to all persons liable to be affected by a major-accident. These are that off-site emergency plans should provide for alerting the population and managing the evacuation of persons from accident areas.

The problems are quite naturally much greater when the accident happens in connection with a transport on land or at sea and not at an installation. A transport accident can happen anywhere. It is therefore, in most cases, difficult to define beforehand the threatened area and the persons liable to be affected. Also the chemicals that may be involved in the accident will not, as a rule, be known. The capacity and ability of the emergency response to deal with situation will differ from one place to another. The situation is quite different when you are dealing with installations. There, most of the risks or threats are known and can be taken into consideration in the contingency planning. The situation can to a certain extent be quite simular in a port or certain other marine areas. Generally speaking it is, however, quite clear that the risks for and consequences of chemical accidents at sea are not known. The same applies to the risks for people in such cases. A better knowledge in this respect is needed to protect and save human life.

When an accident happens at sea, the chemical spill can threaten people both at sea in different vessels and on land. The threats to the environment, that involve the emergency response organisation directly, are those that ca affect for instance cattle and salmon breeding installations. There is therefore a need for effective communications to the land and sea organizations and between these. As these organizations generally have completely different main tasks, different traditional backgrounds, different training etc. you will find that there are substantial difficulties in getting an immediate and efficient response. Far-reaching measures have to be taken to obtain this.

There is a general feeling that quite few accident happen at sea and the risks therefore are small. A statistical analysis has been made of incidents with hazardous materials, which entailed more than five deaths and which are recorded in the TNO FACTS data bank. There were in the 24 OECD member countries over the time period 1974-1985 eight major accidents per year causing a total of 130 deaths per year. The total number of deaths during the period was 1 562. When a major accident happens, the average number of people injured is 30 in

addition the the number of people killed. 45 per cent of the major
accidents are linked with the transport sector, which causes over
half the total numer of deaths. In 200 major accidents over the
period 1980-1986 just over 700 people were killed and 5 700 injured.
Almost 300 000 persons were evacuated. It is therefore quite clear
that there is a need for efficient means of taking care of the affected population in case of chemical accident.

INFORMATION BEFOREHAND

Threatened people can only be expected to take the most appropriate
measures in case of an accident if they are well informed beforehand.
In many countries the affected population are considered to have a
right to know what risks that are associated with the installation
based on a risk analysis which allows an evaluation of the consequences of potential accidents. In the United States of America this is
part of Titel III in the Superfund legislation. There are similar
obligations in the so called Seveso-Directive for the Member States
and the manufacturers in the European Community. The information
beforehand about what to do in case of an accident is in many countries considered as part of the emergency preparedness obligations
for the manufacturers or the Authorities.

The question of which is the affected public or which are the threatened people that should get information is not solved in a simular
way in different countries. In most countries much remains to be done
in this field. The obligations often only apply to very large installations where dangerous substances are manufactured. But the
risks can be just as high or even higher where such substances are
stored or handled for instance in ports or terminals. IMO has taken
an initiative to establish co-operation with the OECD Expert Group on
Chemical Accidents concerning preventive measures and emergency response for chemical accidents in ports and terminals, which is an
important step. The OECD work is at present concentrated on large
installations but is expected to cover a wider scope in the future.
So far, information has mostly been given to the population in the
vicinity of an industrial installation. There is, however, a clear
tendency to expand the information activities to communities or even
geographical regions. This is necessary if as many as possible of the
people who really might be threatened in case of an accident are to
be reached.

The forms for giving information to people will, naturally, be different. The persons, who can be directly identified as liable to be
affected by an accident as they are living or working in the vicinity
of an installation, can get information through pamplets, personal
contact etc. The local radio or local papers can be used also to
cover a larger area. In Sweden we are trying different ways of spreading information for instance in connection with sport competitions,
through sport clubs and trade unions, visits to peoples homes and

distribution of video tapes. Exercises that are conducted with participation of the public or which are open to the public and massmedia can also play an import part. An exercise in Malmoe recently had about 50 000 spectators. Investigations, that have been made, show that it is difficult to reach the aims of the information activities and that these have to be improved and repeated regularily.

The responsibility for informing the public lies in many countries on the Authorities. This is the case in the Seveso-Directive. In Sweden this is an obligation for the creater of the risk. The Authorities supervise that it is done and can issue guidelines. In practise there will most probably be little difference. The contents of the information should be a general description of the activities at the installation, storage, port etc. and the risks which are envolved. It is important that the information about what to do in case of an accident is brief, distinct and easy for anybody to understand. The printed information should have such a form that it is natural to keep it easy at hand for use in case of an accident.

ALARM

An accident at sea involving chemicals will course problems concerning responsibility and communications, if the consequences are not limit to the sea area. The responsibility lies in most countries on different authorities, and land Authorities can have responsibility, besides what happens on land, also for ports, certain waterways and along the beaches. There will in many cases be a joint responsibility which calls for close co-operation between the Authorities.

The communications will in many cases be between the ship and the maritime Authorities, which will have to depend on the crew until they have been able to take up command or make investigations on site. But it is important to get the land Authorities into the picture as soon as possible as the time factor plays an important part in many chemical accidents. Communications between Authorities in different sectors are not always good, but the need for efficient communications should be taken into account in the contingency planning and exercises conducted to see to it that they work.

It is vital for the responsible Authorities to be able to assess the accident and its consequences as quickly as possible. There are examples of how difficult this can be, for instance in the Mont Louis accident. In many oil spills the amount of oil, which has been spilled, has been considerably underestimated. But it is not only important to get information about what chemical it is and the quantity, the responsible Authorities need to know how the chemical reacts and spreads in the air and in water. Knowledge is at present limited in this respect and the operational aids to make a correct assessment need to be improved. The methods for getting correct and

enough information to make the assessment and the collaboration between different responsible Authorities in making the assessment and
deciding on action need to be improved. Besides this and measures
which concern ships and their crew, the ability of the emergency
response personnel should be improved through better training and
exercises in co-ordination between different responsible authorities.
It is further necessary to introduce efficient means of communication
and to back up the personnel with computerized information systems in
which you can find, not only information about chemicals, experts and
experience from previous accidents, but also spreading models and
information about consequences for human health and the environment.
All this is necessary to make a proper assessment and give the
correct warning and instructions to the affected population in the
short time that is at hand.

These measures are being put into practise in Sweden. A co-ordinated
high level course for officers as been introduced and is now in its
second year. Regular co-ordinated exercises are conducted in all the
regions with personnel from different authorities. A computerized
information system is planned for introduction in 1990. Research and
development is carried out to get operational spreading modells and
means of getting information about the consequences for human life
and the environment.

WARNING

When the situation and the consequences have been assessed the
responsible Authorities have to decide on what action to take and how
the public is to be warned. It is quite clear that the Authorities
have to reach a mutual decision and give the same message to people
at sea and on land.

Different methods are used for warning the population. In Sweden the
sirens of the Civil Defence are available. Extra sirens have been
installed round certain installations. The telephone system has also
been used for giving signals. This possibility is, however, disappea-
ring as the new computerized exchange systems are beeing introduced.
The possibilities of giving indoor warnings are therefore successively getting worse. Certain hopes are put to the development of radio-
data systems, but it is uncertain if these systems can fill the gap
for many years to come. Another method for warning, especially if it
has been decided to evacuate the population, is to send properly
equiped rescue personnel or police officers to contact the affected
population.

In Sweden the local radio and to a certain extent the National radio
and television can be used in emergency situations. It has been decided that the warning signal means seek cover in the nearest house,
shut doors, windows and ventillation and listen to the radio. Responsible for giving the signal is the rescue commander on duty. There is

an obligation to report to the Authorities all spills that are not insignificant and that can not be dealt with immediately. Risky installations are allowed and should, in accordance with instructions from the local rescue chief and under his responsibility, give the warning signal directly. In these cases the radio station has messages which have been prepared beforehand and which can be sent immediately. The commander of the rescue operation can use the radio whenever needed for official messages, which are called Important Messages to the Public. An instruction has been issued for the use of the radio for such messages. The reporters are trained for their task and take part in exercises. Each time the sirens are tested, the radio sends information about what the signal means. A reporter joins the operational staff in an emergency. He is, besides seeing to it that the official messages are sent, also allowed to report in the radio as he finds fit. The radio is therefore used as an important instrument for protecting and saving people in an emergency.

EVACUATION

The time factor plays an important part in the decision making of the rescue commander on what action to take. The situation is quite different from for instance a nuclear accident where you, because of the possibilities that exist of containing the radioactive substances, will have time to evacuate the threatened population. In Sweden the main principle for chemical accidents is to seek cover as soon as possible indoors. This does not mean that evacuation will not be ordered, but the population will get its basic protection by going indoors until the situation has been assessed. A decision can then be made on the most appropriate action to take, and instructions can be given over the radio. This principle should be even more appropriate when the accident happens at sea as the organizational problems and difficulties with the communications might delay the decision of the commander.

The Swedish principles for evacuating people have been elaborated for the wartime Civil Defence and for the emergency response in case of nuclear accidents. The decision to evacuate is always taken by the rescue commander. The evacuationorders are executed principally by the police. In contaminated areas, however, the personnel of the rescue services will do the job as they alone are properly trained and equiped for such tasks. At sea the Coast Guard will be responsible for all evacuation.

For emergency response in case of nuclear accidents, extensive planning of evacuation has been carried out. Different principles have been developed. Elderly people and others who are difficult to evacuate are generally evacuated at an early stage. The same applies to children in schools and play-schools. As the people who will be evacuated often live in the vicinity of and perhaps work at the plant, it has been felt that the knowledge that the children are evacuated early will improve the possibilities of keeping people at the important job of running the plant in a difficult situation. The rescue services naturally have plans for evacuation also in case of chemical accidents. These also contain measures to take care of the evacuated.

One important function in connection with accidents involving evacuation is the social and medical care which might be needed. The medical care has a difficult task in connection with chemical accidents. Respirators etc. are limited in number even in greater city areas. In recent years much interest has been devote to the psychological side. This service is need not only by the injured but also by relatives to killed or seriously injured persons and by rescue personnel. Such service is most probably in many cases also needed for evacuated persons who are not injured.

RESEARCH AND DEVELOPMENT

There is a clear need to develop the means for making decisions in case of chemical accidents which has been mentioned before. It is also essential to get more efficient warning systems. In Sweden attemts are being made to improve the methods of communicating information to the public and making sure that it is really picked up. Investigations that have been made show that the knowledge among people of what to do in an emergency situation is in fact quite poor. Studies are also being carried out to see how the public reacts in emergency situations, i. e. if they evacuate immediately or if they follow the advice given by the authorities. The issues which have been dealt with in this paper will also be dealt with in the OECD Expert Group.

INFORMATION COLLECTION/
INFORMATION SYSTEMS

HAZARD ASSESSMENT/DECISION MAKING:
COLLECTION OF RELEVANT INFORMATION:

ROGER KANTIN

C.E.D.R.E.
Centre de Documentation, de Recherches et d'Expérimentations sur les
Pollutions Accidentelles des Eaux.
P.O. Box 72
29263 PLOUZANE
FRANCE

Abstract

        In the case of an accident involving the spillage of dangerous
substances at sea, the "collection of relevant information" is a fundamental step in bringing to a successful conclusion the reflections
which will lead to the final decision-making.

This paper explains what kind of information should be sought, and how
to do so. It gives a series of recommendations concerning the documentation currently available, computerised or not. It also proposes
documentary research procedures which should lead to the required
objectives: sound knowledge of product behaviour, the most accurate
assessment possible of the hazard, and recommendations concerning methods
and means of intervention.

The addresses and telephone numbers whereby the reader can obtain this
specialised documentation are also given.

Introduction

When an accident occurs involving a ship transporting dangerous substances, the rapidity and the efficiency with which the information-
collecting operations are led are key factors in the success of the
intervention operation. It is this information which will direct the
final decision-making.

P. Bockholts and I. Heidebrink (eds.), Chemical Spills and Emergency Management at Sea, 209—221.
© 1988 by Kluwer Academic Publishers.

Immediately the accident is known to have occurred, 2 levels of information must be taken into account:

- firstly enquiries made to the captain, the owner, the ship's manager, administrations, and port or community authorities in order to obtain information about the ship, the nature and volume of the cargo, etc.

- secondly, and once the nature of the chemicals in the cargo is known, investigation into:

. the properties of the substances concerned

. the danger stemming from these substances

. the emergency procedures to be set up.

The personnel collecting this information must be trained to interpret it rapidly and to convert it into operational terms.

It is essentially the second aspect which will be discussed here.

A recapitulation of these different stages is shown in the sequence below:

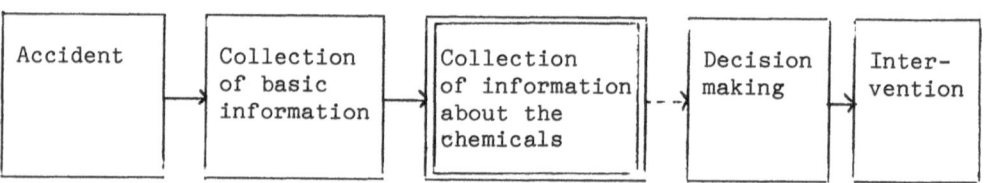

Where and how can information be found on the hazards presented by a dangerous cargo?

It is often said that information on physico-chemical characteristics, risk-profiles etc. are found in 'specialised literature'. This consists of reference books, guides and data-banks, but also of forecast models and   decision-making systems. A team of experts can also intervene at this level to provide additional information.

Faced with the abundance of existing sources of information, we would soon be overwhelmed if we did not preselect those sources of information which are truly complementary and operational.

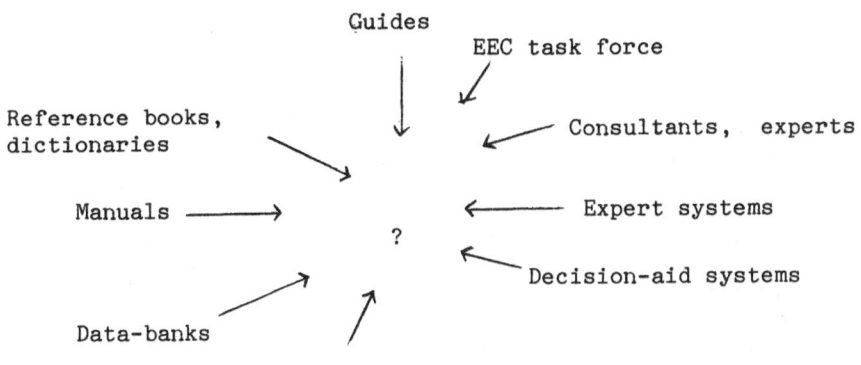

Collection of Information

In order to avoid time-wasting or the collection of non-priority information for decision-taking, it is advisable to have an investigation procedure which could be different in each country. The method used by the C.E.D.R.E. is the one presented here.

But before explaining this tactical approach, it is perhaps advisable to remind the reader what kind of information should be sought.

What information should be collected?

The information to be gathered should permit an assessment of the nature, duration and extent of the hazard. It concerns first and foremost human life (the crew, the rescuers, neighbouring populations), but also property and the environment.

This information concerns the physico-chemical behaviour of the substance, its reactivity, its toxicity, the nature of the products formed in case of fire, etc.

212

The box below summarises the main information to be sought, once the
concerned chemicals have been identified.

What must be ascertained about the product:

. If the product ignites or explodes, and in what conditions.
. If the product is toxic in air, with or without fire hazard. Do not
  forget insidious and long-term effects.
. If the product is toxic in water, and if it persists in the en-
  vironment.
. If the product evaporates, floats, sinks, and its distribution
  between the different compartments air/water/sediment/biota.
. If the product is reactive to: water, air, itself (polymerisation)
  and to other products
. If the product is radioactive.

But 'specialised literature' also provides another type of information
which it is also useful to gather before the decision-making phase. This
information does not concern the characteristics of the product and
the danger it represents, but the approach to the vessel and intervention
procedure.

The main fields covered in specialised literature are given in the box
below:

What must be ascertained for the intervention:

. Emergency measures in the case of intoxication by inhalation, contact
  or ingestion
. Emergency measures in the case of fire: means of extinguishing or
  reducing a fire
. Emergency measures in the case of leakage or spillage
. Approach to the vessel in difficulty (excluding meteo-oceanic
  conditions):

            - protective clothing, breathing apparatus

            - measuring and sampling apparatus
. Means to combat pollution
. Compatibility of products and material (cargo transfer)
. Stocking, transport, processing and elimination of waste.

Note also:

. Compendium of intervention capacity (means which can be used) in
  national, and if necessary foreign stocks.

```
┌─────────────────────────────────────────┐
│  Information research procedure           │
└─────────────────────────────────────────┘
```

Given the multiplicity of documentation available worldwide, computerised
or not, on the dangers of a product and the appropriate measures to be
taken, it is advisable to have a proven information-collecting procedure.
Of course, this is only one approach among others, for each organisation
has its own network of relations which should be maintained.

Be that as it may, the sources of information should not be too many;
they should be complete, complementary, easily accessible, and should
have already been tested in the case of an emergency.

Approach adopted:

- First and foremost, consult the works which give synthesis data:

. GESAMP - The Evaluation of the Hazards of harmful substances carried by
          ships, 1982 (currently being updated). Reports and Studies
          N° 17. Accessible through the IMO or other United Nations
          Offices (FAO, UNESCO, WMO, WHO, IAEA, UNEP).
          This volume gives, in codified form, synthesis information on
          the toxicity of a product for man, for the environment and for
          attractive sites.

. CHRIS - Chemical Hazard Response Information System, 1984. Developed
          by the American Coast Guard , available in all scientific
          bookshops or direct from the United States at the following
          address: Commandant (G-WER-2), U.S.Coast Guard, Washington
          DC 20593, USA. Phone (202) 426-9568. This document consists
          of about 1000 index cards containing information on the
          chemical designation, classification, characteristics, risks
          to health, reactivity, water pollution and the implementation
          of initial emergency measures. The document was created from
          several American information sources, notably the EPA
          (Environmental Protection Agency) which possesses its own card
          index file: OHM-TADS, and CHEMTREC

. IMDG code of the IMO (Publications Section, 4 Albert Embankment, London
          SE1 7SR, U.K. Phone: 01-735-7611. Telex 23588).
          Consult the General Index at the end of volume 5 to find the
          name of the chemical, which is cross-referenced with the page
          in the IMDG Code and the EMS (Emergency Schedule N°) in the
          "Emergency Procedure for Ships carrying Dangerous goods" or
          with the MFAG table N° (Medical First Aid Guide for Use in
          Accidents involving Dangerous goods).
          The IMDG code gives data on packing, stowing and segregation.

214

. ENVIRONMENT CANADA has published a volume in English and in French
        entitled "Manual for Spills of Hazardous materials", 1984
        (Environment Canada, Technical Services Branch, Environmental
        Protection Service, Ottawa, Ontario, Canada).
        It comprises 220 index cards, each with the indispensable
        basic data about the product, hazards, emergency measures and
        measures to protect the environment.

- Although these works give relevant information concerning initial
  emergency measures to be taken, it is perhaps necessary to analyse the
  initial emergency guides already available and thus confirm the
  different information provided. Among them, apart from the MFAG and
  the EMS from the IMO, should be mentioned the following:

. Hazardous materials Emergency Response Guidebook, 1980 (U.S.
  Department of Transportation, Research and Special Programs
  Administration, Materials Transportation Bureau, Washington DC 20590,
  U.S.A.).

. Dangerous goods guide to Initial Emergency Response from CANUTEC, 1986,
  in English and in French (Canadian Government Publishing Center, Supply
  and Services Canada, Ottawa K1A 059, Canada).

. Dangerous chemicals Emergency First Aid Guide and Dangerous chemicals
  Emergency spillage Guide (Wolters Samson Ltd., Gower House, 173
  Kingston Road, New Malden, Surrey KT3 355, U.K. Phone: 01-942-8966).

. Le Guide des sapeurs-pompiers Genevois (SPG), 1985 (Service d'Incendie
  et de Secours, Rue du Vieux Billard, 11; 1205 Geneva, Switzerland.
  Phone: 213144-310700).

- Additional information can be obtained from many other documents or
  data bases. There are 3 in particular which, in our opinion, seem to
  be important in guiding the decision to intervene. Whilst the sources
  previously mentioned consist of index cards and are rapidly consulted,
  the following documents contain twenty or thirty pages for each
  chemical:

. ENVIROTIPS (Enviroguides in French) gives useful information about
  fifty chemicals in order to help decision-making when intervention is
  necessary (obtainable from Publications Section, Conservation and
  Protection, Environment Canada, Ottawa K1A OE7, Ontario, Canada).

. HSDB - Hazardous Substance Data Bank, a very comprehensive American
  data base (22 pages on acrylonitrile).
  In addition to information about characteristics and hazards found in
  basic reference works, it also gives operational information concerning
  anti-pollution methods; protective clothing and emergency measures for
  different situations, comprehensive data on toxicology and ecotoxicology
  as well as other biochemical information , can also be found.
  Obtainable in France through the CNIC (Centre National de l'Information

Chimique, Tel.: (1) 45 51 37 40) or the National Library of Medicine, National Institute of Health, 8600 Rockville-Pike, Bethesda MA 20894, U.S.A. Phone: 301 496 11 31.

. IRPTC - International Register on Potentially Toxic Chemicals (listing of approx. 70 pages on acrylonitrile), with very comprehensive data from public and private sources. Among the most items: environmental transformation, chemical biocinetics, various studies of toxicity, sampling, processing the hazardous substance, management of waste, legal procedure.
This data bank, which belongs to UNEP, can be contacted at the following address: IRPTC/UNEP, Palais des Nations, 1211 Geneva 10, Switzerland. In an emergency, phone Madame Laurijssen: (41) 22 98 84 00, or telefax to Switzerland, N° (41) 22 98 39 45

- Specialised dictionaries or handbooks should also be consulted. These give succinct information, but cover a great number of products. There are about 10 works, of which 3 are listed here:

. Dangerous properties of Industrial Materials, 6th edition 1984, by Irving Sax, edited by Van Nostrand Reinhold Company. This is a condensed version in one volume of approx. 3000 pages (18000 entries). English editor's address: Molly Millars Lane, Workingham, Berkshire, RG11 2PY, U.K.

. Condensed Chemical Dictionary, 10th edition 1981, same editor as above. More succinct than the above-mentioned work.

. The Merck Index, 10th edition, 1983 published by Merck & Co. Inc., probably the most widely distributed dictionary in its field. 10 000 entries, 55000 synonyms.

- If pesticides are among the products being investigated, then more specialised works should be consulted. Two complementary works are:

. The Pesticide Manual, a world compendium, 8th edition 1987 edited by the British Crop Protection Council, 20 Bridport Road, Thornton Heath CR4 7QG, U.K.

. The Agrochemical Handbook, 2nd edition 1988, edited by the Royal Society of Chemistry, Information Services, Nottingham N97 2RD, U.K.

- Parallel to these investigations, there should also be someone to study the forecast models, which permit the assessment of hazards according to well-defined input data (circumstances of the accident, environmental data). Several models exist. When an accident occurs, they do not all need to be tested; it is advisable to use only one, the performance and limits of which have been proved.

216

- HACS - Hazard assessment computer system, designed by the American
  coastguard. This system enables the user to obtain details on the
  type and the extent of the hazard; it allows, for around 900 chemicals,
  an estimation of the hazards in terms of distance and times for which a
  toxic or flammable concentration of a given chemical may exist in water
  and air, an evaluation of the minimum safe distance between the spill
  site and people or combustible materials, and to predict whether the
  chemical could ignite and cause a fire or explosion.

  A version for the microprocessor (micro HACS) is being developed. In
  order to make the HACS system more operational in France, the C.E.D.R.E.
  has made a translation. 2nd Street S.W. Washington DC, 20593, U.S.A.
  Tel.: (202) 267 04 21. Telex: 89 24 27.

- SEABEL, whose module II (effect diagnosis module) contains a model for
  the drifting of slicks, is being prepared by T.N.O. for - and in
  collaboration with - the Dutch North Sea Directorate and the European
  Community. It contains four modules which give information on hazard
  identification, threats for man and the environment, and countermeasures to be undertaken. The hazard identification is based on the
  classification of chemicals according to their physico-chemical
  behaviour (and reactivity) and concerns substances coming from the nine
  IMO categories. Distribution and evolution of the chemical in water
  and/or in the atmosphere, and other relevant information concerning thei
  hazards in case of a spillage at sea are given. The system also
  provides indications for measures to be taken in case of an emergency
  response.
  PO Box 342 - 7300 Apeldoorn NL. Tel.:(055) 77 33 44. Telex: 36395.

- CAMEO - Computer-Aided Management of Emergency Operations, conceived by
  NOAA and the Seattle Fire Department. This prototype computer program
  and datebase allows the user to assist first responders in chemical
  accidents. Maps of the city of Seattle can be visualised including
  shorelines, city boundaries, street names, parks and major landmarks,
  and it is possible for the decision-maker to "zoom" onto a location of
  interest. Personnel protective measures, fire explosion hazards,
  human health threats, evacuation recommendations and spill cleanup
  procedures are also contained within the database.
  U.S. Department of Commerce, 3300 Whitehaven Street N.W., Washington DC.
  20235, U.S.A.

Several other models exist and concern more particularly the diffusion of
chemicals in the atmosphere (ATMOS, CHARM, DOURY models, EFFECTS, OCDISP,
OPIA, RISKCURVES, UNAMAP, etc.)
The existence of abaci should not be forgotten (see introduction manual in
Canadian ENVIROTIPS. These help to obtain results, sometimes faster than
with computerised models. Indeed, it is to that end that RIJKSWATERSTAAT
is preparing a "slide rule" for chemicals which dissolve, similar to the
oil spill slide rule for hydrocarbons, or for chemicals which evaporate
(gas clouds hazard estimation handbook), documents given out at the RISC
training course. (Rotterdam International Safety Centre sponsored by EEC)

- Useful additional information can be found in more specific fields:

- <u>firefighting</u>, in for example:

. Fire protection guide on hazardous materials, 9th edition
1987, edited by the National Fire Protection Association
(NFPA) of the United States.
Available from the NFPA, Batterymarch Park, Quincy, MA 02269,
U.S.A. or from INFONORME London Information, Index House,
Ascot, Berks SL5 7EU, U.K. Telex 849 426; Phone 0990 23377.

. Incendie et produits chimiques (chemicals fire), available
from the Société Alpine de Publications, 48 Cours de la
Libération, 38100 Grenoble, France.

- <u>reactivity</u> between different chemicals:
(most important when the damaged ship is carrying several
products)

. Handbook of reactive chemical hazards by Bretherick, 3rd
edition 1985, in the Butterworths, U.K. Collection.

. Réactions chimiques dangereuses, 1st edition 1987, by the
INRS (Institut National de Recherche et de Securité), 30
rue Olivier Noyer, 75680 Paris Cedex 14, France.

If the chemical is not listed in any of the aforementioned works or
banks, a <u>telephone enquiry</u> may be made (the answer will be given by
telefax) through one of the following organisations:

- Canadian Minister of Transport, Transport Canada, Canadian
Transport Emergency Centre (CANUTEC), Place de Ville, Ottawa,
K1A ON5, Ontario, Canada.

. by phone - emergency: (613) 996 6666

- information: (613) 992 4624

. by telex: 053-3130(DOT-OTT)

It possesses a data bank called CANCHEM. CANUTEC provides a permanent
24-hour service (in English or in French).

- Ministère des Transports Américains (CHEMTREC), Chemical Manufacturer's
Association (CMA), 2501 M Street N.W., Washington DC 20037. The main
role of CHEMTREC is to give immediate operational advice.

CHEMTREC may be contacted by telephone: (202) 887 1255

or by telex: 89617 (CMA WSH)

Generally, this information is sufficient to reach the required
objectives. It should nevertheless be pointed out that other data
banks may be consulted to supplement or replace other sources.

Thus, in France, in addition to the technical card index files of
the IRCHA and the INRS, we consult three French data bases, through
the MINITEL system :

. SECURICI, a data bank set up by the Ministry of the Interior
  (Sécurité Civile)

. FDE (Fondation de l'Eau) which contains amongst others information
  concerning methods of sampling and analysis.

. SECURLINE (Société Alpine de Publications)

Details of addresses etc of these three French data banks can be
obtained through the C.E.D.R.E

Other computerised sources exist : OHM-TADS, RTECS, CHEMLINE, CIS,
(Chemical Information System), CAS, EXIS, CIRUS, CHEMDATA, CHEMSTOR,
HAZFILE, DABAWAS, BIG, BVINFOTANK, etc

Among works, we have to add :

KIRK Othmer Encyclopedia, Gardener's chemical synomims and Trade
Names, Handbook of poisoning, Niosh Pocket Guide, Chemical hazards,
Hommel Handbuch Der Gefahrlichen Güter, Janssen Chemica catalog of
Fire chemicals, LAMY Transport Marchandises dangereuses, Dictionnaire
de la Chimie ...

Finally, in order to offset deficiencies in seeking information, it
would be useful to contact specialists from the universities and
from industry, ships captains, consultants, service companies, etc,
who have more detailed knowledge,

              - either on the behaviour and hazards of a specific chemical
              - or on the methods of intervention and anti-pollution to
                be adopted.

It is with this in mind that the Community Information System (C I S)
provides in section 2 Part F an "Inventory of Resources for Intervention" which gives a list of human and material resources at European
Community level. It should also be noted that the EEC can help in seeking information on the nature of a cargo, for example by making enquiries with the port authorities or at the ports where the cargo was
loaded .

The EEC can be contacted

        - during office hours by telephone : 32 2 235 5475
                          or by telefax   : 32 2 235 0144

        - outside office hours, by telex : 21877

Each message must be preceded by the expression "URGENT POLLUTION-ALERT SECTION"

Furthermore, by consulting the work entitled "Major chemical and petrochemical companies of Europe 1987-88" published by Graham + Trotman (Sterling House, 66 Wilton Road, London SW 1 V 1DE U.K) the names, addresses and activities of the main industries in 17 European countries can be obtained. At industrial level, certain countries even have a team of experts for each chemical.

REMARKS AND ADVICE CONCERNING RELATIONS

THE NEWS MEDIA

---

1 - Accept that an accident concerning dangerous cargo is justifiable news, and do not dispute the right of journalists to be on the scene of the accident.

2 - During your journey to the scene of the accident, think about the questions you could be asked. In your office, as soon as you learn that there has been an accident, expect a call from journalists and be prepared to answer their questions.

3 - Reply frankly, without hiding anything.

4 - Reply briefly and comprehensibly.

5 - Give the most important facts first.

6 - Give facts, and not opinions. Never answer a question which begins ; "And if ... ?" Deal only with facts.

7 - Always speak in the interest of the public.

8 - If you don't know an answer, don't invent it. If you promise to phone back later with the answer, do so.

9 - Don't adopt a defensive or antagonistic position.

10 - Don't try to explain the cause of the accident.

11 - Don't ask the journalist to show you his article before he sends it to the newspaper.

12 - Call your public relations department if you need aid.

13 - Don't try to conceal the danger from the product, or the gravity of the accident : the risks are that nobody will believe you, and that you will complicate the situation.

FROM THE OUTSET, IT IS YOU WHO MUST ESTABLISH THE TONE OF THE INTERVIEW WITH THE MEDIA. DO NOT BE CONTENT TO SIMPLY REPLY TO THEIR QUESTIONS.

In the case of accident, it is the duty of the carrier to provide the media with the necessary information. Often it is the manufacturer of the product involved in an accident who is expected to provide information about the dangers of that product.

The media will ask you five types of question : when, where, why, who, and what ?

Extract from the Conference on
Emergency Intervention
"Transporting dangerous goods"
TRANSPORTS CANADA

CONCLUSION

Information collecting is essential in order to prepare decision-making. The information must be gathered as rapidly as possible, but also with the utmost precision, by trying to define the hazards (their nature, duration and extent) for man and the environment.

The information must be gathered by trained personnel : data-bank enquiries, interpretation of results (vapor pressure may be given in mm of mercury, in kilopascals or in many other units), in a position to synthesise the data and communicate it to the decision-making authority.

It is, in fact, thanks to sound information-collecting, but also to the best transmission of information, that the best decision can be taken.

SEABEL: HAZARD IDENTIFICATION AND DECISION SUPPORT SYSTEM FOR CHEMICAL
SPILLS AT SEA.

Ingrid Heidebrink
Netherlands Organization for Applied Scientific Research TNO
Division of Technology for Society
Department of Industrial Safety
P.O. Box 342, 7300 AH Apeldoorn, The Netherlands

1. INTRODUCTION

The use of chemicals and oil products has grown progressively
Incidental spillages of hazardous substances may lead to very serious
consequences for people and the environment. Unfortunately incidents in
the past made the world aware of the risks involved. This awareness has
led to various steps. These steps deal either with the prevention of
incidents or with the reducement of the consequences, in case an
incident has happened. To be mentioned are for instance amendments of
the respective legal frame work; preventive regulations; hazard
assessments of production, storage and handling facilities and of
transport of hazardous materials; adjustments of construction
regulations; formation of emergency management and response teams and
drawing up emergency plans. The emergency management organization of
the North Sea Directorate of Rijkswaterstaat is commissioned to respond
in case of chemical incidents on the Dutch North Sea Shelf. This
organization liked to dispose of a tool for supporting them in taking
decisions where various elements had to be weighed! The European
Community was interested to support new development in this direction
as well. After a feasibility study, leading to a description of the
possible functions of such a tool, the development of the so called
Seabel system was started in September 1986. This paper describes
amongs others the actual functions of Seabel, the databases that are
incorporated and archive facilities.
The signification of the Seabel system for other areas than The Dutch
Shelf is presented as well.

2. APPLICATION OF SEABEL

Seabel is a tool for decision makers who have to come into action when
oil and/or chemicals are spilled at sea. With regard to chemicals; bulk
chemicals, liquefied gases and packaged goods are all incorporated.
With respect to oil, the unrefined as well as the refined products are
meant.

P. Bockholts and I. Heidebrink (eds.), Chemical Spills and Emergency Management at Sea, 223—232.

At sea, three kinds of pollution sources are foreseen: ships, pipelines and platforms. Seabel does not take into account the pollution of the sea as a consequence of polluted river water or for instance dumping of waste for which the authorities gave permission.
In addition to the registration of incidental chemical and oil pollution one may register, using Seabel, reports of persons in danger; of lost objects; of undefined spills; of oiled birds and coastal pollution.
Seabel runs on a IBM AT compatible Personal Computer and is also available as a Handbook. However, the flexibility of the handbook with regard to the calculations of the behaviour of chemicals is not as great as the computerized Seabel version.

3. FUNCTIONS OF SEABEL

The Seabel system as it is to be delivered to the Dutch North Sea Directorate and the European Community on September 1st 1988 has four mainfunctions:
- accident diagnosis (Module I);
- effect diagnosis (Module II);
- hazard identification (Module III);
- decision support on selection and evaluation of measures (Module IV).

3.1 Accident diagnosis

The first step in case an incident is reported to the responsible authorities, is to collect relevant data on the incident. The response to an incident will be more efficient if a good picture exists of the actual situation as well as of the possible consequences in the (near) future.
The data collection procedure is based on two principles that may be summarized in two words: completeness and rapidity.
The data collection should be as complete as possible. This means that all relevant questions should be posed. Therefore checklists should be available that cover all relevant aspects in such a way that an emergency management team can do something with the data. One can imagine however that checklists may differ depending on the kind of pollution, the source of the pollution, specific circumstances such as fire etc. Therefore the data collection starts with a number of keywords. The user of Seabel makes a selection of these keywords.
On the basis of the selected keywords a checklist is formed.
On the other hand the data collection procedure should be quick. Therefore the user has the disposal of the complete checklist immediately after the selection of the keywords, that are characteristic for the reported incident. Except for the data collection function, module I has the possibility to look in various databases such as a chemical database, an address and telephonebook and a database with data on ships.

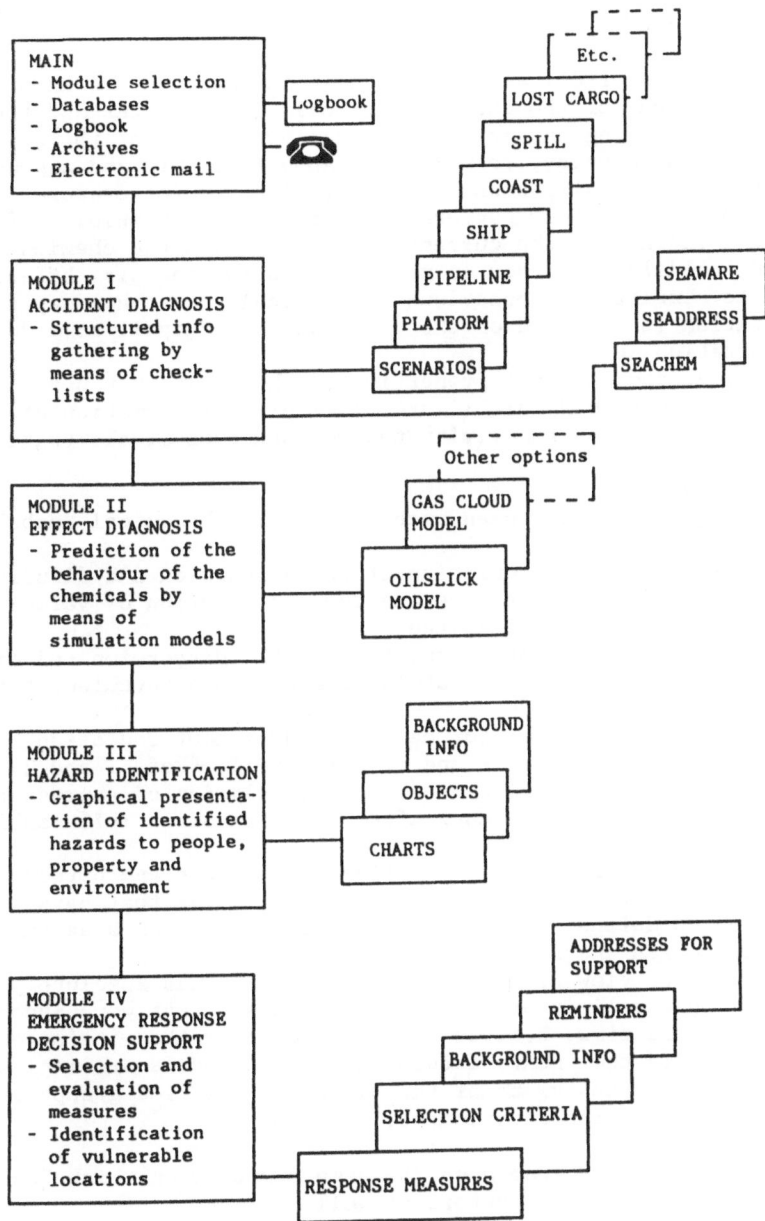

Figure 1: Structure of the Seabel system

3.2 Effect diagnosis Module II

In case of spills of oil and chemicals it is important to know the physical behaviour of the chemical (or oil) in the water and the atmosphere. Depending on certain physical properties of the spilled material such as density, solubility, vapour pressure and aggregation state, the chemical will sink, dissolve, float and/or evaporate. Combinations of behaviour patterns are of course possible. (Ref. 6) Under the influence of the current and/or the wind the chemical will be displaced and diluted either in the water and/or the air. Effect models are used to simulate the behaviour of chemicals in such a way that a representation is obtained of the position of the spill as a function of concentration and time.
Other physical and chemical properties of chemicals that have to be taken into account in Module II are flammability; formation of explosive mixtures and reactivity towards other chemicals (water included!).

The following models have been incorporated in the actual Seabel version:
1. Oil spill model for calculation of the dispersion and dilution of an oil slick as function of time and the calculation of various aging processes of the oil slick. (Ref. 3)
2. The Dispersion model for calculation of the dispersion and dilution of gases and vapours in the atmosphere due to a turbulent airflow. (Ref. 1 and 2)
3. The Heat Radiation model for calculation of heat radiation as a consequence of a pool fire and of a fireball. (Ref. 1 and 2)
4. The Vapour Cloud Explosion model for calculation of a pressure wave as a function of the distance after ignition of an inflammable gas or vapour. (Ref. 1 and 2)
5. The Search and Rescue model for floating objects and persons. This model comprises the total set of activities that have to be performed to define a locus and a size of a search area and to allocate search units to that area. (Ref. 4)
From the point of view of project-management, it was absolutely necessary to put a limit of the number of models that were incorporated in Seabel at this time. However it might be clear that it is very important that in the (near) future simulation models are implemented for describing the behaviour of the various classified groups of chemicals in the marine environment. (Ref. 5)

One may run simulation models on the same computer on which Seabel is installed and on remote computers as well.
For implementation of simulation models in the Seabel system without modifications in the computer program of the model itself, it is required that the model runs on an IBM AT compatible computer.
In this case an interface program is written to collect the necessary data. Data is collected from the completed checklists, or by presenting the relevant questions on the screen. The interface program transfers the data into the format known by the simulation model. After running

227

the simulation model, the interface program transfers the output in a
format, that can be used for graphical presentation in the hazard
identification module of Seabel.
In fact the same procedure is followed for simulation models, running
on external computers. The interface program prepares a file with input
data for the "external" simulation model and after running the model,
the file with ouput data, to be interpreted in Seabel, is transferred
to the Seabel computer.
For communication with other computers - this means transfer of files -
the RELAY package is implemented.

The reason for using simulation models on remote computers lies in the
fact that these models may be too large to run on a PC or they do not
run on PC-AT compatible machines.

3.3 Hazard identification - module III

This module may be used in two ways:
1. immediately after the report of the incident; to verify the position
   of the incident in relation to the surrounding area. In other words
   are there any vulnerable locations - such as platforms, municipalities etc. in the neighbourhood of the incident?
2. after running of the simulation models; the contours of the spilled
   chemical are shown on the nautical chart and all objects and
   activities at sea may be highlighted.

Examples of graphical representations are shown in figure 2 and 3.

Features of module III are:
- zooming in on nautical charts in order to present the contours of
  chemical spills and possible vulnerable areas in detail;
- locating objects on the nautical chart and querying for background
  information on the selected object.

Example: one may put the pointer on a particular platform and withdraw
the respective data on that platform.
The number of crew members, kind of equipment, communication
frequencies are then presented.

228

Figure 2: Graphical presentation of the displacement of an oil spi

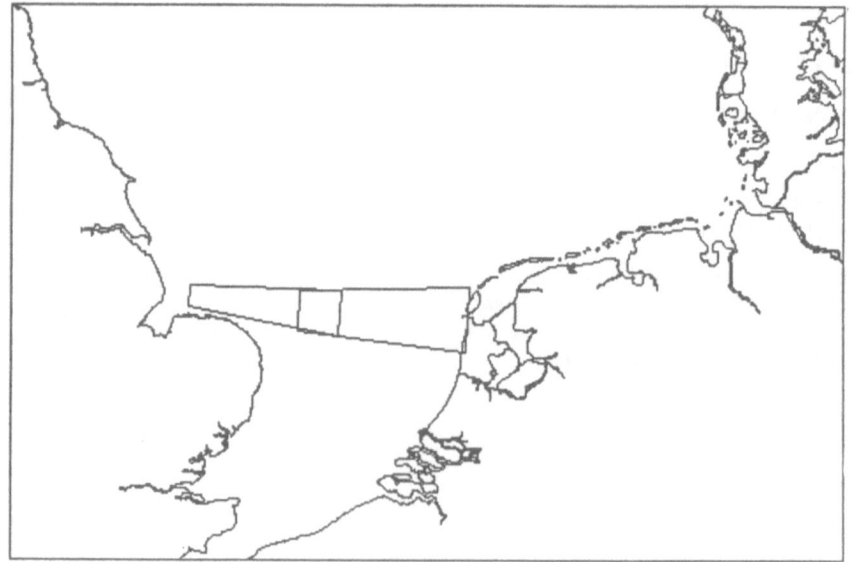

Figure 3: Graphical presentation of the search area in SEAFLOAT
           A typical DATUM-LINE search area

3.4 Decision support-module IV

The main objective of module IV is to present applicable response
measures to the decision maker who has to decide "what to do" in case
of a chemical (oil) incident at sea. In the actual Seabel version the
response measures are related to chemicals, packaged or unpackaged, and
oil.
With regard to the source of pollution most response measures refer to
ships.
Response measures are described in the following terms; category-
method-submethod.
Six categories of response measures are distinguished:
- repressive response actions
- protective response actions
- warning response actions
- response actions to be taken on board
- response actions to be taken on land
- miscellaneous response actions.
A category exists of a number of methods and a method exists of the
individual actions, also called submethods.

An example is shown in figure 4.

```
CATEGORY: PROTECTIVE RESPONSE ACTIONS

NR. METHOD - SUBMETHOD                        SCORE 1 2 3 4 5

  1. MEASUREMENTS - MOBILE LABORATORY           24  *
  2. MEASUREMENTS - GAS TUBES (TOXICITY)        24  *

                                              OPTIONS

                                              1. background
                                              2. N.I.S.
                                              3. score
                                              4. ranking
                                              5. notebook
Select an option (1- 6), e for exit, or h for  6. reminders
help
```

Figure 4   Submethods of one category are ranked on the basis of their
           score

Scores are awarded to each action with regard to
- human risc (with regard to people involved in an action);
- environmental risc;
- nuisance;
- costs (of the actions itself).
The presented score to the user of Seabel is a weighed score.

Selection parameters for response measures are:
- hazardous characteristics of chemicals and behaviour in the marine environment (classification schemes);
- meteorological conditions, such as windforce, mist;
- phenomena like fire and endothermal reaction;
- type of transport.

Other functions within the decision support module are:
- Retrieving background information on the action itself.
- Looking at the reminders; they do not belong to the selection parameters for response measures. However, they should be taken into account by the expert, who takes the final decision.
- Retrieving information on organizations that may supply expertise and equipment.

3.5 Archive facilities

Using the various modules of Seabel "data" is produced.
It is of great importance that the results of runs can be saved or retrieved.
This makes it possible to evaluate incidents. Lessons are learned with respect to the precision of calculations, the correctness of the score of response actions, omissions of whatever nature, etc.
A systematic analysis of all the facts of past incidents supports the expert in taking decisions in the future. The difference with the actual situation without a decision support system is the systematic approach and the fact that the knowledge of the expert can be incorporated in the system. The expert can train his new colleague by presenting him an 'old case'. He can show him the deviations in actual response actions if the circumstances change and make him aware of all the facts that might play a role in incidents.
'Archives' has 3 functions:
1. saving results of running the various functions of Seabel;
2. restoring previous results;
3. deleting results.

4. DATABASES IN SEABEL

A review of databases belonging to Seabel.

SEACHEM
The chemical database contains approximately 400 chemical substances with their relevant data and synonyms.

SEADDRESS
The address database with a national and an international section. The
international section is filled with the addresses of the participants
in the Bonn Agreement.

SEAWARE
This database contains sets of average data on ships, and is used for
estimation of amounts of cargo and spill.

CHARTS
This database contains the coordinates of coastlines. The data are
given in geographics and they are derived from admiralty charts.

QUERYPOINTS
This database contains background information on platforms,
municipalities and pipelines.

SEAVALUATION, METHODS DESCRIPTION
The response measure database contains approximately 200 response
measures. The database consists of 4 datasets:
- description and characteristics;
- selection criteria;
- background information (textblocks);
- reminders (textblocks).

NIS
National Information System; the database for addresses for
consultation, support, assistance, provision of equipment, etc.

5. EPILOGUE

TNO is convinced that the Seabel system is an excellent tool for the
experts who have to come into action when chemicals and/or oil are
spilled at sea.
Because of the fact that the logic is firm and the data can be adjusted
to the actual needs, Seabel may be build for any country in the world.
For this reason a national start-up programme is available describing
amongst other hard- and software requirements, data that should be made
available by the principal and training facilities, offered by TNO.
With respect to the training facilities. TNO has developed a course of
$2\frac{1}{2}$ days to make the future users familiar with the system and its wide
abilities.
At the end it should be emphasized that because of the modular
structure of Seabel the hazard identification and decision support
system may be extended with other modules such as a communication
module, input of traffic quidance systems, etc.

REFERENCES

1. Methods for the calculation of the physical effects to the escape
    of dangerous material (liquid and gases), (Yellow book).
    Directorale General of Labour;
    Ministry of Social Affairs, Voorburg, The Netherlands

2. EFFECTS Manual
    TNO, Division of Technology for Society, Apeldoorn, The Netherlands

3. Theory of the advanced Transspill model
    Ministry of Transport and Public Works, North Sea Directorate,
    Rijswijk, The Netherlands

4. L. Schrijnen, H. Bekkering
    Product Information and User manual of the Search and Rescue
    Decision Support System SEAFLOAT
    TNO, Division of Technology for Society, Apeldoorn, The Netherlands

5. C.J.H. van den Bosch
    'The use of mathematical models in accident consequence assessment
    for transport of hazardous materials at sea'.
    Proceedings First International Conference on Chemical Spills and
    Emergency Management at Sea, November 1988 (Amsterdam).

6. W. Koops
    'Classification of chemicals for situation analysis'.
    Proceedings First International Conference on Chemical Spills and
    Emergency Management at Sea, November 1988 (Amsterdam)

THE USE OF INFORMATION SYSTEMS FOR SUPPORTING DECISIONS ON EMERGENCY
RESPONSE AT SEA

G. Wagenaar
Netherlands Organization for Applied Scientific Research
Division of Technology for Society
Department of Industrial Safety
P.O. Box 342, 7300 AH Apeldoorn, The Netherlands

Abstract
Today's knowledge for combating chemical spills at sea is complex.
A lot of information concerns criteria and methods that are applicable
to reduce or eliminate the effects of a spill. The use of information
systems to assist in the decision-making process offers new challenges,
because such a use brings about advantages as well as disadvantages;
both categories are addressed in this paper. A current tendency in the
use of information systems is the use of expert systems. Examples of
this tendency are presented and handled in some detail. This concerns
CODA, the chemical oil dispersant advisor system, PECOS, the expert
system for oil spill emergency and SEABEL, the hazard identification
and decision support system for emergency response of chemical spills
at sea.

1. INTRODUCTION

Chemicals are among the most commonly used products in today's life.
Their application is wide-spread, resulting in an extensive transport.
Throughout the world large quantities are being transported over land
and water. Especially the transport over oceans, seas and other waterways involves large amounts of chemicals. This transport brings about
certain dangers to people and environment. Therefore, many countries
have agreed upon international regulations with the aim to reduce risks
to a minimum. However, preventive measures reduce the number of accidents; they do not exclude them! As an example some outstanding cases
from the last decade are presented in table I.

Emergency response facilities are therefore required in order to
minimize the consequences of an accident. Such consequences are for
example: pollution of the sea and/or shore, harm to the health of
people, effects on fishing grounds or shipping trade, etcetera. In
order to deal adequately with the consequences, many countries have
established emergency management organizations.

P. Bockholts and I. Heidebrink (eds.), Chemical Spills and Emergency Management at Sea, 233–247.

234

Table I: Accidents at sea (source: TNO accidents database FACTS)

TYPE OF ACCIDENT	NAME OF SHIP	DATE	CHEMICAL	MAIN THREAT	MEASURE(S) TAKEN
Run aground	Amoco Cadiz	1978	Oil	Oil pollution of sea and shore	Mechanical removal
Polymerization reaction in tanker	Stolt Busan	1979	Styrene	Fire and explosion; human health	Cooling with water; addition of toluene and inhibitor
Loss of cylinders	Sindbad	1979	Chlorine	Burst of cylinders; inhalation	Tracing and destruction
Collision	Stanislaw Dubois	1981	Caustic soda; Calcium carbide	Explosion	Ship was scuttled
Collision of ferry and vessel	Mount Louis	1984	Uranium-hexa-fluoride	Toxic gas; human health	Salvage
Collision of fishing vessel and tanker	Orleans	1986	Oil	Fire on board	Natural dispersion
Sinking of ship after collision	Olaf	1986	Fly ash	Bio-accumulation of heavy metals	Salvage of ship and cargo

The organizations, responsible for a proper handling of calamities, have experts in the field. The decisions of an expert are based on his best knowledge and judgement. However, the process of decision-making is a complex one where many different aspects play a role such as, for instance, behaviour and properties of the chemical involved, the variety in actual situations at the location of the accident and many more. Since calamities do not occur frequently, decision-making never becomes a routine procedure.

Information systems (ISs) represent a development in the field of decision-making, because they allow more different aspects to be

considered than an expert can. They also allow predictions and conclusions to be drawn more accurately, more consequently and most of the
times faster. The use of ISs for supporting decisions on emergency
response at sea is the primary subject of this paper, but other
functions are touched upon as well. A current tendency in the use of
ISs, namely the use of expert systems (ESs), is presented, tailored to
decision-making for emergency response at sea. Some examples, like
SEABEL, the hazard identification and decision support system for
emergency response of chemical spills at sea, and CODA, the chemical oil
dispersant advisor, are discussed.

2. INFORMATION SYSTEMS AND EMERGENCY RESPONSE AT SEA

As stated above, the process of decision-making for emergency response
at sea is a complex one. The support of this process by an IS offers new
and challenging opportunities. Nowadays, automation is common property
in many disciplines. The field of emergency response at sea is no
exception. Some five years ago, computer applications were foreseen
(Cormack, 1983); now they are a reality. The use of ISs for emergency
response at sea (from now on the expression IS refers to an IS for
emergency response at sea, unless explicitly stated otherwise) can in
general take two shapes, namely strategic use and operational use. This
distinction is not typical for the use of ISs; it also applies to the
field of emergency response at sea in general (Psaraftis, 1987).

Strategic use of an IS is defined here as the use of an IS for purposes,
which precede the occurrence of an accident. A typical example of such a
use is the planning of optimal placing of equipment for recovery. This
category is not treated in this paper, although some types of ISs, which
are used operationally, can also be used strategically.

Operational use of an IS is defined here as the use of an IS during or
shortly after the occurrence of an accident. One important implication
of operational use is the need for a nearly real-time performance.

ISs may perform several operational functions during an accident. These
functions include:

- gathering of information;
- modelling of spill behaviour;
- decision support on execution of measures;
- presentation of relevant information;
- availability of information (databases);
- exchange of information.

Most of these functions also apply for the strategic use; only the first
and the last seem to be exclusive for operational use. The coherence
between these functions is depicted in figure 1.

236

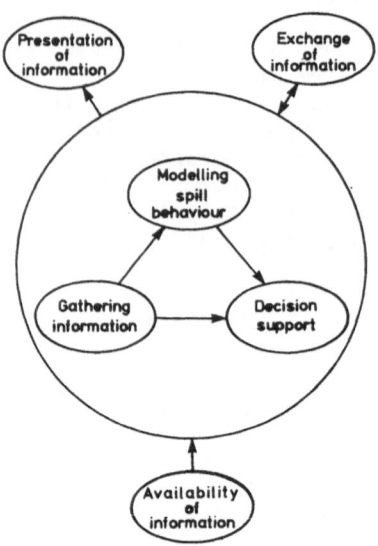

Figure 1: Coherence between functions in an IS

The rest of this section briefly describes the lion's share of the functions. The function of decision support is described in detail in section four since it represents the most interesting development.

2.1. Gathering of Information

The function of information gathering is the process of collecting relevant data on an accident. The keyword here is "relevant"; the collection of too much data has to be avoided, the collection of too little data has to be avoided even more.

An IS in itself cannot gather information. There is one exception to this statement, namely the system can exchange information with another system in one or another network configuration. This aspect is treated in paragraph 2.5.

The foregoing thus implies that a person is responsible for the gathering of information, whereby the system supports the gathering. Here lie possibilities to compose lists with questions to be asked, to supply default and/or standard information to questions (think, for example, of date and/or time of reports), to provide access to appropriate (external) information sources (think, for example, of databases) and others.

2.2. Modelling of Spill Behaviour

The function of modelling of spill behaviour is an attempt to simulate the behaviour of a chemical (or oil) as it is spilled in an aquatic environment. Traditionally, this function has drawn much attention, resulting in numerous implementations in ISs. However, most of the time the function is used on a stand-alone basis; that is, the function is not integrated with any of the other functions mentioned above. Such an integration could, for instance, allow the transfer of information from one function to another.

Since the modelling of spill behaviour is such a well-known function, which is also well described elsewhere, it suffices with the mentioning of some examples plus a reference to their description. Some examples of ISs that include models are HACS (Pascoe,1987), SCICON (Cormack, 1983), EFFECTS (Bockholts and colleague, 1987) and many others.

2.3. Presentation of Information

The function of presentation of information is inherent in any IS; it is not specific for ISs for emergency response at sea. This is why this function is not studied in depth here. Nowadays, two major trends can be distinguished, namely:

- textual presentation of information; until recently this trend was dominant.
- graphical presentation of information (including menu-driven presentation); this trend is rising sharply.

Two examples of presentation of information in ISs for emergency response at sea are depicted in figures 2 and 3.

2.4. Availability of External Information

The function of availability of information offers the possibility to use data that are not incorporated in the IS itself. Of course, frequently used data should be incorporated in the system, but it goes without saying that the function has considerable potential, because it offers entrance to a world of information far bigger than can be incorporated in any system. Any database containing, for instance, information on chemicals is suitable at this point and a large number of such databases exist. The ways of contacting an external database are discussed in the next paragraph.

```
┌──────────── SEACHEM 1 - identification & verification ────────────┐
│                                                                    │
│  chemname   VINYLIDENE CHLORIDE                    tno#            │
│  subname    INHIBITED                              380             │
│                                                                    │
│  un#            imoclass    transport code         physical state │
│  1303           3.1         B,G,P                  LIQUID          │
│                                                                    │
│  gi#   si#      imopage     packaging group        temperature (C) │
│  339   1303     3050        1                      AMBIENT         │
│                                                                    │
│  iata#          label       type of ship           density (kg/m3) │
│  1866,1867      3           2                      1210.0          │
│                                                                    │
│  cas#           sub risc    physical behaviour     pressure (Pa)  │
│  75-35-4        -           substance /  package   UNDER PRESSURE  │
│                             S            PS                        │
│                                                                    │
│  family name                colour                 odour          │
│  VINYL HALIDES              COLOURLESS             UNPLEASANT ODOUR │
│                                                                    │
└────────────────────────────────────────────────────────────────────┘
```

Figure 2: Example of textual presentation of information in SEABEL

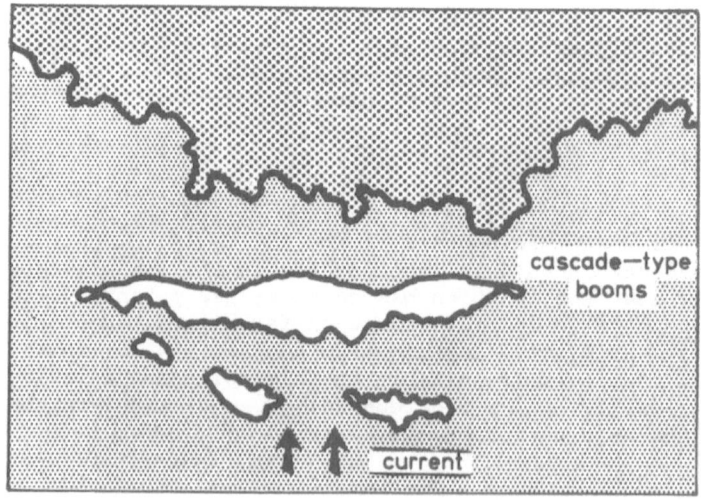

Figure 3: Example of graphical presentation of information in PECOS
(adapted from (Baldassarre and colleague, 1987))

2.5. Exchange of Information

The function of exchange of information deals with the possibility of a fast and reliable communication between ISs. This communication can be static or one-way, such as the retrieval of information from databases as mentioned above. It can also be dynamic or two-way, for instance, the merging of two complementary descriptions of an accident from two different sites. Especially the latter one is an important feature.

Communication between ISs is communication between computers. Several options are available to accomplish communication between computers:

- temporary or permanent communication by means of a telephone line (modem connection by (auto-)dialing);
- permanent communication by means of a network connection; this ranges from a local area network (LAN) to a wide area network (WAN).

These options are equal in the sense that they all allow transfer of any kind of information; they differ, among others, in speed and reliability.

3. SOME EXAMPLES OF INFORMATION SYSTEMS

3.1. CODA

The Chemical Oil Dispersant Advisor (CODA) system (Bradford Lowe, 1986) is designed to provide advice on whether or not chemical oil dispersants should be used as a cleanup strategy following an oil spill. It uses data available at the time of the spill to conclude whether or not dispersants are recommended. Explanations are given along with the conclusions to clarify how they were reached. CODA uses knowledge from three interrelated areas:

- chemical knowledge; the first question is whether dispersants are effective or not on an oil spill with particular chemical and physical characteristics.
- logistical knowledge; a second question is whether or not the resources exist to mount a cleanup operation with dispersants.
- environmental knowledge; the final question is whether or not dispersants alleviate any potential negative environmental impact. An oil spill model is included in this knowledge.

3.2. PECOS

The goal of PECOS system (Baldassarre and colleague, 1987) is to identify the best intervention level strategy with regard to the action to be taken to manage an emergency plan in case of an oil spill at sea. In order to solve this problem, answers to the following questions are addressed:

240

- Is there oil on the sea surface, that requests immediate action of recovery or dispersion?
- Are the operations of mechanical recovery and dispersion of the oil applicable and what are the most appropriate techniques in each case?
- What are the most sensitive resources of the coast and the most appropriate techniques to protect them?
- What are the techniques to cleanup the oil impacted shores (and what disadvantages do they have)?

PECOS is subdivided into four phases, which deal with the focal point of a emergency plan, namely:

- determination of the need for immediate intervention;
- evaluation of the applicability of mechanical recovery and chemicals;
- determination of coast protection conditions;
- criteria for reclamation of the impacted coastal areas.

3.3. SEABEL

The goal of the SEABEL system (Heidebrink and colleague, 1987) is to give a decision maker a clear view of the impact of an accident, in which dangerous chemicals are involved, and to enable him to evaluate possible applicable measures. This goal is derived from three basic questions, that should be answered before a decision maker will be able to respond adequately to emergency situations:

- What are the actual and predicted hazards?
- Are immediate actions required or is there still time available to investigate the accident in more detail?
- Which measures are the most suitable in terms of success and effort required?

To answer these question SEABEL is divided into four modules. Module I deals with the gathering of information; it allows the composition of relevant checklists with questions and it supports the filling in of a checklist. Module II deals with the simulation of the effects of a spill; models for the behaviour of chemicals and oil are included. Module III deals with the presentation of the results of the previous modules; the results are merged with geographical data.
Module IV deals with the decision to be made on emergency response; first, all applicable measures are selected and second, the measures selected are evaluated.

4. INFORMATION SYSTEMS FOR DECISION SUPPORT

After the description of three ISs in the previous section, it is time to turn to the remaining function of decision support (see figure 1). All systems described have this function and this section goes into detail of its implementation. The function is new in the sense that relatively few investigations into this area are described. One of the early results is from five years ago; it concerns an ES for emergency management of inland oil and hazardous chemical spills (Johnson and colleague, 1983); this ES is not treated here. It is equally impossible to treat the fundamentals of ESs here; reference is made to (Hayes-Roth and colleagues, 1983) and (Waterman, 1986).

4.1. CODA

CODA is an expert system, which uses a standard rule based knowledge representation scheme and a backward chaining inference mechanism. Figure 4 illustrates the important parts of the architecture of such a system. Only the knowledge base and the inference engine are described here.

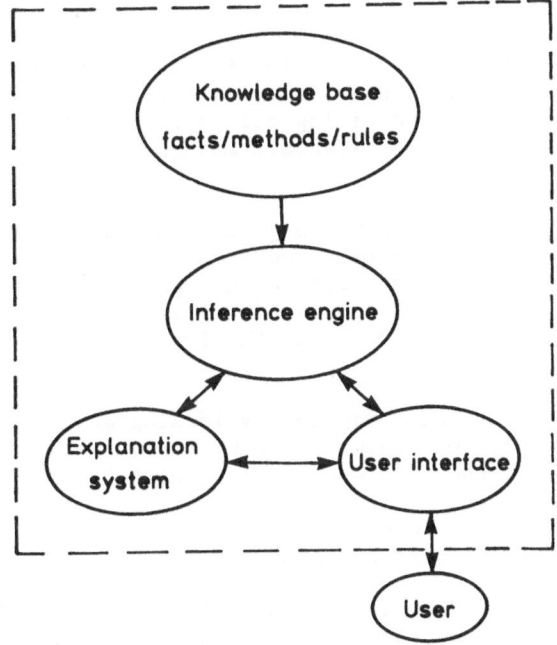

Figure 4: Architecture of CODA (adapted from (Bradford Lowe, 1986))

4.1.1. Knowledge base. The knowledge base stores all domain specific data (including the chemical, logistical and environmental knowledge). It is stored in the form of facts, rules and methods.

Knowledge base facts are data, pertaining to the particular situation at hand. Facts consist of objects attributes and values. An object attribute is a static parameter, which the system tries to instantiate with a value. A value is dynamic and varies based on the current oil spill situation. As an example, an object attribute could be oil-specific-gravity; its corresponding value in a certain situation could be 0.83. The combination of object attribute and value is a fact.

Knowledge base rules are the rules of thumb (heuristics) that are used to make a decision. They use facts to update other facts, for which the values are not yet known. A rule consists of a premise and a conclusion; if the premise is true then it can be inferred that the conclusion is also true. An example of a rule is:

    IF   oil-specific-gravity = 0.83
    THEN oil-classification = type 2

Through the use of rules the system is able to make powerful inferences, thus using various combinations of known facts to infer conclusions about unknown facts.

Knowledge base methods are similar to rules in that they use known facts to instantiate unknown facts. However, they differ in that they use mathematical calculations instead of logical inferences to find these values. Methods are represented by standard mathematical functions (formulas, etcetera).

4.1.2. Inference engine. The inference engine of an ES is responsible for sifting through the knowledge base and combining the rules, methods and facts to make conclusions. It exists as a separate module and uses inference procedures, that are independent of domain specific knowledge. The inference engine in CODA is of the depth-first backward chaining type.

The inference engine begins its execution by looking at the end goal of the system, represented by some uninstantiated fact. Since there is not yet a value for this fact the inference engine tries to find one by:

  - looking for a method that calculates the value;
  - a rule that concludes the desired fact.

If such a rule is found, the premise of the rule is checked. If it contains further unknown facts, the process repeats itself, thus continuing in a backward fashion deeper and deeper along the inference chain until a premise is found with a known value. If that premise is true then the inference engine instantiates any fact value concluded by that rule and starts to look for other facts values needed to make a high level conclusion. If the premise is false, then the present line of reasoning is dropped and a next rule making the desired conclusion is tried. At some points it is possible for the system to ask the user to supply a value thus using basic data to build higher level conclusions. An example of this inference mechanism is shown in figure 5.

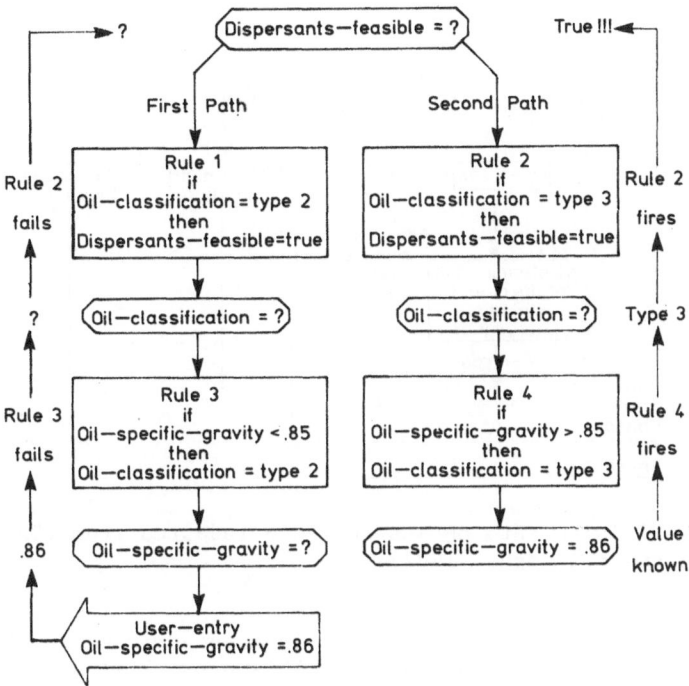

Figure 5: Example of inference mechanism in CODA (adapted from (Bradford Lowe, 1986))

4.2. PECOS

PECOS is an ES divided into four phases. These phases are structured like the flow chart in figure 6 (only the first phase is described in detail).

244

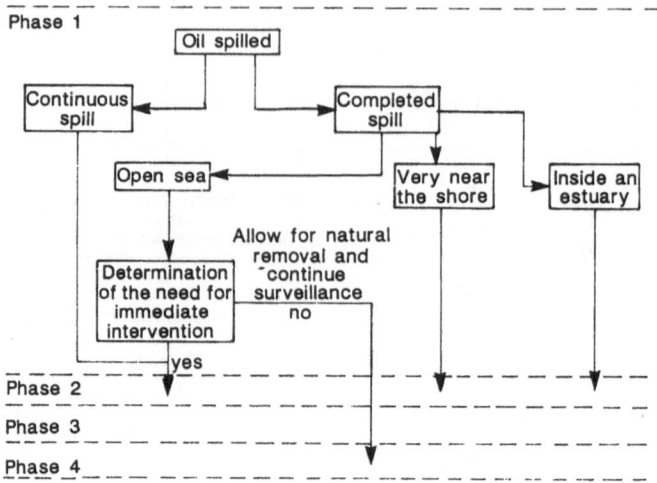

Figure 6: Division into phases in PECOS (adapted from (Baldassarre and colleague, 1987))

This flow chart is elaborated further in a semantic network; part of this network is depicted in figure 7. This network illustrates PECOS' line of reasoning.

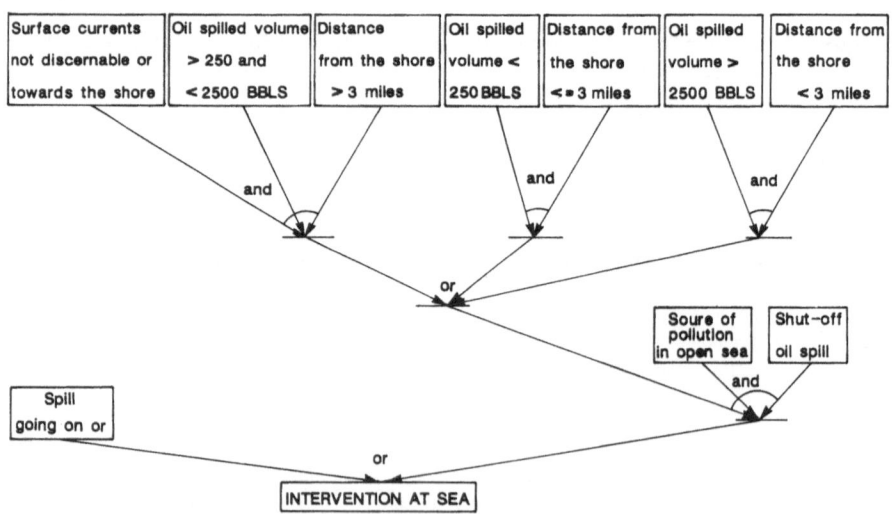

Figure 7: Part of the semantic network in PECOS (adapted from (Baldassarre and colleague, 1987))

Boxes at the top of figure 7 represent input data from the user. These data tell the system which path to follow, what rules to infer and what further data to request. The technique used by PECOS is to try to solve the final problem by using the solution of problems that gradually scale down in size, so as to obtain the necessary information. This process is called problem decomposition; it is applied here again in a backward reasoning fashion.

4.3. SEABEL

Because the scope of SEABEL is much wider than both systems described above, the description of SEABEL is limited; to be precise, the description is limited to module IV, the decision support module. Decision support in SEABEL is a two-step process. First, measures are selected and second, measures are evaluated.

The selection of measures is based on:

- the chemical involved in accident (for instance, its behaviour);
- the location of an accident (for instance, near a harbour);
- meteorological data (for instance, wind-speed);
- the general description of an accident (for instance, is the cargo inside or outside the ship?).

These characteristics are refined to some eighty criteria. These criteria are plotted against the measures (some two hundred). The result is the decision matrix; part of it is depicted in figure 8.

	TYPE OF SHIP		IMO CLASS			WIND-FORCE		
	BULK CARRIER	PACK CARRIER	1	...	8	1	...	12
ADD INERT GAS	1	1	0	...	0	1	...	0
COOL THE CARGO	1	1	0	...	0	1	...	1
DO NOTHING	1	1	1	...	1	1	...	1

Figure 8: Decision matrix in SEABEL

A '0' in the decision matrix means that if the criterion in that column applies, then the measure in that row is not suitable for execution given the accident on hand. It is now obvious that, given a set of instantiated criteria, this process results in a set of measures.

Each measure selected is then evaluated. Evaluation is performed on the basis of four aspects:

- risk for the people involved;
- risk for the environment;
- nuisance to parties involved;
- costs.

These aspects are rated (numerically) and they are combined in two overall scores; a linear score (all aspects added) and an incremental score (all aspects added with weighting factors). This results in a ranking of all applicable measures. Furthermore, each measure has its corresponding set of information such as information on limitations to its use, addresses of organizations that supply equipment to execute the measure and other information.

5. CONCLUSIONS

It is obvious that ISs are penetrating in the discipline of emergency response at sea. One does not have to be gifted with second sight to foretell that this penetration shall increase significantly over the next few years. Although much of these systems shall be "conventional" of nature, part of them shall also use technology from ESs. As shown, some of today's ISs already use this technology; they were described in this paper.

Two systems, CODA and PECOS, are ESs. They are both aimed at emergency response at sea for oil spills. In this restricted area they seem to function adequately. The SEABEL system on the other hand has a wider applicability. It supports the greater part of the functions, enumerated in section two, partly by using conventional programming techniques and partly by using techniques related to ESs.

Particulary this last development, integration of conventional ISs and Ess, reflects the future. The power of a conventional IS in data processing combined with the power of an ES in symbolic (knowledge) processing opens new perspectives. The integration is already taking place in other disciplines and there is no reason to believe that emergency response at sea is an exception.

6. REFERENCES

Baldassarre, E. and L. Ceffa (1987). *PECOS: an expert system for oil spill emergency*. Rel. nr. 1217, AGIP S.p.A., San Donato Milanese, Italy.

Bockholts, P. and I. Heidebrink (1987). 'SEABEL: hazard identification and response decision for chemical spillages at sea'. In: *Proceedings "Workshop on risk assessment of accidental pollution related to maritime transport of harmful substances"*, pages 409-417. 25-27 March 1987, Brest, France.

Bradford Lowe, T. (1986). *'CODA: A knowledge based system for advice on the use of chemical oil dispersants'*. Report number 86529-3, Senter for Industriforskning, Oslo, Norway.

Cormack, D. (1983). *Response to oil and chemical marine pollution.* Applied Science Publishers Ltd., Barking, England.

Heidebrink, I. and P. Bockholts (1987). 'SEABEL - hazard identification and decision support system for emergency response of chemical spills at sea'. In: *Proceedings "MariChem87"*, paper 2.12. 20-22 October 1987, Hamburg, Germany.

Johnson, C.K. and S.R. Jordan (1983). 'Emergency management of inland oil and hazardous chemical spills: a case study in knowledge engineering'. In: Hayes-Roth, F., Waterman, D.A. and D.B. Lenat (editors). *Building expert systems.* Addison-Wesley Publishing Company, Reading, Massachusetts, USA.

Hayes-Roth, F., Waterman, D.A. and D.B. Lenat (editors) (1983). *Building expert systems.* Addison-Wesley Publishing Company, Reading, Massachusetts, USA.

Pascoe, D. (1987). 'Hazard assessment computer system (HACS)'. In: *Proceedings "Workshop on risk assessment of accidental pollution related to maritime transport of harmful substances"*, pages 229-240. 25-27 March 1987, Brest, France.

Psaraftis, H.N. (1987) 'Oil spill risk management: the strategic response case'. In Kuiper, J. and Van den Brink, W.J. (editors), *Fate and effects of oil in marine ecosystems*, Martinus Nijhoff Publishers, Dordrecht, The Netherlands.

Waterman, D.A. (1986). *A guide to expert systems.* Addison-Wesley Publishing Company, Reading, Massachusetts, USA.

AN ANALYSIS OF HISTORICAL EVENTS DURING TRANSSHIPMENT OF DANGEROUS GOODS
IN HARBOURS

Ing. L.J.B. Koehorst
Netherlands Organization for Applied Scientific research TNO
Division of Technology for Society
Department of Industrial Safety
P.O. Box 342, 7300 AH Apeldoorn, The Netherlands

1. PREFACE

From time to time people are startled by major disasters where hazardous
materials are involved. This often induces detailed accident investigations, in order to discover the cause of the accident, and by this the
ways to prevent repetition of such an accident. In this respect,
analysis of historical data of accidents is a very important tool. Not
only data about major accidents is useful for such an analysis but also
data on accidents with minor consequences can be used. The reason is
that the mechanism for major accidents as well as for minor accidents is
for a great deal the same.
In this paper a specific type of accident, namely accidents during
transshipment is analysed in order to trace important characteristics
from which it may be useful to learn what went wrong in the past.

2. SCOPE OF THE ANALYSIS

Every way of transport starts and ends with the transshipment of cargo,
implicating that during two occasions (loading and unloading) the cargo
is handled. During handling there is a dynamic interaction between
cargo, people and equipment which induces potentially dangerous situations that may cause accidents.
Cargo can be transported over sea, inland waterways, rail, road and
through pipelines. Figure 1 illustrates the share of each type of transport for 10 basic types of cargo. The data is related to the situation
in 1985 for the harbour of Rotterdam.

The type of cargo is a factor which may contribute to the potential
risks of transshipment. In particular for the categories crude oil, oil
products and chemical products a greater potential risk exists, not only
because the relative great amounts that are handled (49,9%), but also
because their physical and chemical properties mean a much higher risk
than the in other categories.

P. Bockholts and I. Heidebrink (eds.), Chemical Spills and Emergency Management at Sea, 249–258.
© *1988 by Kluwer Academic Publishers.*

Type of transport Type of cargo	sea	inland water-ways	rail	road	pipe-line	Total	%
	numbers x 1000 tons						
Agriculture products	11,375	3,222	231	1,292	-	16,120	4.2
Foods	29,290	6,336	371	1,864	-	37,861	10.0
Solid fuels	14,113	5,317	119	285	-	19,834	5.2
Crude oil and oil-products	111,107	21,773	250	677	34,887	168,694	44.4
Ore	40,258	36,300	2,216	258	-	79,032	20.8
Metals	6,049	2,829	179	1,440	-	10,497	2.8
Building materials	4,677	3,175	243	426	-	8,521	2.2
Fertilizers	3,925	2,203	169	125	-	6,422	1.7
Chemical products	14,217	3,657	747	2,108	224	20,953	5.5
Other products	8,375	608	446	2,789	-	12,218	3.2
Total	243,386	85,420	4,971	11,264	35,111		
%	64,0	22,5	1,3	3,0	9,2		

Figure 1: Amounts of supplied and discharged cargo in 1985 in the harbour of Rotterdam
Source: CBS, statistics supply and discharge of cargo per transport type.

In this paper the transshipment of dangerous goods in harbours such as crude oil, oil products and chemical products is discussed. Based upon historical data of accidents resulting from these activities an analysis has been made which gives an insight into the situations that lead to an accident.
The accident data, used for this analysis, is obtained from the TNO database FACTS[*]. For this reason a short explanation of FACTS and the type of information that is stored in this database is given below.

3. THE TNO DATABASE FACTS

FACTS is a database with technical information about accidents with hazardous materials that happened during all types of industrial activities (processing, storage, transshipment, transport, etc.). At this moment FACTS contains information about 15,000 accidents that happened all over the world. Most information concerns accidents during the last 30 years.

[*] Failure and Accidents Technical information System.

The information stored in FACTS is derived from several sources such as:
- literature;
- periodicals;
- technical reports;
- environmental and labour inspectorates;
- industrial companies;
- fire brigades;
- police.

There is always a discrepancy between the number of accidents that actually have happened and those that are recorded. Accidents having minor consequences may not have been recorded at all, while accidents with more consequences may be recorded incidentally. Only accidents where severe damage or danger is involved will be publicized, analysed and documented. The discrepancy between events that actually have occurred and those that have been recorded, is shown in figure 2. The quality of the available information of recorded accidents is also related to their seriousness. The most serious accidents are also those of which good and detailed information is available.

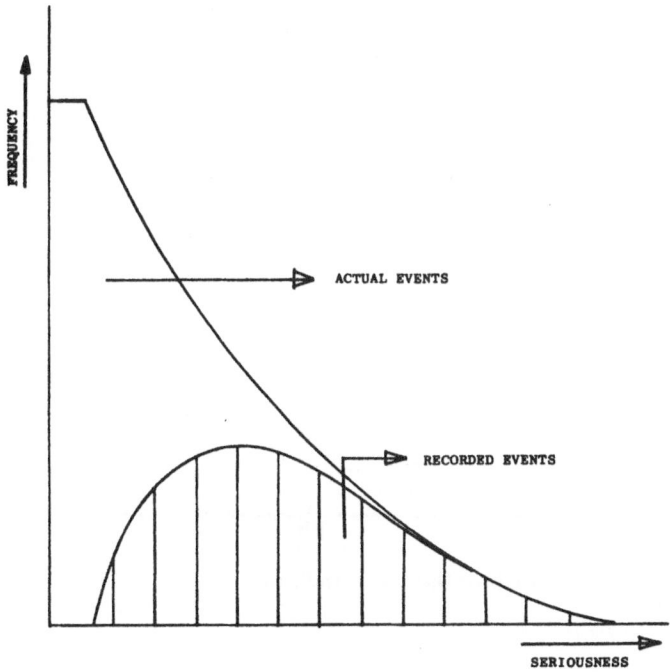

Figure 2: Comparison of actually occurred incidents and recorded incidents.

4. ACCIDENT DATA FROM FACTS ON TRANSSHIPMENT IN HARBOURS

396 Accidents during transshipment of dangerous goods in harbours are selected. For analysis the following aspects of the selected accidents are necessary:
- injuries and fatalities
- chemicals involved
- accident cause

4.1. Injuries and fatalities

Figure 3 represents the amount of the selected accidents during transshipment with more than a certain number of injuries and fatalities. The figure shows that in 38 accidents fatalities occur and injuries in 75 accidents.
From the reviewed 396 accidents it can be concluded that in 20% people got injured and that people were killed in 10%.

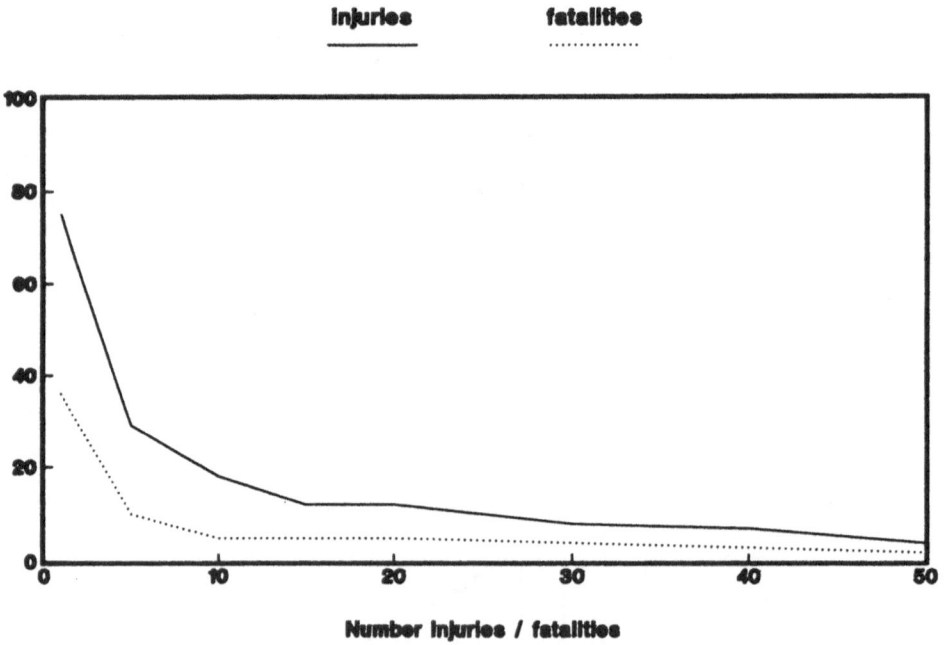

Figure 5: Accidents with more than a certain number of injuries and fatalities

Injuries can be characterized as follows:

type of injury	number of accidents	%
burns	32	36
toxic inhalation	28	31
physical impact by blast waves	20	22
unknown	10	11
total	90	100

4.2. Involved chemicals

The types of chemical and their state during transport vary considerably. Fertilizers, crude oil and oil products for example are transported in bulk. For relative small quantities of chemicals, most of
which are very aggressive or toxic like for instance toluenedi-iso-
cyanate and pesticides, the cargo can also be transshipped in drums and
containers. All these types of cargo require a specific way of handling.

Irrelevant to the way dangerous goods are transshipped, the first consequence of almost any accident will be a release of at least a part of
the cargo. It depends on the physical properties of the released
chemical how the chemical will behave. In only 2% of the accidents no
release took place but people got injured when they inspected a tank, or
when for example decomposition of the cargo took place in the cargo deck
of a ship.
Moreover, after a release of a chemical, it depends on several factors
if and when a fire or an explosion will take place. The availability of
an ignition source with sufficient energy, the presence of oxygen and
the concentration and type of the released chemical are such determining
factors.

In figure 4 the number of accidents that resulted in a release and
eventually in fire and/or explosion are presented.

The chemicals that were involved in the selected accidents are classified; see figure 5.

254

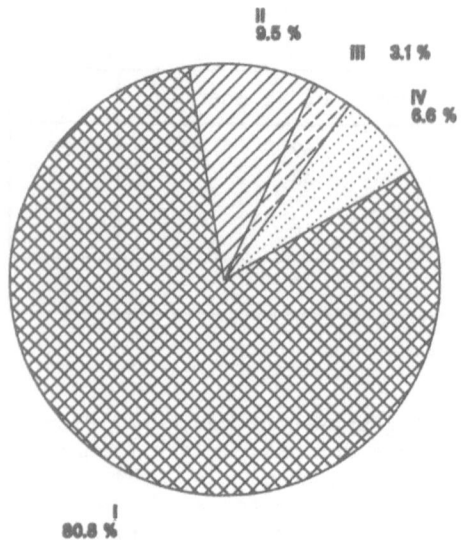

Consequences	Number of accidents
I Release	315
II Release + fire	37
III Release + explosion	12
IV Release + fire + explosion	26

Figure 4: Accidents during transshipment, resulting in a release and eventually in fire and/or explosion

type of chemical	number of accidents	%
ammonia	11	3
chlorine	6	1
anorganic acids	10	3
light hydrocarbons	15	4
aromates	39	10
toxic organic compounds*	29	7
crude oil	45	11
diesel, fuel, gasoil	83	21
benzine, kerosine	52	13
naphta	10	3
lpg, butane, propane	16	4
rest**	80	20
total	396	100

* for example acrylonitrile, aniline, pesticides etc.
** for example anorganic salts, complex organic compounds.

Figure 5: Involved chemicals

Looking at figures 4 and 5 it is remarkable that, although a great deal
of the involved chemicals is highly inflammable, only in 20% of the
accidents the release of chemicals resulted into a fire and/or
explosion. Earlier studies (2) show that almost in every situation an
ignition source is present, and oxygen is sufficiently available due to
the fact that the transshipment takes place in the open air.
A possible explanation might be that the amount of the released chemical
is so small that the mixture of vaporized chemicals and air does not
result in an explosive vapour cloud.
Figure 6 presents the amount of spill of the involved chemicals. In
about 65% of the selected accidents during transshipment, data about the
amount of spill is available.

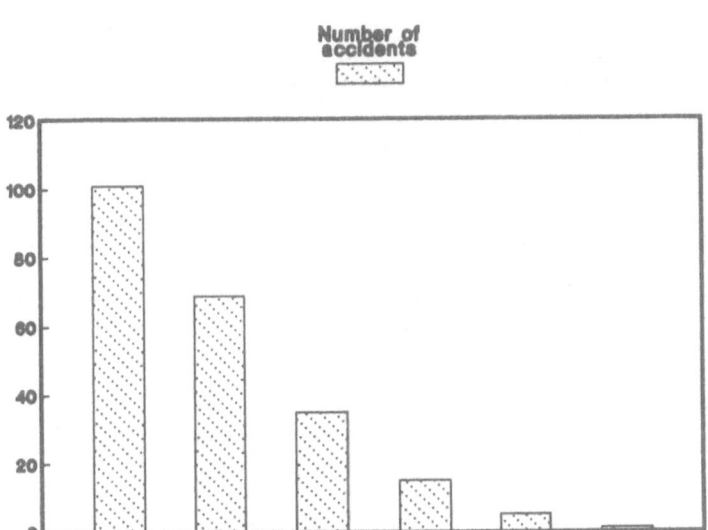

Figure 6: Spills of involved chemicals

In 45% of the accidents there is less than 1 m^3 spilled and in even 75%
the spill is limited to 10 m^3.
For those accidents where the amount of spill is unknown, the actual
spills are probably relatively small because no serious damage was
reported.

4.3. Accident causes

Looking to the cause of the accidents resulting in a release, the
following general causes may be identified:

cause	%
human failure	41
technical failure	25
other failures	4
- domino effects	
- sabotage	
- natural causes	
unknown causes	30
total	100

Figure 7: Cause classification

Human failure and technical failure are the most important causes, but these descriptions are so general that it is necessary to have a more detailed look into these two categories. Technical failure is a general description which implicates the malfunctioning or demolishing of equipment. In most cases the reason why the equipment failed is not known or investigations show that wrong human decisions concerning the choice of material, the calibration of equipment or the injudicious use of equipment caused the malfunctioning of the equipment which resulted into an accident.
In about 100 cases the accident occurred due to failure of some type of equipment. Figure 8 presents the frequency of failure of equipment which caused an accident.

equipment	number of accidents	%
drums	20	19
hoses/lines	18	17
valves	17	16
pumps	8	8
welds	5	5
tanks, containers	8	8
measure devices	6	6
small equipment	11	11
- coupling		
- flange		
- fitting		
- packing		
abnormal process conditions	7	7
elevators, loading arms	4	3
total	104	100

Figure 8: Failure of equipment which caused accidents.

The relative high number of accidents with drums are caused by damaged or leaking drums. Accidents with drums that got loose from a crane are not listed in this figure, but are classified as operation failures; see figure 9.
Human failure can be divided into the following classes. The first three classes will be illustrated with some examples of characteristic failures that frequently occur.

1. Manipulation failures: failures during the handling of single pieces of equipment such as valves, hoses etc.
   Examples: - opening or closing the wrong valve
             - valve set in wrong position
             - wrong switching
             - disconnecting wrong hose
             - start-up wrong pump

2. Operation failures   : failures during the preparation and the execution of loading and unloading activities.
   Examples: - collision of fork-lift truck with drums
             - wrong storage of containers or drums
             - wrong hoisting
             - wrong connection of loading/unloading facilities.

3. Organization failures: failures caused due to acting against proce-dure regulations or by contradictions in these procedure regulations.
   Examples: - fell asleep
             - unpermitted smoking
             - no supervision during loading/unloading
             - repair activities during transshipment

4. Inspection failures
5. Maintenance failures
6. Construction failures

Figure 9 presents the frequency of the several types of human failure.

human failures	number of accidents	%
manipulation failure	76	48
operation failure	38	24
organization failure	23	14
inspection failure	6	3
maintenance failure	7	4
construction failure	11	7
total	160	100

Figure 9: Classification of human failure.

5. CONCLUSIONS

The conclusions of this analysis, which is based on historical data, are influenced by the limitations caused by the incompleteness of the data. On the other hand, the relative great number of accidents (396) that have been analysed, quarantee that frequently occurring specific cause types and accident circumstances, are noticed.

1. The release of chemicals during transshipment causes only in 10 à 20% of the occasions injury to the directly involved people (operaters, workers).

2. Although the major part of the handled chemicals is inflammable, in 20% of the accidents a release of chemicals results into a fire or explosion. A possible explanation is that the amount of the released chemicals is relatively small, 75% less than 10 m^3 and 45% less than 1 m^3.

3. The cause of the release often is the wrong handling and manipulating of drums, valves and hoses. This is not specific for accidents during transshipment, but earlier analysis (3) show that relatively simple actions cause a great deal of the accidents. Not only the fact that these simple actions are executed very frequently, but also the daily practice of these actions which makes that the operators underestimate the dangers of the chemicals is a contributing factor.

6. LITERATURE

1. 15th International TNO Conference.
   Trade, Transport and Technology, march 1982.

2. P. Bockholts and L.J.B. Koehorst.
   Accident analysis of vapour cloud explosions, 17 june 1987.

3. L.J.B. Koehorst.
   Analyse van ongevalsinformatie naar het ontstaan en verloop van ongevallen in de chemische industrie. Draft-report, july 1988.

THE COMMUNITY INFORMATION SYSTEM FOR THE CONTROL AND REDUCTION OF POLLUTION BY THE SPILLAGE OF HYDROCARBONS AND OTHER HARMFUL SUBSTANCES AT SEA

Claus Hagebro, the EEC Commission
Directorate-General, Environment, Consumer
Protection and Nuclear Safety. Rue de la Loi 200
B-1049 Brussels - BELGIUM

ABSTRACT

The Council of Ministers of the EEC adopted Decision 86/85/EEC on the 6th March 1986 establishing a Community Information System for the control and reduction of pollution caused by the spillage of hydrocarbons and other harmful substances at sea.

The paper gives a short introduction to the background of the Decision and explains the aims of the system, which forms a part of the Community action plan.

The information system when fully developed will include :

- a list of national and joint plans for combating pollution caused by the spillage of hydrocarbons at sea, comprising a brief description of the content of the plans and naming the authorities responsible for them;

- an inventory of resources for combating pollution of the sea by hydrocarbons;

- a compendium of hydrocarbon properties and their behaviour and of methods of treatment and end uses of mixtures of water-hydrocarbon-solid matter recovered from the sea or along the coast;

- an inventory of resources for intervention in the event of a spillage at sea of harmful substances other than hydrocarbons.

In addition, the Commission will gradually compile, in the light of experience, a compendium of information concerning the properties and behaviour of harmful

259

P. Bockholts and I. Heidebrink (eds.), Chemical Spills and Emergency Management at Sea, 259—268.
© *1988 by Kluwer Academic Publishers.*

substances or groups of harmful substances other than
hydrocarbons.

The different parts of the system are described as well
as the state of development of the different parts.

1. INTRODUCTION

The policy of the Commission of the European Communities
for dealing with sea pollution has been conceived mainly
as a preventive and protective approach. In particular,
the key element of this policy remains a number of
directives which aim to reduce the quantity of pollutants
which reach the sea every day either directly or through
rivers and estuaries. This paper on the Community
Information System, however, deals with questions related
to pollution as a result of accidents.

In the late seventies massive accidental pollution
incidents such as the Amoco Cadiz wreck and the blow-out
of the Bravo Platform in the Ecofisk field provoked great
public concern.

Following these events, the Council adopted an action
programme on the control and reduction of pollution
caused by hydrocarbons discharged at sea (1). This
programme was in fact a mandate for the Commission to
undertake studies and to make the appropriate proposals
in different fields and resulted in several important
initiatives which constitute the so-called Community
Action Plan. Today this Community Action Plan for dealing
with spills of oil and other harmful substances at sea
consists of four main parts : the Community Information
System (CIS), a training programme, studies and pilot
projects and the Community Task Force. The latter is
available to support authorities confronted with critical
situations. For instance in the case of an emergency on
request from a Member State the Commission can provide
experts to attend the scene of the incident to advise the
national authorities involved.

Before the Community Information System is described in
detail it should be noted that Western Europe is a region
of the world where the greatest number of agreements are
in force to ensure the most efficient cooperation between
public authorities for dealing with accidental marine
pollution.

There are in particular several bilateral agreements
(e.g. Manche Plan and Denger Plan), agreements to which

four Parties subscribe (e.g. Copenhagen Agreement) and
multilateral agreements such as the Bonn Agreement or the
Barcelona Convention, the last one going far beyond the
Western Europe limits.

The Community Action Plan is superimposed on most of
these agreements (figure 1). It is seen that Portugal and
Ireland are covered only by the EEC framework for
international cooperation in case of major accidental
pollution at sea. It should also be mentioned here that
the aim of the Community action is only to complete and
extend the actions achieved at international level and
not to duplicate existing work.

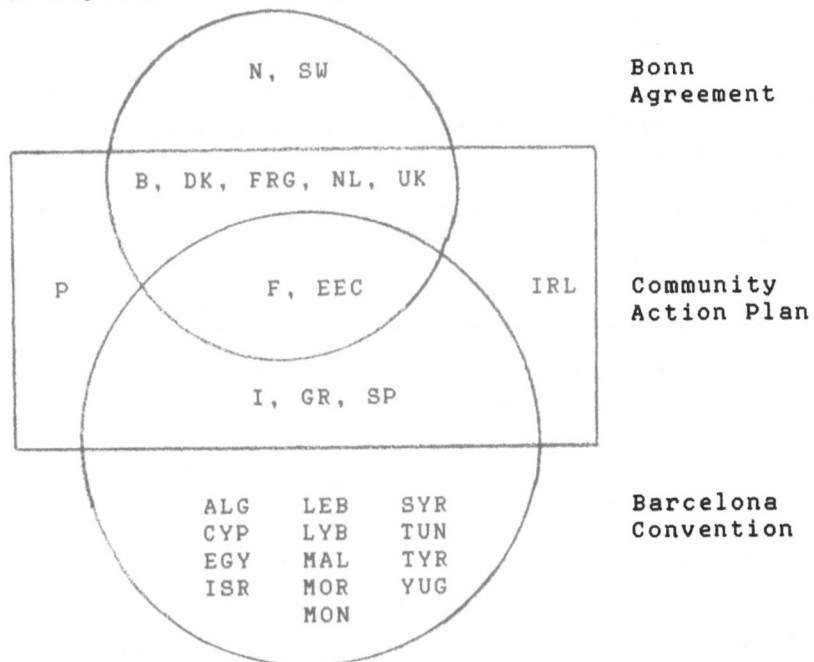

Figure 1 : Main frameworks for cooperation in western
          Europe

ALG:Algeria,   B:Belgium,   CYP:Cyprus,   DK:Denmark,
EEC:European Economic Community, EGY:Egypt, F:France,
FRG:Federal Republic of Germany, GR:Greece, IRL:Ireland,
ISR:Israel, I:Italy, LEB:Lebanon, LYB:Libya, MAL:Malta,
MOR:Marocco, MON:Monaco, N:Norway, NL:The Netherlands,
P:Portugal, SP:Spain, SW:Sweden, SYR:Syria, TUN:Tunisia,
TYR:Turkey; UK:United Kingdom, YUG:Yugoslavia.

2. THE COMMUNITY INFORMATION SYSTEM

The Community Information System (CIS) for the control and the reduction of pollution caused by hydrocarbons discharged at sea was established by Council Decision of 3 December 1981 (2). This has since been repealed by Council Decision of 6 March 1986 (3) which extended the Community Information System to other harmful substances.

The Commission presented the first operational version of the system in June 1985. The system has the character of a handbook or manual giving a lot of useful information on national contact points, telex and telephone numbers, etc. as well as lists of equipment for combating pollution at sea.

The System is updated annually. The updating in 1987 contained a preliminary version of the section on "other harmful substances".

For the moment the system consists of the main parts shown in table 1.

TABLE 1 : the Community Information System

Section 1 : Hydrocarbons

Inventory of resources
for combating pollution
of the sea by hydrocarbons
(Part A)

Catalogue of resources
for combating pollution
of the sea by hydrocarbons
(Part B)

Properties, behaviour and
impact of hydrocarbons
(Part C)

Storage and use of oily
debris (Part D)

Other aspects (Part E)
E1 : customs questions.

Section 2 : Other harmful
substances

Inventory of resources
for intervention in the
event of a spillage at
sea of harmful substances
other than hydrocarbons
(Part F). Preliminary
version.

2.1. Inventory of resources (hydrocarbons)

The inventory contains a list of resources for dealing with pollution which can be made available to affected authorities by other authorities or other interested parties.

The resources are described very briefly, classified by country and by categories such as equipment, material, ship, aircraft, storage facilities, cargo transfer, people, strike teams,etc. The complete address of the owners of the aforementioned resources is recorded. In this way, the authorities wishing to obtain equipment etc. during an incident can have their dialogue with the owners of the equipment on the basis of a common text of reference.

An example of the information presented in the inventory is shown in Table 2.

Table 2 : Example of information in the inventory.
        Netherlands. Categorie = equipment Boom

```
MANUFACTURER          : SANERA
PRODUCT NAME          : EXPANDI 3000
ORGANIZATION          : RIJKSWATERSTAAT BENEDEN RIVIEREN
TELEPHONE             : NL-70-132266 - NIGHT:1747-3791
TELEX                 : NL-29230 - NIGHT : 33028
MOBILIZATION TIME     : 4 HRS
OWNER                 : RIJKSWATERSTAAT
LOCATION              : HOOK OF HOLLAND
DESCRIPTION           : CURTAIN BOOM-SELF INFLATING-HARBOUR
QUANTITY              : 2 930 m
COMMENT               : 15 m SECTIONS-COMPLETE WITH END
                        CONNECTIONS
NUMBER OF OPERATORS: 12
REFERENCE             : 3
```

Unfortunately, for the moment, Portugal and Spain are not included in the inventory as no data has been recieved by the Commission.

2.2. Catalogue of resources (hydrocarbons)

The Catalogue consists of a collection of sheets. Each sheet gives a succinct but detailed description of one of the resources recorded in the inventory.

The description of the resource contains all the information essential for utilisation; the product name, the manufacturer, the technical characteristics, the size and the weight of the containers suitable for their transport, indications of the possible utilizations, the limits of such uses and instructions for maintenance.

An example is shown in Table 3.

2.3. Properties, behaviour and impact of hydrocarbons

This section gives details of the physical/chemical properties and behaviour of a hydrocarbon when released into the sea. Also included are apects of the impact of hydrocarbons on marine flora and fauna.

It has, however, to be mentioned that this part of the CIS has been withdrawn for the moment in order to permit a further examination. The revised version will try to convey in the best possible way scientific knowledge to the operational people in charge of combating pollution. At this stage only two analytical sheets have been retained : an updating and extension of a report on characteristics of petroleum and its behaviour at sea and experimentally determined parameters of crude oils and their emulsions.

2.4. Storage and use of oily debris

This part has been withdrawn until further notice.

2.5. Other aspects

This part aims to cover all information of a diverse character which could concern the combating of pollution by spillages at sea. For the moment part E1 contains a note on customs problems liable to arise in the event of marine pollution control operations i.e. the international movement of equipment.

```
----------------------------------------------------------------
```

Table 3 : Example of information in the catalogue
          Categorie : Boom

```
----------------------------------------------------------------
```

PRODUCT NAME   : EXPANDI 3000
MANUFACTURER   : SANERA
DESCRIPTION   : SELF-INFLATING FLEXIBLE CURTAIN FAST
                    DEPLOYMENT HARBOUR BOOM. COLOUR ORANGE
CHARACTERISTICS:
  FREE BOARD   : 0,28 m
  DRAUGHT   : 0,47 m
  OVERALL HEIGHT : 0,75 m
  BUOYANCY   : 53 kg/m Internal bonded pliable
                    frames of polypropylene plastic
                    which are pre-sprung with rust-
                    proof springs. Nylon reinforced
                    polyurethane PVC. Each 25 m boom
                    contains 18 sections, with
                    airtight spacer between sections.
                    Diamond shaped floats.
  SKIRT   : Nylon reinforced polyurethane PVC
  BALLAST   : Bottom tensioning ballast chain
  TENSION LINE   : Bottom tensioning ballast chain
  TENSILE STRENGHT : 2 590 kg
  MOORING-ANCHORING : Incorporates anchoring rings and
                    grip handles at 8 m intervals
  WEIGHT PER METER : 2,48 kg
  STANDARD LENGTH : 25 m
  CONNECTION   : Male/female connectors using
                    locking bar and bottom tension
                    ballast chain
  COST   : Available upon request
PACKAGING
  TYPE   : On wooden pallet or on reel or on
                    EXPANDI ROTO PACK or in a package
  DIMENSIONS   : 0.9 x 0.7 x 0.3 m or 0.25 M3 for
                    25 m in a package
UTILISATION
  AREA OF USAGE   : Harbour
  USAGE   : Can be deployed at laying speeds
                    of up to 5/min.
                    Personnel required : 1 to 3
  TOWING   : Up to 18 kt
  SWEEP   : 1 kt
MAINTENANCE
  CLEANING   : Industrial detergents and water or
                    with the EXPANDI BOOM CLEANER
  REPAIRS   : Patch kits of adhesive and fabric
                    are available

```
----------------------------------------------------------------
```

2.6. Inventory of resources (other harmful substances)

A preliminary version of the inventory was distributed at the end of 1987. So far information from Greece, Italy, Portugal, Spain and United Kingdom has not been included due to missing information or too late arrival for the 1987 updating of the system.

It must be underlined that the inventory is not exhaustive. As the amount of information related to harmful substances is very large, the Commission has decided to structure the inventory as a Directory. In this way the inventory has a structure different from that of hydrocarbons.

The inventory contains :

- List of resources : outlines the various resources likely to be found in the inventory. Divided in human resources (expertise, advisors), material resources (equipment, storage, measurements), combined resources (aircraft, ships, strike teams) and other resources (databases, evaluation methods).

- List of national contact points : operational centres for contact on 24h/24h basis. They constitute a great network of information which can be used in case of emergency in order to request assistance.

- Summary of resources : synoptic presentation for each Member State of the resources included in the inventory and actually available in the State concerned.

- Inventory of resources for each Member State : divided into the four types of resources mentioned above.

FUTURE DEVELOPMENTS

The Council Decision instituting the CIS states that "the Commission will gradually compile, in the light of experience, a compendium of information concerning the properties and behaviour of harmful substances or groups of harmful substances other than hydrocarbons" (3).

The first analysis carried out by the Commission clearly showed that a great quantity of information is already available in numerous data banks, in manuals and other various works. As the Commission does not want to duplicate already existing work it is now being considered if the compendium should be developed as a sort of indexed "guide" to already existing information sources.

Besides the current work on the compendium and the problems mentioned under item 2.3 and 2.4, the main task is to obtain information from those Member States which are not yet included in the CIS.

The present state of the CIS displays several weak points, due to the fact that the CIS was developed in two steps - first involving hydrocarbons and from 1986 harmful substances other than hydrocarbons - and because it is a question of large amounts of information of complex and different nature organized and presented in different ways by the Member States.

The information contained in the system is rather detailed and due to the stepwise development exhibits a certain degree of overlapping. What is intended to be a handy manual which can easily be brought to the place of the incident now consists of two big files. It will therefore now be considered whether the two major sections on hydrocarbon and harmful substances can be combined together.

In the Inventories and Catalogue of resources are listed equipment and materials which might be found in rather small amounts or where the transfer time is so long that in practice the resource is not available to other Member States. This data might be removed from the system in the future.

Therefore, the Commission is now starting a project to consolidate the CIS with the main aims to integrate, simplify and shorten the data contained in the System, thereby making the system more operational to those directly involved in the pollution combating actions. This project will also consider whether the system should be computerized, which would be very useful in view of the annual updating of the system.

Finally, it should be mentioned that the management of all activities in the framework of the Community action plan is carried out by the Commission with the invaluable assistance of an Advisory Committee of high level governmental experts. This Committe has been involved in the development of the CIS and will also play an important role in the activities required to simplify and consolidate the Community Information System making it an

important practical tool for dealing with pollution
incidents.

ACKNOWLEDGEMENTS

The author would like to thank Mr A. BARISICH for his
advice on the CIS. He is also grateful to Mr P. WOOD and
Mr C. BYRNE for their comments on the manuscript as well
as Miss A. DE TORNACO for typing the paper.

REFERENCES

(1) Council Resolution of 26 June 1978, Official
    Journal N° C 162, 8.07.1978, p. 1.

(2) Council Decision 81/971/EEC of 3 December 1981,
    Official Journal N° L 355, 10.12.1981, p. 52.

(3) Council Decision 86/85/EEC of 6 March 1986
    establishing a Community Information System for
    the control and reduction of pollution caused by
    the spillage of hydrocarbons and other harmful
    substances at sea, Official Journal N °L 77,
    22.03.86, p. 33.

State-of-the-Art Emergency Information Management and Communication
at the Chemical Transportation Emergency Center

Claudette Cofta
Chemical Manufacturers Association
2501 M Street, N.W.
Washington, D.C.

ABSTRACT

The Chemical Transportation Emergency Center (CHEMTREC) has
recently implemented state-of-the-art digital-optic technology to
manage its vastly expanded chemical information base. The new information system can store millions of pages of information and still
retrieve any single record within seconds. The information is stored
in the CCITT group 3 fascimile standard, opening the door for providing information in a hard-copy format as well as verbally. Combined
with interpreter service from the United States Agency for International Development, CHEMTREC hopes to greatly enhance its ability to
assist in regions outside of North America.

CHEMTREC, the Chemical Transportation Emergency Center, is the
official hazardous materials emergency assistance in the United States
(U.S.). CHEMTREC has assisted in more than 50,000 incidents involving
chemicals since the Center opened its doors in 1971. For situations
ranging from leaking gallon containers to major derailments, CHEMTREC
has performed the two services for callers: immediate basic information is provided about the material and its hazards, and contact with
the product manufacturer is provided for expert assistance and cleanup
instructions.

Fulfilling the first segment of our service, that of providing
fast, chemical-specific information to the caller, has become an increasingly complex task. In recent years the U.S. has experienced
an explosion in the proliferation of published chemical information.
Today I will tell you about CHEMTREC's unique approach to handling
this massive amount of technical information while satisfying the
hotline's requirement for rapid, random access to individual records.
But first, some history to help put the information management demands
into perspective.

CHEMTREC began its service by providing information for the
high-production-volume commodity chemicals, such as sulfuric acid,

269

P. Bockholts and I. Heidebrink (eds.), Chemical Spills and Emergency Management at Sea, 269—273.
© 1988 by Kluwer Academic Publishers.

chlorine and ammonia, from emergency action reports written at the
CHEMTREC headquarters. In these reports the information pertinent
to emergencies was distilled from a variety of sources, including
manufacturer information bulletins and chemical reference manuals.
It was then synthesized into a single set of emergency instructions
for dealing with spill, leak, fire or exposure to the commodity,
along with chemical-specific information such as physical data, decomposition products and reactivity. Because this is a relatively
small file, the records were managed as a paper library.

This approach sufficed for the first years of the program, when
the call volume was fairly low and information on commodity chemicals
was not widely available. For materials not covered by our information base, our emergency contact at the company provided the primary
information.

During the 1970's, our members began to produce Material Safety
Data Sheets (MSDSs) for their employees. They agreed to also submit
them to CHEMTREC because of the emergency data contained in several
sections, such as Fire and Explosion, and First Aid. This was the
beginning of the CHEMTREC MSDS library, and remains today the largest
library of information on trade name and commercial products in the
world.

CHEMTREC now offered a much more powerful information base:
explicit instructions for emergency response for commodity chemicals,
and MSDSs which contained important information relevant to emergencies involving the trade name products and commercial mixtures.

We estimate that there are roughly 3,000 chemicals that fall into
the category of "commodity chemical." For these materials we have
continued to develop and refine our own reports for emergency responders, now maintained on a relational database. The form, appearance,
odor, and any outstanding hazards of the material are reported first.
The pertinent physical data: flash point, vapor density, boiling
point, and flammable limits, are also reported at the beginning, in
an easy-to-read tabular format.

The body of the report is then presented in six clearly marked
paragraphs. The product is characterized according to human health
and reactivity hazards in the first two paragraphs. The third provides precautions to be taken for any incident, and recommends a level
of protective clothing. The next three are incident-specific, for a
spill, leak, fire or exposure. Finally, the decomposition products,
reactivity, container types, and common uses of the material are
described in the last paragraph. These reports can be transmitted
from our microcomputer to firefighters that have computer equipment
available to them in about five minutes. Otherwise the information
is read to them over the telephone.

However, the Material Safety Data Sheet has always been communi-

cated verbally over the phone. Verbal communication of the information
has been necessary because this data is generated and submitted by the
CHEMTREC company registrants as a paper document. Hence, the information is not resident in our computer system. Therefore, the information can neither be retrieved for the hotline via a computer terminal,
nor transmitted to emergency responders.

The obvious answer may seem to be to key the data from the paper
record into our own computers. However, there are a number of com-
pelling reasons that make this impossible.

The first step we would have to take is to create a standard
Material Safety Data Sheet, as none of our 3,800 CHEMTREC registrants
format their MSDSs the same way. For a simple commodity such as
chlorine, we have data sheets ranging in length from two to fifteen
pages, in format from textual letter-style to a highly structured
form document, and in reading level from high school to post-doctoral.
In order to accurately translate this information to our own database,
we would require a PhD industrial hygienist and toxicologist.

But even if we could supply the qualified translators, we still
have serious problems. The paper documents presented to us have
graphical features such as bold-face type, all sizes of type fonts,
molecular sketches or even a skull and crossbones. If we omit any
one of these features, which occurs when a paper record is "computerized," we have altered the communicability and content of the
original document. And therefore assumed the liability for the document content.

Finally, in the United States, the sheer scope of the project is
overwhelming. There are more than 600,000 commercial and trade name
products manufactured in the U.S., for which the MSDS is the ONLY
source of emergency data. It would be impossible for us to transcribe
this data at a rate that would keep it up-to-date with the manufacturer's latest issue.

By 1985 the MSDS library at CHEMTREC had grown to be so large,
200,000 records, that it presented extremely serious problems for
record retrievals during emergencies. We had created an index to the
records on a computer database. The communicator would first query
the database, to confirm that the record was in our library. Then,
the communicator would go to that company's record area, and pull the
appropriate binder. This process took as long as five minutes,
assuming that the record had been filed correctly. It also required
the communicator to vacate one of the hotline stations during the
search period. Difficulties in record retrieval was certainly the
most important problem we were experiencing.

However, there were also serious problems in attempting to manage
and maintain the MSDS file. It was very labor-intensive to file new
records or update existing records, and we began to fall behind. The

filing itself is also very difficult because of the many special characters often found in trade names and chemical names.

Therefore, we recognized that we had to automate our file. Yet, for the reasons discussed previously, the usual method of moving the information to an internal computer system was not an option open to us.

We satisfied our requirement for rapid information retrieval, preservation of the original record format, and rapid uploading of new information through the use of digital-optical technology to write images with lasers onto optical disks. We use the FileNet Corporation image management system, marketed in Europe by Olivetti, to manage all of our information as image-records.

We now take the paper MSDS record that is submitted by a chemical company and scan, or make an electronic snapshot, of each page. The MSDS image record is then indexed according to the product name, manufacturer, date of issue and product code. The images and index record are then written by a high-powered laser on to 12" optical disks. These disks are stored in a special cabinet (known as a jukebox) and retrieved when needed by a robotic arm. The appropriate record is read by a low-powered laser, and the images are transmitted to a high resolution terminal. Two full page 8½" x 11" records can be displayed simultaneously on the extra-wide terminals. The images are displayed in windows, which can be manipulated according to the size and shape desired by the user, in a manner intended to simulate a desk top. The terminals also use a mouse and soft keys, to provide a user-friendly environment.

The most important result of implementing this system is that a record that used to require a manual search of five minutes or more, is now brought to the communicator's terminal automatically in 20 seconds or less following a query. The MSDS images are displayed at the communicator's work station, or can be printed if hard copy is preferred. The data sheet that appears on the terminal is an exact duplicate of the paper record, so that we have not lost any of the information or presentation features of the original record.

We also address the problem of rapid uploading through this system. Because we are simply making a picture of the original record, rather than re-keying the data, the data sheets are loaded very rapidly. We scan documents at a top rate of 24 pages per minute, more than 1,000 times faster than character entry. The cradle-to-grave processing time for an average four page document is about 2 minutes. We are currently loading about 700 MSDSs daily into the system.

Our new method of information management has also opened the door for the first time to electronic communication of the MSDSs. The digital MSDS images are stored on the optical disk in the CCITT Group

III facsimile format. Therefore, the images are "fax ready." We are currently considering faxing the records from the hotline directly to emergency response organizations at the scene of an incident.

In the United States, more and more of our emergency response teams, fire, police and medical, are adding portable faxes and cellular phone equipment to their mobile units. In the U.S., the datalink and mobile fascimile can be purchased for $1500, within the budget of at least medium to large-size fire departments.

Our experience at CHEMTREC leads us to believe that imaging is the future in crisis information management. It circumvents the difficulties with inter-computer compatibility, and can communicate large amounts of information to remote sites without exotic equipment or a high level of computer expertise required by the user receiving the transmission.

It also allows us to cross international borders easily. With the institution of this new service, CHEMTREC hopes to provide more assistance in international incidents. CHEMTREC has also arranged for 24 hour interpreter service in 18 languages to facilitate international assistance. We appreciate any feedback from the international community on this project or other ways in which we can help to provide effective assistance during chemical emergencies.

ECOLOGICAL EFFECTS OF CHEMICAL SPILLS

EVALUATION OF ENVIRONMENTAL HAZARDS

Klavs Bender
Water Quality Institute
11, Agern Allé
DK-2970 Hoersholm
Denmark

ABSTRACT. The paper describes various sources of chemical, physical, biological and toxicological information on substances. It describes classification systems of shipborne chemicals and the grouping of chemicals according to their primary spreading capabilities when discharged to the environment. It refers to the ICES research activity guide to be used after oil pollution incidents and suggests a similar guide be developed covering all aspects of chemical pollution in emergency situations. This guide should be developed in the regime of an international body, i.e. ICES, The European Community or the Bonn Convention.

1. Introduction

In the event of a spill of chemicals at sea, the immediate response of the authorities concerns adverse effects on human health. Depending on the nature of the substance which is discharged to the sea, various types of adverse effects on human health can be visualized. On a longer term the effort is concentrated towards avoidance of environmental impact on a broader scale with emphasis on species which provide food for human consumption. But in order to be able to evaluate long term environmental impact after a certain spill of chemicals at sea, it is necessary to carry out research which includes the entire ecosystem of the impacted areas.

It is important in pre-spill situations to evaluate the potential harm to the marine environment of a discharge of a certain substance. By doing this, necessary precaution can be taken, and in case of a spill, proper action towards combating the spill can be carried out. Several classification and information systems concerning chemicals are in the following described briefly.

For the purpose of defining pollution, The United Nations joint group of experts on the scientific aspects on marine pollution (GESAMP) have formulated a definition, which is reflected in many international conventions.

The definition reads:

"Pollution" means introduction by man, directly or indirectly, of substance or energy into the marine environment, including estuaries, resulting in such deleterious effects as hazard to human health, harm to living resources and marine life, hindrance to legitimate use of the sea inclu-

P. Bockholts and I. Heidebrink (eds.), Chemical Spills and Emergency Management at Sea, 277—283.
© *1988 by Kluwer Academic Publishers.*

ding fishing, impairment of the quality for use of sea water, and reduction of amenities".

This definition is highly valid in connection with evaluation of damage after a spill of chemicals at sea.

2. Pre-spill situation

2.1 Classification of chemicals

Chemicals are classified by several systems. In order to fulfil the goals for an environmental classification taking into account the potential for pollution, criteria such as "harmfulness to living resources" and "fate in sea by spreading" are applied.

In the MARPOL convention noxious liquid substances which are carried in bulk are classified into four categories. Beside these classified substances, the convention provides a list of other liquid substances carried in bulk /1/.

In order to make this classification operational, the MARPOL convention provides guidance and interpretation of the criteria used to classify the various substances in the four categories. These criteria are developed by GESAMP /2/.

In accordance to the definition of pollution the following criteria were taken into consideration when giving the various substances a hazard profile:

- Bioaccumulation

- Damage to living resources

- Hazards to human health ⟨ by oral intake
  by skin contact and inhalation

- Reduction of amenities

- Interference with other uses of the sea, i.e. fishing and corrosion of structure

In addition, carcinogenic properties were taken into account. All shipborne substances with the exception of oil and radioactive substances have been considered by the group.

Generally speaking, high acute toxicity, high potential for bioaccumulation and low degradability in the marine environment increase the environmental hazard, as well as the hazard to human health, when a certain compound is discharged to the sea.

In the latest revision of the MARPOL list of noxious liquid substances carried in bulk, approximately 460 compounds are classified. However, GESAMP has developed a hazards profile for more than fifteen hundred substances or groups of substances.

Another classification system which covers a large amount of substances is also developed within the International Maritime Organization, IMO. The name of the code is:"The International Maritime Dangerous Goods Code (IMDG)". Specifically in relation to the evaluation of potential environmental hazard the system is of more limited use. It is mainly developed with a view to working conditions on shipboard.

In addition to the above mentioned classification systems, different sources of open literature and databases can be used in order to evaluate the potential environmental hazard by discharge of a certain substance. One title, reference /3/, is of outstanding importance due to the comprehensive coverage of the subjects. The volume contains physical, chemical, human toxicological and ecotoxicological data of almost 20,000 different substances. The compounds are listed alphabetically with a synonym-index containing approximately 50,000 names.

2.2 Databases

The most commonly used databases for physical, chemical and biological information of substances are OHMTADS, ECDIN and EXIS. OHMTADS is a part of US-EPA's database system, Chemical Information System, CIS. ECDIN is produced and updated in ISPRA, Italy, on behalf of the Commission of European Communities, and EXIS is British. EXIS contains several modules and individual databases, among these a module with the International Maritime Dangerous Goods Codes.

2.3 Fate of Chemicals in Environment

Besides basic knowledge of toxicity, bioaccumulability and degradability of a given compound, physical and chemical data is of importance in order to predict the fate in the environment of a certain compound in emergency situations. Figure 1 reflects the primary spreading in environment based on physical and chemical characteristics of the various substances.

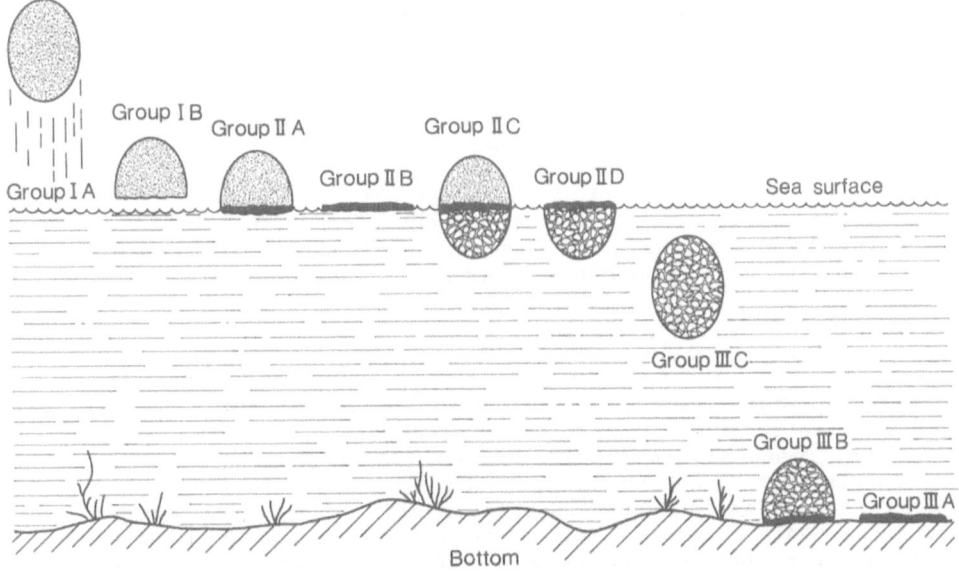

Figure 1   Primary spreading of various groups of chemicals when spilled in sea. Se text for further explanation. From /4/.

As can be seen from the figure, the division of the chemicals is based on three main groups, each of which is divided into several subgroups /4/.

Group I: Substances which form gas- and vapour clouds

                A: Density of substance smaller than air, cloud rises.

                B: Density of substance greater than air, cloud remains suspended

Group II: Substance which continues to float on the water

                A: Substances which float on water and form a slick over which a gas cloud is suspended, their reaction with water being negligible

                B: Substances which float on water and simply form a slick, evaporation and reaction with water being negligible

                C: Substances which float on water and form a slick over which a gas cloud is suspended, a reaction with water also occurring

                D: Substances which float on water, form a slick and react with water, virtually no evaporation occurring.

Group III: Substances which sink or remain suspended.

                A: Substances which sink and react with water

                B: Substances which sink without reacting with water

                C: Substances which remain suspended since their density is similar to that of seawater.

Besides the primary spreading of a certain compound in sea, which can be predicted based on information as mentioned above, wind, tides, current and waves in the specific emergency situation add to the description of the total fate of a chemical discharged in the sea.

2.4 Sensitivity Mapping

In order to complete the evaluation of potential environmental hazards by a certain substance in pre-spill situatuions, it is important to identify key resources in specific areas. The resources in question could be: Special fishing grounds, spawning and nursery area for commercially important species, aquaculture area, marine sanctuaries, amenity beaches and areas where coastally located industries depend on clean water.

In order to get a comprehensive view and compile this type of information, mapping of sensitive resources is a proper tool. Figure 2 from /5/ is an example of mapping of spawning area of commercially important

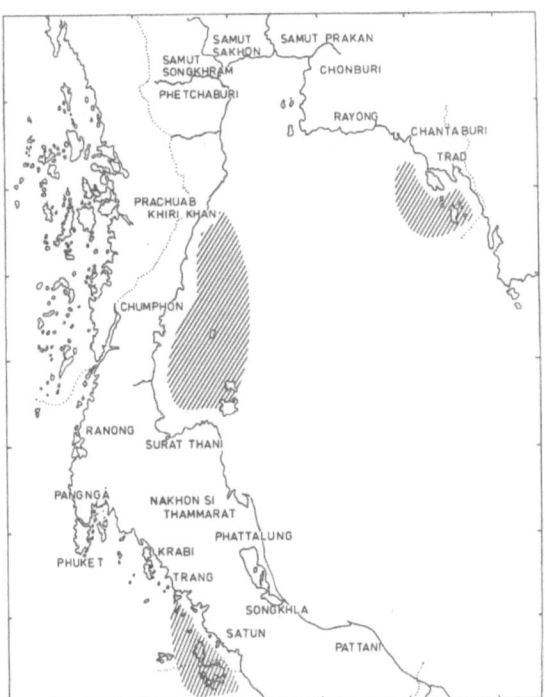

Figure 2   Spawning area for the Indo-Pacific mackerel in Thailand,<u>Rastrel-</u>
           <u>liger brachysoma</u>. Two spawning peaks have been reported: (1)
           February-March, and (2) June August. From /5/.

species in Thailand. This sensitivity mapping was carried out as part of
oil spill contingency planning. In the North Sea, for example, reference
/6/ can be used as a basis for a more detailed mapping of resources in
specific areas.

2.5 Mathematical Modelling

In order to process all the information gathered both in pre- and post-
spill situations, mathematical models are useful tools. In pre-spill situations, mathematical models can be used in spill scenarios for the purpose of training of the emergency response team. In post-spill situations, mathematical models are useful as a tool for the emergency authorities by assisting in the decision-making process in order to combat the
spill or observe and describe the potential effects.

3. Post-Spill Situations

In order to evaluate the environmental hazard during and after a spill of
chemicals at sea, an environmental impact assesment has to be carried
out. Figure 3 visualizes the possible transport routes and the potential
fate which chemical compounds can pursue before they finally either are

degraded to inorganic compounds, deposited in the sediment bottom or combated. Chemical, physical and biological information about the various compounds which are spilled in the actual situation will have to be collected and processed before further steps can be taken as described above.

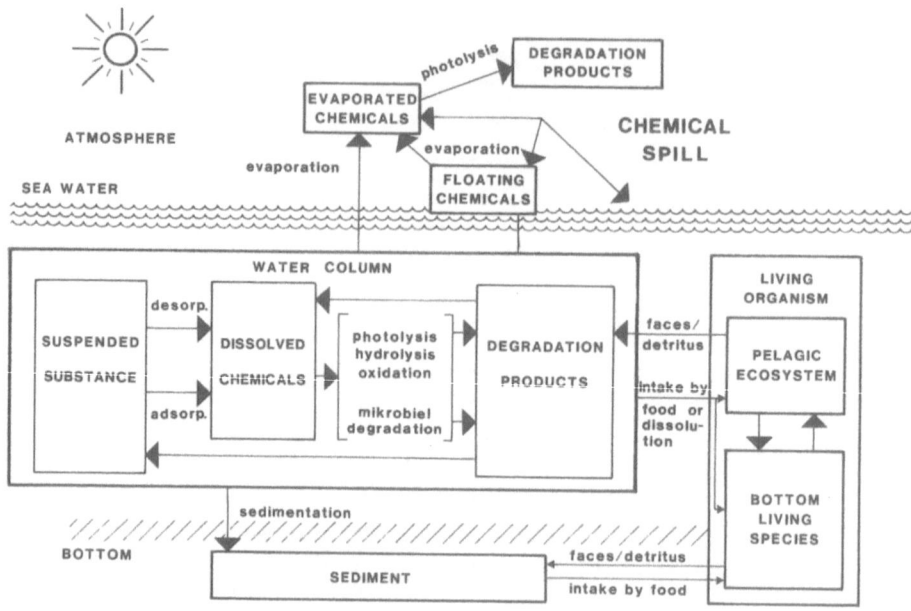

Figure 3   Fate of chemicals when spilled in the sea.

According to this information, the combating action and the exact nature of the study evaluating the actual environmental impact can be decided upon.

If a given compound is a very fast precipitator and at the same time a potential danger to human health, scientific efforts should be concentrated in order to be able to answer questions such as: "Should a town 50 km downwind from the point of discharge be evacuated" or "Is any long term effect likely to occur after a dilution of third order?" Contrary to this, if the compound spilled in the sea is dissolved in the water column or floating on the surface, research efforts will have to be concentrated on effects on pelagic and bottom living organism, and question like: "What is the immediate damage to sensitive area" and "How do we evaluate the long term impact in a specific region", should be addressed.

On a short term basis emphasis should be concentrated on species which are used for human consumption. On a longer term basis, depending on the seriousness of the spill, efforts will have to be directed towards the total ecosystem. Reference /7/ is a guide to be used in scientific research relating to oil spills. It has been developed by the International Council for the Exploration of the Sea (ICES). The guide covers several aspects to be taken into account, both in pre-spill and in post-spill situations. A similar guide should be developed to cover all aspects of chemical pollution in emergency situations.

4. Conclusion

It is of great importance for authorities to be in a position where a proper response can be taken in case of chemical spills at sea. The physical and chemical behaviour of many substances transported at sea makes combatting actions extremely difficult and sometimes even impossible. In these situations the responsibility of the authorities should be concentrated in order to minimize and avoid unnecessary environmental damage to monitor the situation and give proper advice. In order to fulfil this goal pre-event preparations will have to be carried out.

In relation to the particular problem of assessment of environmental hazards in case of chemical spill at sea, a guide for research activities should be developed as described above. This should be pursued on an international basis, either within ICES, the European Community or the Bonn Convention.

5. References

/1/ MARPOL convention. International Convention for the Prevention of Pollution from Ships.
1973. Ammended 1978. IMO. London.

/2/ GESAMP. The evaluation of the hazard of harmful substances carried by ships.
Report and Studies. No. 17. IMO. London.

/3/ Sax, J., B. Feiner, Y.Y. Fitzgerald, T.Y. Aley:
Dangerous Properties of Industrial Materials.
Sixth Ed. Van Nos Reinhold, New York, USA.

/4/ Spaans, L.:
Methods of combatting pollution resulting from chemical spills.
North Sea Directorate. 1981. The Netherlands.

/5/ Bender, K. and R. Taylor:
Oil Spill Contingency Planning in Thailand.
Proceedings of the 1987 Oil Spill Conference. American Petroleum Institute. Washington D.C. USA.

/6/ Atlas of the Seas around the British Isles.
Ministry of Agriculture, Fisheries and Food. 1981. London. Great Britain.

/7/ Research activities related to oil pollution incidents.
Cooperative Research Report No. 107. ICES. Copenhagen Denmark.

ECOLOGICAL CONSIDERATIONS WITH RESPECT TO ACTIONS TOWARDS HAZARDOUS SPILLS

W. Zevenboom
North Sea Directorate - Rijkswaterstaat
P.O. Box 5807
2280 HV Rijswijk
The Netherlands

ABSTRACT. This paper discusses the ecological aspects that are to be taken into account in the various stages of actions towards accidental hazardous spills at sea. The salvage operation of the coaster MS OLAF with her cargo, which formed a threat to marine life, is given as an example. The action scheme comprised of: notification, situation analysis, action plans, execution, and a continuous evaluation and feed back of results. Questions involved were: the possibility of leakage, the amount, behaviour, chemical-physical-ecotoxicological aspects of the cargo-components, environmental situations and the location of the wreck, in order to assess its impact in space and time to the marine environment (water, sediment, biota). Answers to these ecological aspects were provided at relative short notice, using results of analysis of the cargo, transport models, and ecotoxicological guidance values, and led to the decision to salvage the OLAF with her cargo. Seven sampling campaigns conducted during and after the salvage operation indicated that no significant pollution of seawater had occurred and that sediment-pollution was within the acceptable range-level. Recommendations are made regarding improvements in notification and knowledge of ecotoxicological profiles (in seawater) of oversea transported cargo-compounds.

P. Bockholts and I. Heidebrink (eds.), Chemical Spills and Emergency Management at Sea, 285—302.
© *1988 by Kluwer Academic Publishers.*

1. **INTRODUCTION**

Overseas transport of hazardous substances (chemicals, complex wastes compounds) involve certain risk for the marine environment. In case of a ship-accident, hazardous cargo may leak into the sea, and, depending on its amount, behaviour, chemical-physical-ecotoxicological features, may thus threaten marine life over a wide area (transport) and over a long period of time (redistribution of sediment, bioaccumulation).

In such cases, action is required at <u>a short notice</u> in order to prevent the possible large scale threat, in space and time, to marine life.

At the 2nd North Sea Ministerial Conference, in November 1987, it was internationally agreed to reduce pollution of the North Sea and to follow "the principle of precautionary action". The latter implies that immediate prevential action can be taken towards hazardous substances that are likely to threat marine life, without awaiting further scientific evidence of proof of their harmful effects in the marine environment.

In the Netherlands, a precautionary attitude towards hazardous substances-inputs to the sea has been adopted for quite some years. This paper describes shortly a general approach which is designed to deal with a broad spectrum of accidental spills of hazardous substances at sea, and discusses in more detail the ecological considerations that were taken into account in the various decisions and actions concerning the salvage operation of the coaster MS OLAF and her hazardous cargo.

2. GENERAL APPROACH TOWARDS HAZARDOUS ACCIDENTAL SPILLS/ CARGO

Shortly, the approach comprises three major phases: one before, one during, one after decision making (see also Koops, 1986).

Phase 1: - notification
(basic information at the spot, classification)
- situation analysis
(define problem, assess environmental and human impact)
- action plans
(solve problems)

Phase 2: - decisions
(best solutions for various problems)

Phase 3: - execution of action plans
- continuous evaluation and feedback to phase 1, 2.

During phase 1 questions regarding the nature and amount of the spilled cargo compounds, the environmental (local and faraway) situations need to be answered, and required measures/actions should be taken and appropriate tools be used, all in a short notice, in order to assess the environmental impact and risks involved for personal in action.
This is schematically shown in TABLE I.

TABLE I. Phase 1 questions, measures, tools, impact

Question	Answer	Impact	Measures/tools
1. risk of escape?	yes	yes	study at spot repair?
2. - physical behaviour? - fase of spill?	- evaporator - floater - dissolver - sinker	- air, water, a,b,g - water, a,b,c,d,g - water, sed., a-f - water, sed., a,f	- classification, - handbooks, - analysis of original sample
3. - chemical composition? - amount?	- black-grey list - ,	yes yes	- classification, - handbooks, - analysis of original sample - leaching experiments
4. - marine <u>ecotoxicological profile?</u>	- $Lc^t_{50} < 1000$ ppm - $NOEC^t_{50} < 100$ ppm - no biodegradation - bioaccumulation - risk quotient < 0.1	yes yes yes yes yes	- handbooks - when time available: tests with indicator species
5. - scale of effects in space and time? - vulnerable areas?	large long-time involved	yes yes	- transport models for relevant compartments combined with toxicity data

a= mammals, b = birds, c = phytoplankton, zooplankton, d= macrophytes, e = pelagic fish, eggs, fishlarvae, f = bentic fish, macro-meiobenthos, g = (wo)men (personal in action), sed. = sediment, Lc^t_{50} = Lethal Concentration, $NOEC^t_{50}$ = No Effect Concentration (assumption: $NOEC^t_{50} = 0.1 \ Lc^t_{50}$).

Phase 2 is the decision-making phase, and deals with questions regarding salvage actions; is it necessary?, is it possible?, and how should it be done to guarantee minimum risks for the marine environment and human life? The answers depend on the results of phase 1.

In phase 3 the best solutions and tools are implemented in the execution of the action plans. Control and specific sampling campaigns are required in order to follow the salvage operation and to judge whether additional actions are required.

3. CASE MS OLAF

3.1. PHASE 1: SITUATION ANALYSIS

On July 7th 1986 the coaster MS OLAF sank some 40 nautical miles west off the coast of Den Helder as a result of a shift in the cargo. There were no reasons from a navigational point of view to salvage the ship.

The OLAF's red/purple coloured cargo consisted of 3555 tonnes of fly ash produced from pyrite combustion and contained large amounts of heavy metals that represented a significant proportion of the total yearly input into the North Sea (Pb: 444 tonnes (24,6%), As: 114 tonnes (23,5%), Cd: 0,5 tonnes (0,3%), Cu: 20,6 tonnes (1,4%)). It should be noted that international treaties prohibit the dumping/discharge of Cd at sea since it has been put on the "black-list". The others, being "grey-list" substances, require special exemptions.

In order to assess the potential risks of the cargo to the marine environment, series of questions were raised, according to the scheme shown in TABLE I.

Question 1: Risk of escaping of cargo?
Based on divers-reports it was concluded that some leakage of cargo from the ship via a hatch opening occurred.

Question 2: Physical behaviour and fate?
- Insufficient information was available to classify the cargo. A sample of the original material was used to determine the grainsize distribution. 50% of the cargo consisted of grains smaller than 90µm. It was expected that this fraction would behave similar to silt (sus-pended matter) and would leave the observed opening relatively easily, thereby producing a plume of increased turbidity that would be transported in a north easterly direction by residual current.
- Sedimentation of the large, heavier particles, close to the vessel was also expected. This would be easily recognized, by its red/purple colour, and distinguished from the normal sandy coloured clean sediment. The polluted sediment was also expected to move in north easterly direction under influence of watermovements and residual current.

Question 3: Chemical composition, amount, leaching?
- The specification supplied by the manufacturer of the fly ash agreed more or less with the results of ana-lysis of a sample of the original material. TABLE II shows a summary of relevant data.

- Most of heavy metals present in the cargo were
  soluble in seawater (TABLE II, Pb, Cd, Cu, Cr, Zn).
  These results of leaching experiments confirmed the
  suspicion that fly ash from pyrite combustion had
  different chemical (and physical) properties than fly
  ash from coal-fired power stations.
- It was important to know whether the low pH (1,52) of
  the cargo would pose a threat to divers, personal in
  action. Results of dilution experiments (with seawater pH = 8,2) showed that in immediate distance of
  the cargo low pH levels in seawater were to be expected (TABLE II).

TABLE II. Relevant data of cargo of MS OLAF.
          () = data supplied by manufacturer.

Element	Concentration in fly ash mg/kg	Total amount tonnes	Max. proportion leached out %
Pb	125000 (103000)	444	21
As	32000 (21000)	114	0.1
Cd	140 (140)	0.5	17
Cu	5000 (5790)	18	5
Cr	16 (7)	0.06	25
Fe	345000 (406000)	1226	0.03
Zn	5000 (10000)	18	11

pH cargo after 9 times dilutions : 2.12

Question 4: Toxic to marine life?

- This "simple" question hits the gaps in our present knowledge! "Idealy", one would like to use ecotoxicological profiles, in seawater, of the elements/ compound and results of toxicity tests with marine species of various trophic level. The relevant parameters to consider are: LC^t_{50} (Letal Concentration) at various exposure times (t), $NOEC^t$ (No Effect Concentration), biodegradation rate, bioaccumulation rating, and Risk Quotients (RQ, the ratio of in situ background concentration and the lowest concentration at which sublethal effects on various species, in laboratory, are observed) (see also TABLE I). It is expected that results of current research will fine-tune present rather simple schemes like: highly (LC50 $<$ 1ppm), moderate (1-10ppm), slightly (10-100ppm) and non-toxic ($>$ 100ppm) substances. Quite recently a scheme has been developed in which LC^t_{50}, $NOEC^t$ and biodegradation rate values of certain compounds have been combined in order to assess their ecotoxicological spectrum from which their possible risk for marine life can be estimated (Anonymous, 1987).
  In case of OLAF's cargo, being a complex mixture, available data on the elements were used to assess its risk for marine life.

- The majority of heavy metals in the North Sea have RQ values of $<$ 0,1. Based on the background concentrations at the spot, and RQ = 0,1, "maximum permissible" concentrations were calculated. The data are given in TABLE III.

- With regard to biodegradation and bioaccumulation it was concluded that heavy metals are not broken down, and do accumulate in organisms. As a result of this they form a long-term risk to marine life.

Question 5: Scale of harmful effects in space and time;
          Vulnerable area's.

a) Water

The scale of pollution of the waterphase was assessed by using transport models for continuous or instantaneous release of the cargo (TABLE III).

TABLE III. Measured and computed concentrations of heavy metals, around MS OLAF.

Element	Background concentration $\mu g l^{-1}$ a)	"Permitted" concentration $\mu g l^{-1}$ b)	Predicted concentration $\mu g l^{-1}$	
			c)	d)
Pb	0.8 - 0.1	8	444	111
As	0.9	9	114	28.5
Cd	0.04	0.4	0.5	0.13
Cu	0.1	1	18	4.5
Cr	0.5	5	0.06	0.02
Fe	0.01	0.1	940	235
Zn	1.0	10	18	4.5

a) Data from WAMONO-monitoring network, July 1985 (see also Fig. 1).

b) Based on RQ values, WKP part 2B.

c) Continuous release of total cargo within one month; maximum concentration at the centre of the plume, at 1 km distance from OLAF (Spanhoff and Suylen, 1986).

d) Instantaneous release of total cargo with gradual diffusion, calculated for water column of 30 m in depth; maximum concentration one month after release is shown here. The patch length of 1 ton, instantaneously released, is after one month 16 km with a maximum concentration of 0,25 $\mu g l^{-1}$ (Spanhoff and Suylen, 1986).

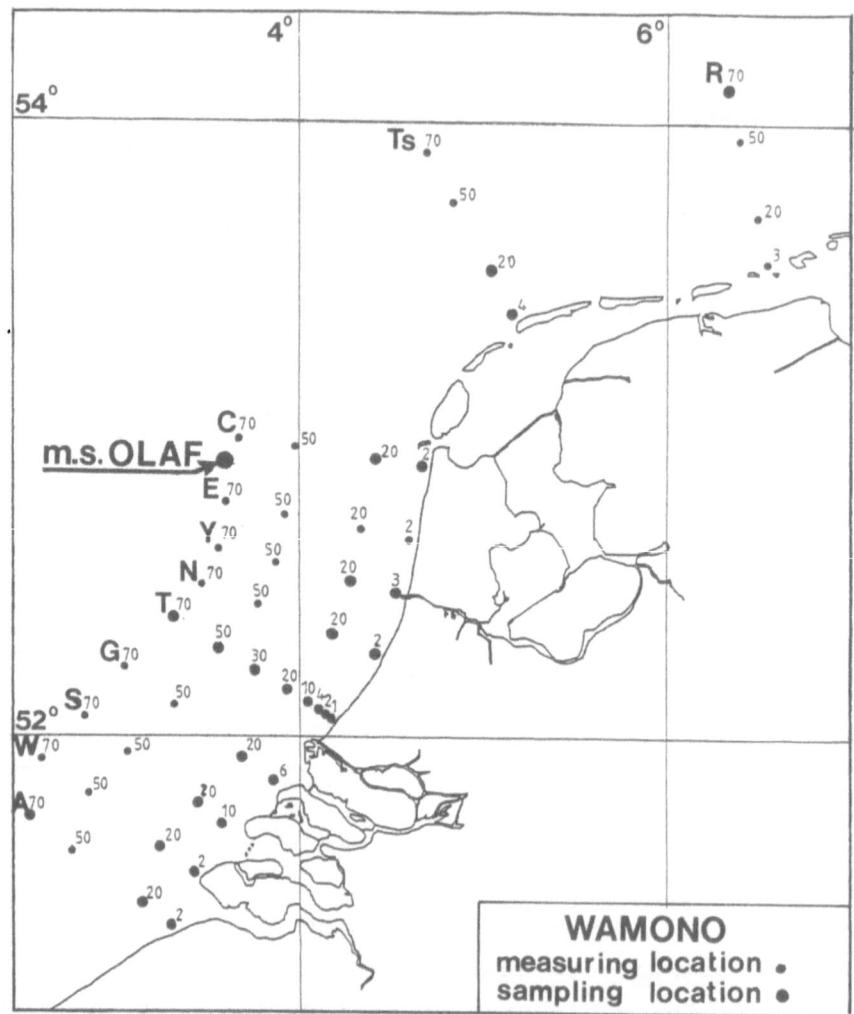

Figure 1. Locations of OLAF and the measuring/sampling locations of the WAMONO network (Water Monitoring North Sea), July 1986.

In case of no salvage of OLAF, a situation of a continuous release of cargo would occur. The concentrations
would have increased by a factor of 10 (for Cd) to 4000
(for Pb) relative to the normal background values (TABLE
III). Due to rotation of the plume under influence of
tidal motion, an area of serveral tens of kilometers would
have been exposed to high, toxic concentrations of heavy
metals for over a month. Moreover, the pollutants would
have been transported to vulnarable areas (Wadden Sea,
Germain Bight) by prevailing currents.

b) Sediment
The levels of the contaminants would increase by a factor
of 10 (for Pb), in the few upper cm over an area of some
30 km².

c) Biota
As already mentioned above, marine pelagic and benthic
life would be threatened for a long time over a large
area.

3.2. PHASE 2: CONCLUSIONS ⟶ DECISIONS

The following ecological considerations, summerized below, led to the decision to salvage MS OLAF with her
cargo, as quickly as possible in a manner which would be
least detremental to the marine environment.
1. The cargo was able to escape from the wreck.
2. The presence of significant quantities of toxic heavy
   metals, in particular those of Pb, which were soluble
   in seawater (TABLE II).

3. Part of the cargo would leave the wreck as suspended matter, which would have serious consequences for waterquality and marine life over a large area (TABLE III).

4) Part of the cargo would show sedimentation, would act as a long-term source of pollution, would affect both water- and bottomquality and would threaten marine life dependent on these environments.

5) Considering the point of discharge and direction of current, the pollution would form a threat to vulnerable area's such as the Wadden Sea.

6) Heavy metals are not broken down, they do accumulate in organisms and, as a result, cause long-term damage to the marine ecosystem.

3.3. PHASE 3: CONTROL- AND SPECIFIC SAMPLING COMPAIGNS

- At least three critical phases during the salvage operation were identified: preparation to tilt and raise the vessel, and transport of the vessel. A total of seven campaigns of control- and specific monitoring of seawater and sediment were performed in order to take immediate action to any detoration that might occur.

3.3.1. seawater monitoring

From MS HOLLAND (and Razende Bol), watersamples were taken, with CTD equipment, at various depth and various distances from the OLAF-location, before, during and after the salvage operation.

During campaigns IV, V, VI, VII, <u>in situ</u> turbidity measurements were performed to trace the plume. The samples were analysed for pH, suspended matter, Pb, Cd, Cu, Ni, Zn, As. The results for Pb (target element, since it was present in highest concentration in the cargo, (see TABLE II) are shown in Fig. 2 (see further results by Zevenboom, 1986).

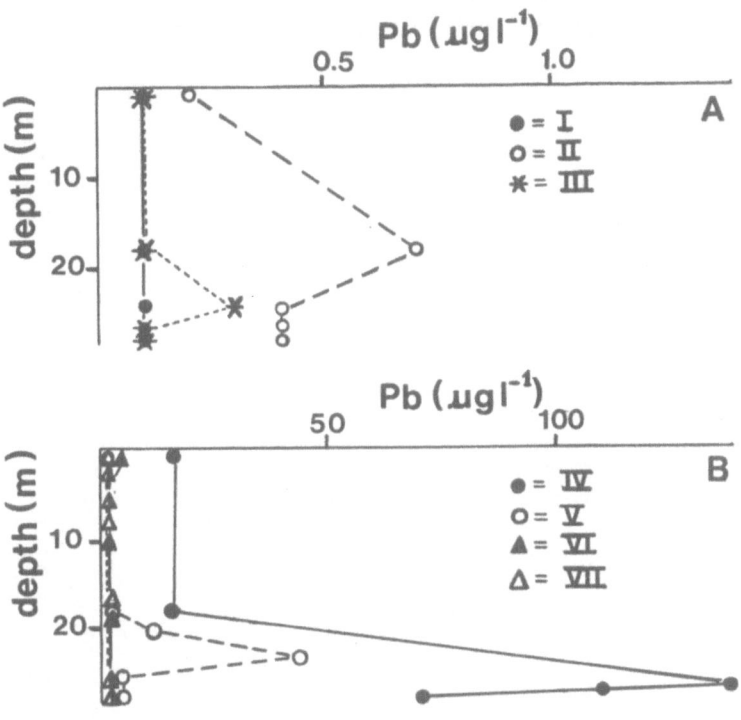

Figure 2. Results on Pb in watersamples taken at P100 location at different depths, during seven campaigns: A) I = 17 July (●), II = 19 July (○), III = 25 July (✳), B) IV = 30 July, preparations for tilting (●), V = 8 August, tilting operation (○), VI = 12 August, raising operation (▲), VII = 13 August, survey after OLAF's departure (△)

298

3.3.2. <u>sediment monitoring</u>

Sediment samples were taken with the Van Veen Crab
(various campaigns), and by divers with perspex tubes,
(campaign IV, Fig. 3) in order to assess the amount and
distribution pattern (horizontal and vertical) of leached
cargo on/in the sediments. The purpose of this systematic
sediment-mapping was to decide, whether cleaning up of
contaminated sediment was required.

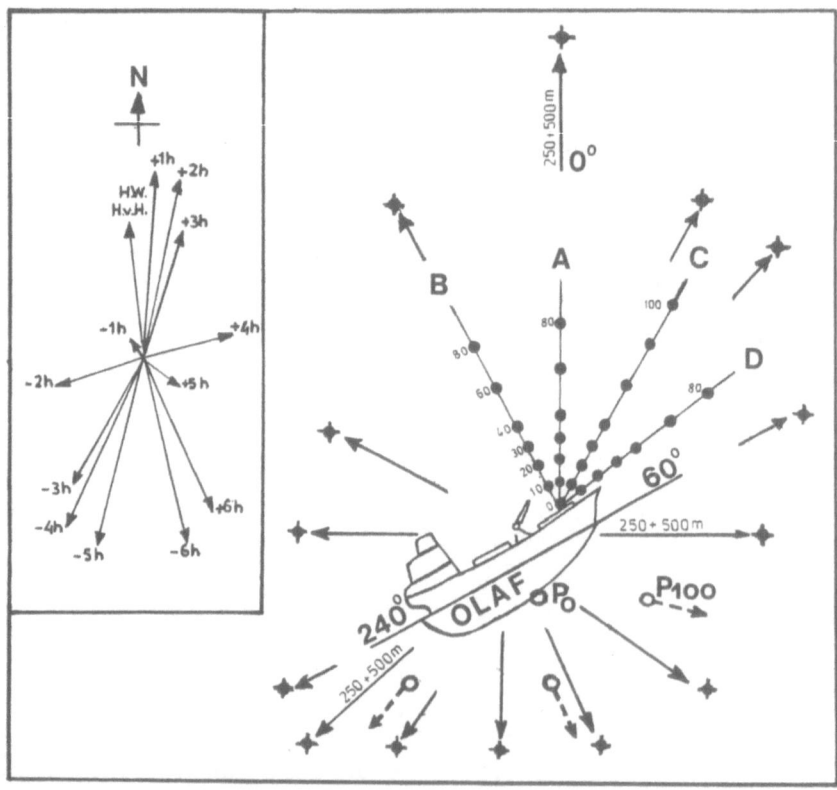

Figure 3. Sample locations of sediment monitoring, Perspex
Cores (taken by divers, campaign IV) (●), and Grab sample
locations (-◆-) at 250 m and 500 m. Also shown are Po and
(variable) P100 (100 m) locations (watersamples), and
flowchart (inserted).

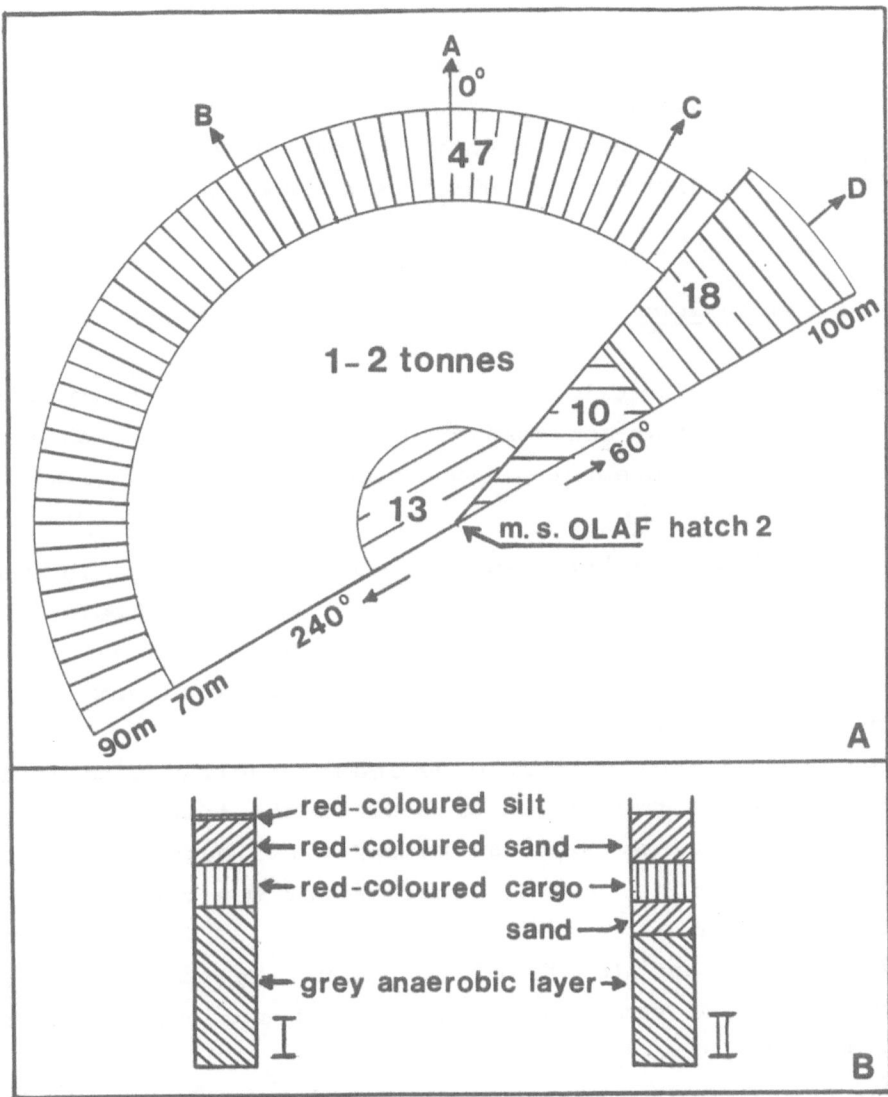

Figure 4. Summary of results of analysis on the amount of escaped cargo found on/in the sediment.
A) Distribution of amount of cargo.
B) Vertical distribution of cargo and sediment charac-teristics in sampling perspex tube cores (30 cm - 40 cm sediment depth).

Using Pb as tracer, calculations were made, prior to the measurements, on the "permissible" range-level of spilled cargo. Values in the range of 1-3% of the total cargo were regarded as an environmental acceptable amount. The results of sediment measurements are shown in Fig. 4.

3.3.3. conclusions seawater – sediment monitoring

1. Apart from the tilting and rising operation (campaigns IV, V, VI) the contamination of seawater by released heavy metals present in the cargo did no give rise to concern (Fig. 2).

2. Based on the highest concentration of Pb found in sea-water samples (138 μgl^{-1}, 100 m. distance, campaign IV, Fig. 2) it was estimated from the transport model (TABLE III) that only ca. 3% of the total amount of cargo had escaped.

3. Results of the systematic sediment sampling also indicated that a proportion of 2-4% of the total amount of cargo had escaped and was patchy distributed on/in the sediment (Fig. 3 and 4). This amount was considered to be permissable.

4. No significant levels of contamination were found in seawater and sediment after the salvage operation was completed.

5. No evidence of cargo leakage could be detected during transport of MS OLAF to the port.

4. CONCLUSIONS AND RECOMMENDATIONS

- Based on ecological aspects it was decided, in the case of OLAF, to salvage the wreck with the hazardous

cargo, as quickly as possible in a manner least deter-
mental to the marine environment and personal in
action.

- It is quite clear that knowledge on ecotoxicological
profiles of hazardous elements, compounds complex substances, wastes, in seawater, should be increased in
order to be able to assess their risks for marine life,
during the situation analysis (phase 1).

- Likewise, ecotoxicological information should be incorporated in transport models, designed for relevant compartments air water, sediment, and biota), in assisting
to assess the spatial and temporal scale of the risk.

- The necessary information (on the cargo, and the situation at the spot) should be available at short notice.
Adequate measures can then also be taken during phase 3
(execution of action plan).

- Although each "case" (incident) may have its own specific problems, it is advisable to have a first-aid con-
tainer-equipment for environmental control measurements.

- Gained experiences should be incorporated in the
general approach -"combatting plan" towards hazardous
accidental spills.

- More specific control on loading, transport and registration of cargo is advisable, and a quick notification
of all relevant data (amount, chemical-physical-
ecotoxicological behaviour) is required.

- A joint approach in handling hazardous cargo by
transport-industry, incident combatters and ecotoxi-
cologists is advisable.

- During the three major phases of the "combatting plan",
good coordination, cooperation, and information exchange
is required. The OLAF case can be regarded as an excellent example.

302

5. ACKNOWLEDGEMENTS

This report summarizes the collaborative efforts of TAUW
Infra Consult BV, The Institute for Inland Water
Management, The Tidal Waters Division, and the North Sea
Directorate. All colleaques, including the salvage company
are greatly acknowledged for their efforts, and excellent
cooperation.

6. REFERENCES

ANONYMOUS, 1986. Water Quality Management Plan North Sea.
Ministry of Transport and Public Works. The Netherlands.
(part 2B, Dutch version).
ANONYMOUS, 1987. Criteria and standards for substances
used in marine offshore mining Working Group Toxicolo-
gical Aspects of OM-MRCP (Ed. J. Boon). (paper in Dutch).
ANONYMOUS, 1987. Second International Conference on the
protection of the North Sea. Ministerial Declaration.
London, 24-24 November 1987.
KOOPS, W. This journal. Policy in the Netherlands with
respect to response of chemical spills.
KOOPS, W. This journal. Classification of chemicals for
situation analysis.
SPANHOFF, R. & J.M. SUIJLEN, 1986. Bijdrage AOG aan DGW
advies inzake Olaf.
Note: GWAO - 86.312 (Dutch paper).
ZEVENBOOM, W. 1986. Salvage of the MS OLAF: Considera-
tions given to environmental concerns. Report NZ-N-86.23
(Dutch and English version).

ANALYSIS OF HISTORICAL EVENTS

EXPERIENCES AND FINDINGS IN CONNECTION WITH THE CASUALTY INVOLVING THE
SHIP CASON

J.A. Madiedo,
Director General de la Marina Mercante,
c/Ruiz de Alarcon, 1
28004 MADRID, Spain

1. INTRODUCTION

At 04.55 hours on 5 December 1987, the Panamanian ship <u>Cason</u>,
navigating close to the Spanish coastline on a voyage from Antwerpen to
Shanghai, sent a distress message reporting that there was a fire on
board and requested assistance. At 05.55 hours, it was reported from
the ship <u>Cason</u> that the fire was out of control and that the ship was
being abandoned.

The rescue service was immediately mobilized involving the intervention
of three helicopters, one tug, rescue launches and ships navigating in
the vicinity. Eight survivors and twenty-three bodies were recovered.

In the first few hours following the incident, the class of goods being
carried by the ship was unknown and it was only possible to ascertain
in general terms, from the hazard labels of the IMDG Code observed by
the rescuers, that it was carrying flammable and toxic substances.

2. SALVAGE PLANNING

After recovery of the survivors and bodies, the rescue tug tried to
salve the ship but the bad weather and the fire which continued on
board made the operation difficult and the wind drove the ship ashore.

Through the EEC Task Force and the maritime authorities of the ports of
Hamburg, Rotterdam and Antwerp, the Spanish Administration received
full information on the ship's cargo and its stowage to enable it to
recover the dangerous goods and in particular the pollutants. The ship
<u>Cason</u> was carrying the following dangerous goods which were stowed on
board in accordance with the attached diagram:

305

P. Bockholts and I. Heidebrink (eds.), Chemical Spills and Emergency Management at Sea, 305—313.
© 1988 by Kluwer Academic Publishers.

Quantity	Substance	Packaging	IMDG-class
125.8 tons	Sodium metal	11 containers (1,430 drums)	4.3
254 tons	Xylene	1,123 drums	3.3
228 tons	N-Butanol	1,285 drums	3.3
110 tons	Ortho-cresol	500 drums	6.1
109.4 tons	Aniline oil	434 drums	6.1
0.7 tons	Diphenylmethane-di-isocyanate	3 drums	6.1
86 tons	Butyl acrylate	430 drums	3.3
10.4 tons	Formaldehyde	46 drums	9
1 ton	Halon	cilinders	2.2
135 kg	Aerosols	2 cases	2.1
5.6 kg	Aerosols	1 case	9
3 tons	Ortho-phosphoric-acid	24 cans	8
50 tons	Phthalic anhydride	2.000 bags	8
2.3 tons	Sulphuric acid	6 units	8
12.8	Gas-oil	55 drums	3.3
75 kg	Resin solution	2 cases	3.2
1 ton	Paint	37 cans	3.1
2.6 tons	Paint	29 cases	3.3
4.7 tons	Cola flavours	86 drums	3.2
3.6 tons	Pentanol	20 drums	3.2
245 kg	Propylene aldehyde	70 drums	3.3
9 tons	Cyclohexanone	36 drums	3.3
750 tons	Bunker fuel	bulk	-

After identification of the cargo and its location on board, an analysis was made of the situation and the salvage and pollution prevention operations were planned. The findings were as follows:
The ship was aground at a distance of about 100 m from the shore in a rocky area and almost the whole of the ship's bottom was damaged letting in water to the holds and the engine room space; as a result it would have been very difficult to refloat.
Smoke was observed coming out from holds 1 and 2, although there were no visible fires, and small quantities of fuel were spilling into the sea.
The main threat of chemical pollution came from the aniline oil, ortho-cresol and diphenylmethane di-isocyanate which are classified by GESAMP with the following hazard profiles:

	A	B	C		E
ANILINE OIL :	0	2	2	II	XX
ORTHO-CRESOL:	T	3	2	II	XX
MDI :	0	(2)	(1)	II	XXX

Furthermore, the IMDG Code includes the three substances mentioned in class 6.1 and only ortho-cresol is given with the new characteristic of "marine pollutant".

The remaining substances were flammable or corrosive and, with the exception of the ship's fuel, presented little threat of marine pollution.

The cargo of sodium metal presented the greatest threat to safety.

In the light of the situation, and after having made a survey on board the Cason, the following plan of action was prepared:

1. unloading the sodium from the deck;
2. unloading the sodium from hold No. 1;
3. unloading the dangerous goods located on deck in the following order of priority: ortho-cresol, MDI, formaldehyde, phosphoric and sulphuric acids, remaining goods;
4. transferring the bunker fuel;
5. unloading the goods from the holds beginning with the most dangerous.

In addition, account was taken of the risk of the possible reaction of sodium with water which, in view of the large quantity being carried, might have caused a major fire and the destruction of highly flammable goods.

In order to determine the possible degree of pollution, it was agreed to analyse the seawater periodically by taking samples at various places on the shore or near to it and a network for detection of possible air contamination of land areas close to the scene of the accident was also set up.

In order to carry out the operations, four salvage vessels were available at the outset with pollution prevention equipment and other auxiliary vessels. Given the need for other items, particularly large cranes, arrangements were made for the provision of such equipment.

3. SALVAGE OPERATIONS

While awaiting the arrival of the special cranes for unloading the containers of sodium, the ortho-cresol loaded on deck was recovered; it was possible to unload 204 drums of ortho-cresol as well as 29 of formaldehyde. However, bad weather in the area unfortunately made it necessary to suspend operations.

As was to be expected, during the following days, from 10 December onwards, drums of sodium were broken open by wave action and a violent reaction with the water began, extending to almost the entire ship since the sodium was stowed above holds 1, 2 and 4 and the reaction spread the fire to the flammable goods. During 10 and 11 December, the

sodium reaction continued as did the fire which, at its period of
greatest intensity, turned the ship into a great mass of flames.
The sodium reaction was also observed out at sea, far from the ship,
owing to drums which had fallen into the sea during the accident.

When all the sodium had reacted, the fire was put out by the waves
which swept the entire deck of the ship.

After the fire had been extinguished, the sea state on 12 December made
it possible to go aboard and carry out an inspection to verify that
there was no release of toxic gases, that the holds were flooded with
water, that there was no remaining fire and that all the sodium had
reacted with water.

The plan of action envisaged was then continued; a large floating crane
was by then available in the area as well as a barge for storage of
recovered cargo and two other support vessels.

However, given the season of the year, the bad weather prevented
operations from being carried out normally and there were few days on
which it was possible to work.

During 19 and 20 December, 21 containers and small quantities of drums
and cases were recovered.

Many drums were thrown into the sea by the wave action and floated on
to the beaches; it was therefore necessary to set up a location,
identification and recovery service in case any of the drums contained
dangerous substances.

The bad weather continued and it was only occasionally possible to work
on recovery of the cargo; these periods were also used for recovering
part of the fuel of the Cason.

From 18 February 1988 onwards, it was possible to work more or less
continuously; in accordance with the plan already drawn up, the
dangerous cargo was recoverd although great difficulty was experienced
in unloading the drums stowed below deck, as the fire and the action of
the sea had converted the holds into an enormous and confused mass of
general cargo which blocked access to the dangerous goods.
Consequently, a great deal of time was lost while the general cargo was
unloaded.

In order to recover the dangerous cargo as early as possible and given
the total disorder in the holds, it was decided to use a large grab
crane capable of seizing parts of the ship's deck and sides and tearing
them apart in order to open up a way to the cargo deposited in the
lower holds. Even using these methods, operations proceeded slowly
because of the stowage pattern of the dangerous goods which were mixed
up with general cargo and, in some cases, under the general cargo
practically throughout the floor area of the hold.

When most of the cargo classified as dangerous had been recovered, all
the aniline carried on boad the Cason still remained on board because
it was stowed in the deep tank which was very inaccessible. Finally,
after removal of the tank deck it was possible to recover this cargo
which was highly dangerous to the personnel involved in the operation.

On 12 March 1988, the work of recovering the dangerous goods was
regarded as complete, although the area continued to be monitored in
order to identify any drums which had been swept into the sea by the
waves as well as other effects of pollution due to fuel or engine oil
residues.

It is to be emphasized that in spite of the spectacular nature of the
operations no fatal accident occurred; only three people were slightly
injured and one showed symptoms of slight poisoning as a result of
contact with aniline.

4. POLLUTION

At the start of the salvage operations, a sea and air pollution
monitoring and control plan was set up.

As far as air pollution is concerned, even though this is not within
the competence of the maritime authority, the responsible bodies set up
a network of detectors which did not at any time register appreciable
air contamination levels, even in the area close to the accident.
However, during the days when the sodium was reacting, large and mostly
white plumes were observed; these were largely composed of water vapour
but alarmed the public.

Pollution of the sea at no time reached high concentrations because
both the fuel and the chemicals were spilt sporadically and in small
quantities, for example, a drum of 200 litres.

During the days when the sea was at its roughest, a number of dark
yellow slicks formed in areas not more than one mile from the ship
because of the breakage of drums; these slicks which gave off a smell
of chemicals, sometimes looked like a kind of "lemon mousse" as far as
their colour was concerned.

It is assumed that the slicks described above consisted of fuel and
chemicals, but they did not give off toxic vapours as was confirmed by
the analyses and also by observation of sea birds which flew over the
area without being affected.

It was not possible to recover the spilled substances because at the
time there were severe storms and the boats were prevented from working
by the high waves and the proximity to the breakers on the shore.

It was also ascertained that, because of the strong wave action, the slicks of chemicals, including the "mousse" that had been formed, disappeared within one or two days.

4.1 Results of analyses of sea water

The analysis of the first samples gave negative results indicating that even if a certain quantity of substances was present, they were at such a low concentration that they could not be detected.

Subsequent analyses began to show the presence of substances carried by the Cason, no doubt due to the fact that drums containing dangerous goods had been broken open by the heavy swell.

Other quantitative analyses gave simular results to those summarized below:

DATE	SUBSTANCE	RESULT
18.12.88 (pH=8.1)	Aniline	Negative
14.1.88	Ortho-cresol Xylol Aniline	11-35 ppm 0.01-0.1 ppm 0.01-0.7 ppm
17.1.88	Ortho-cresol Aniline	Negative Negative
23.1.88	Ortho-cresol Aniline	Negative Negative

These results show that only occasionally, as on 14 January 1988, did appreciable concentrations appear in samples taken within a mile of the ship, perhaps due to the breaking open of a drum as a result of heavy weather. In any case, it is to be noted that the above concentrations diminished sufficiently to be undetectable on 17 and 23 January.

Analyses were also made of fish and shellfish taken from the area close
to the ship and the results of these were as follows:

DATE	TYPE OF SAMPLE	SUBSTANCE	RESULT
18.12.87	Mussels Octopus Goose barnacles	Aniline Formaldehyde	Negative
24.12.87	Mussels Goose barnacles	Aniline Ortho-cresol	Negative
11.01.88	Goose barnacles	Aniline Ortho-cresol	Negative
3.02.88	Mussels	Aniline Ortho-cresol	Negative

5.   CONCLUSIONS AND RECOMMENDATIONS

5.1 It was ascertained that although spillages occurred as a result of
packages being broken open by wave action, water pollution never
reached high levels as the spillages took place intermittently and in
small quantities which were neutralized by the action of the sea which
was at most times very rough in the area of the casualty

5.2 The stowage of dangerous goods in the holds and deep tanks adds
considerably to the difficulties of recovering them and because the
cargo remains longer on board, the likelihood of pollution is
increased.

It is recommended that dangerous goods be stowed principally on deck
or, if not, that they should be located in such a way as to facilitate
unloading in the event of an accident.

5.3. The carriage of sodium metal, which apparently caused the
accident, should be reconsidered as far as the packing of drums in the
container is concerned; the drums should have additional securing
devices and more shock-absorbing material should be used. In addition,
it is recommended that a study be made of the desirability of requiring
stronger packaging than that used hitherto.

5.4. The acceptance of goods exclusively under their trade names
always causes problems of identification; for this reason it is
recalled once again that there is an obligation to use the correct
technical name.

DANGEROUS GOODS LOADED ON BOARD THE M/V "CASON"

Paints (3.3)
Orthocresol (6.1)

Paints (3.1)
Formaldehyde (9)
Sodium (4.3)
Resin solution (3.2)
MDI (6.1)
Aerosols (9)
Aerosols (2.1)

Gas-oil (3.3)
Sulphuric acid (8)
Phosphoric acid (8)
Ortho-cresol (6.1

Xylene (3.3)
n-Butanol (3.3)
Sodium (4.3)
Ortho-cresol (6.1)

Sodium (4.3)

ON DECK

CASON

Fire extinguisher (2.2)
Halon (2.2)

G.C.

Phtalic anhydride (8)

G.C.

n-Butanol (3.3)
Aniline oil (6.1)

G.C.

Cola flavour (3.2)
Pentanol (3.2)
Propylene aldehyde (3.2)
M.D.I. (6.1)

Cyclohexanone (3.3)
Ortho-cresol (6.1)
n-Butanol (3.3)
G.C.
Xylene (3.3)

Sodium (4.3)
G.C.

Butyl acrylate (3.3)
G.C.

5   4   3   2   1

MDI: Di-phenylmethane-4,4 di-isocyanate
GC = General Cargo

5.5 International co-operation to combat marine pollution proved very
positive in the case of the casualty involving the Cason. In this
connection, particular mention is made of the co-operation and advice
received from the Marine Environment Division of IMO and from the Group
of Experts of the European Economic Community belonging to the
recently-created "Task Force".

THE HERALD OF FREE ENTERPRISE ACCIDENT: THE ENVIRONMENTAL PERSPECTIVE

T.G. JACQUES
Ministry of Public Health and Environment
Management of the North Sea and Scheldt Estuary
Mathematical Models, I.H.E.
Rue J. Wytsman 14
B-1050   Brussels (Belgium)

ABSTRACT. Five lorries shipped on board the ferry "Herald of Free Enterprise" that sank off the Belgian coast in 1987 contained together over a hundred different chemicals. During the intricate salvage operation that followed the tragedy, the environmental protection activities met with technical, scientific, legal and organizational difficulties. An assessment of the exact nature, harmfulness, and situation of the cargo had to be done while access to the casualty was limited, and information insufficient. The risks for the personnel and for the marine environment were evaluated using both simplified scenarios and sophisticated computer simulations. Environmental contamination was monitored and protective counterpollution measures were implemented under supervision of the authorities. Although more than half the dangerous cargo was lost, environmental damage was kept to a minimum.

Introduction

On March 6, 1987, the British car ferry "Herald of Free Enterprise" of Townsend Thoresen sank off the Belgian harbour of Zeebrugge causing the loss of about 200 lifes. The ship had heeled over just a few minutes after leaving port and was lying on its side in ten meter depth. After the initial rescue operation, which was one of the most dramatic in history, the recovery of vessel and cargo was made both urgent and delicate by the presence on board of five lorries carrying various dangerous substances. While the owner of the ferry retained his rights and remained responsible for salvaging the ship, the Belgian State assumed overall control of the operations. Refloating the ferry took 52 days and represented an major technical achievement. All work was interrupted six times by bad weather, sometimes for several days. The salvage vessels succeeded in uprighting the ship on April 7, holding her dynamically in position, then had to let her go as a storm was building up. The ship was uprighted again later, holes were plugged, water was pumped out, and the Herald of Free Enterprise (HFE) was carried back into the harbour on April 27 without any further loss of life, or environmental disaster.

Like the salvage operation itself, the counterpollution operation presented pitfalls and difficulties that make the accident exemplary. In the context of intensifying maritime transport of chemicals and chemical wastes, growing awareness of widespread

315

P. Bockholts and I. Heidebrink (eds.), Chemical Spills and Emergency Management at Sea, 315—324.
© *1988 by Kluwer Academic Publishers.*

environmental deterioration, and growing publicity given by the media to threats of marine pollution, public authorities have a duty to react responsibly, decidedly and efficiently whenever a ship casualty carries a danger of accidental marine pollution. As soon as the human aspect has been dealt with, the environmental dimension requires their full attention. The problem is by no means an easy one. To an administration responsible for environmental protection it looks like this: how do we (1) evaluate the harmfulness of the cargo and the danger it represents for people and the environment; (2) monitor cargo and actual pollution; (3) take protective measures to reduce potential impact; (4) integrate those activities in the overall response (rescue operation, salvage, recovery of property, judicial enquiry). Each of these aspects has inherent difficulties which the HFE operation brought to light. This paper examines what these difficulties were and how they were overcome. An attempt is made at deriving some general lessons from the experience.

Base of the Information

In the Belgian organization for dealing with disasters in the coastal zone, the overall co-ordinator is the Province Governor of West-Flanders. The Ministry of Public Health and Environment is the department responsible for national environmental policy and for environmental management of the sea. Within the ministry it is the Management Unit of the North Sea and Scheldt Estuary Mathematical Models (MUMM) that deals with marine affairs. In particular, MUMM is responsible for the environmental monitoring of the sea, it co-ordinates scientific evaluations in case of pollution, and it advises the overall command on counterpollution measures. The present description of the HFE operation is based on the information that was made available to the authorities during the operation. More specifically, it draws on MUMM's records and on the experience of the Unit's personnel engaged in the operation. General accounts of the accident and salvage operation have been published (Vandenbussche, 1987) and lie beyond the subject of this paper.

The Facts

1. NATURE OF THE DANGEROUS CARGO

A complete inventory of the dangerous substances carried by the HFE is given in table I. A first list of dangerous goods was made available to the authorities twelve hours after the accident. That list was incomplete and inaccurate. For example, only "cyanide-containing substances" were reported from lorry B, lorry C was not listed, and the load in lorry E was described as "soluble lead compounds", quite an incorrect description since tribasic lead sulfate is very insoluble in water. It was feared initially that large quantities of tetraethyl lead could be on board. Also, the exact amount of cyanide - in reality less than 750 g - remained unclear for some time. This of course, together with the difficulty of identifying recovered packages that had deteriorated in seawater, later led to considerable confusion.

The full manifest with a summary description of the entire cargo was obtained 73 hours after the disaster. Out of the five lorries with dangerous goods it turned out, however, that three had way-bills lacking essential information. For lorry B the description was so inaccurate that even after additional information had been supplied by the shipper it remained impossible to make out what precisely was on board.

TABLE I. Herald of Free Enterprise accident: inventory of the dangerous cargo based on the documents made available to the Belgian authorities and on the salvaged cargo.

Lorry	Substance	Total quantity (kg)	Packaging		Recovered
			Type	Number	
A	TDI[1]	5,450	218 kg drums	25	7 drums
B	Cyanide-containing wastes (<200 ppm CN)	270 (<540 g CN)	200 l drums 120 l drums	2 3	1 drum?
	Cyanide-containing hardening salts	130 (<208 g CN)	200 l drums	1	?
	CN-containing liquid		30 l drum	?	1 drum
	Chlorine trifluoride + fluorine perchlorate		30 l gas tanks 10 l gas tanks	6 2	4 (some empty)
	Twelve chemicals[2] in small gas bottles		0.1, 0.25 and 0.5 l bottles	21	-10
	Chemicals[3] in 0.1 to 2 l bottles packed in buckets		12 l buckets 25 and 30 l buckets	150 30	-80 -15 buckets
	Paint wastes		200 l drums	40	30 drums
C	Hydroquinone	5,000	25 kg sacks[4]	200	none
D	Leatherpaint	3,500	100 kg drums	35	26 drums
	Leatherpaint with MET[5]	1,500	100 kg drums	15	13 drums
	Leatherpaint diluent	50	10 kg drums	5	3 drums
E	Tribasic lead sulfate (88% in granulates)	19,925	25 kg paper sacks[6]	797	-all

[1]Toluene di-isocyanate in solution.
[2]Hydrogene bromide (7 half-l bottles), Chlorine (3 half-l), Ethylene (2 half-l), CO_2, CO, HCl, Fluoromethane, Ether, Chlorotrifluoromethane, Methylamine, Tetrafluoroborate, Antimony pentafluoride.
[3]Fifty-two reported and several tens of unreported chemicals including Br, Na, K, P, Aniline, Acrylonitrile, acids etc.).
[4]Paper sacks with polyethylene layers.
[5]Leatherpaint agent containing Methoxyethanoltoluene.
[6]Twenty pallets of 40 sacks packed in polyethylene sheets.

Table I therefore combines information that was communicated by the shipping companies with the final inventory of recovered cargo drawn by the authorities. A large amount of work was required to verify the nature and the precise state (solid, liquid, solution, etc.) of the substances, their location on board, the type of package or container and its likely behavior in water, and how each package could be identified.

2. ASSESSMENT OF THE HARMFULNESS OF THE DANGEROUS GOODS

In the first hours of the incident some basic information on the properties of the reported chemicals was supplied by the Information Center on Dangerous Substances of the Ministry of Interior (BIG). Thereafter, a number of sources were consulted in addition to the standard IMDG code and GESAMP hazard profiles of the International Maritime Organization. The most useful were found to be the works of Verschueren (1983), unfortunately limited to organic substances, and Sax (1984). These books give information on the physical and chemical properties of the substances and they report available toxicity data, including aquatic toxicities. A first complete evaluation of dangers for people and for the environment connected with the cargo was presented by MUMM on March 8. It was supplemented by the Toxicity Department of the Institute of Hygiene and Epidemiology in Brussels using among other sources the Institute's ISOTOX data base (IHE, 1987). Table II gives the concentrations at which the most worrisome of the substances could be considered hazardous to marine life.

It remained difficult to properly assess the environmental hazard the dangerous cargo constituted, mainly because so little is known of the toxicity of most chemicals to marine organisms. An important factor to be considered was the fact that once released in the marine environment many chemicals react or degrade, so that the reaction products too require attention. In the present case substances such as quinone, the degradation product of hydroquinone, and toluene diamine, the hydrolysis product of toluene di-isocyanate (TDI), were found to be no more toxic than the original chemical. Finally, an assessment of the actual danger could only be arrived at by evaluating the likely behavior of the goods after their release at sea, and by monitoring the environment to detect the presence of the chemicals. These two activities are described below.

3. PREDICTION OF THE BEHAVIOR OF CARGO LOST AT SEA

As soon as the nature of the cargo was known a major concern of MUMM was to evaluate how grave the situation would be if the toxic substances were spilled, and what measures could be taken to minimize impact and maximize the chances of recovery. Lorry A had fallen overboard as the ferry heeled over. It had been pulled out of the water empty: the drums of toluene di-isocyanate were therefore lost at sea. Lorry B was visible on B deck at low tide and only some of its 11 ton chemical cargo had been released. Lorry C with five tons of hydroquinone was deep in the E deck garage and could not be controlled before the ferry was uprighted. Lorry D had also fallen overboard and only the cab had been recovered. Drums of leatherpaint were adrift, but some had quickly been picked up. Lorry E was deep in the hold on B deck and not visible: the condition of the 20 ton lead sulfate was unknown.

Firstly a simplified scenario was used to quickly calculate the extent of the sea area in which concentrations considered lethal for marine organisms could be reached. This scenario assumed immediate and uniform dispersion of the quantity of each chemical that remained unaccounted for over an average depth of 10 m. The toxicities in table II were used as guides in these estimates. Except for the five tons of hydroquinone none of the dangerous substances could cause real problems beyond 165 m of the source. The greatest risk thus came from the hazard drums, tanks, buckets and other containers constituted for the people who would manipulate them. It was therefore important to try and locate those objects, or predict where they could go. For the hydroquinone, the calculation indicated the possibility of a deadly cloud in the water column extending more than 2 km from the source, and therefore more precise risk evaluations were in order.

TABLE II. Herald of Free Enterprise accident: potential environmental hazard presented by the dangerous cargo. All concentrations in mg/l (n.d.= not detectable). Toxicity data from standard reference works and from review by the Institute of Hygiene and Epidemiology. Measured concentrations in seawater as reported by the various State laboratories participating in the monitoring.

Substance	Concentrations hazardous to marine life		Concentrations measured in/around casualty	
	Tolerance limit (sensitive sp.)	Lethal to some fish species	Minimum	Maximum
TDI[1]	1 - 10	164	n.d.	n.d.
Cyanides (CN)	<10	0.05	<0.005	<0.005
Hydroquinone[2]	0.05	0.044[5]	<0.5	1.8
Toluene[3]	10 - 100	24	<0.005	0.040
Lead[4]	0.22[6]	500	<0.001	0.076

[1]Including toluene di-isocyanate (detection limit 0.001 mg/l) and toluene diamine (detection limit 0.020 mg/l).
[2]Measured as hydroquinone + quinone + total phenols.
[3]Taken as representative of unspecified solvents reported in the cargo.
[4]Total lead in water after removing oily phase in oil-contaminated samples.
[5]LC50(96h) for larval fish. Sensitive adult fish reported to be affected at 0.1 mg/l.
[6]LC50(28 days) for rainbow trout.

Hydroquinone is not a very dangerous product for man, but it can be extremely harmful to aquatic organisms. Using the dispersion model MU-DISPER, MUMM ran a series of mathematical simulations on computer in order to study the fate of the substance (UGMM, 1987). The results are reproduced in table III. A standard run studied the effect of the tide on the extent of the contamination and showed that the affected water mass was largest for a spill starting during flood. The isoline corresponding to the LC50 for larval fish contained an area of 7 km^2. This was less than the crude calculation had suggested, but still significant. A real-time simulation using the actual meteorological forecast and wind forcing on marine currents was run on April 7 when the ferry was uprighted, because it was feared that cargo would shift and cause a spillage. It turned out, however, that by that time the entire load of hydroquinone had dissolved away unnoticed, probably during the storms that had affected the area in the previous two weeks.

TABLE III. Herald of Free Enterprise accident: predicted extent of contamination of the sea by five tons of hydroquinone released from the casualty over a period of six hours. The MU-DISPER computer model was used in a standard situation to examine the effect of the tide, and in real time on April 7 during the uprighting operation.

Time spill begins	Maximum extent of the 0.050 mg/l isoline (km^2)	Highest concentration predicted (mg/l)	Time of the highest concentration (hours after begin of spill)
Standard run[1]			
High tide	4.8	0.300	10
Ebb	3.6	0.247	10
Low tide	5.4	0.200	4
Flood	7.0	0.135	2
April 7			
06:00 UT	-5	0.233	11

[1]Using the M2 and S2 tide components in spring-tide regime.

Since the casualty was also loosing oil through vent pipes, an oil-fate model was run to anticipate the drift of possible oil slicks (UGMM, 1987). An attempt was made at adapting the model in order to simulate the drift of floating drums. The model evidenced the dramatic effect of wind on the predicted trajectories, but it was not felt to be sufficiently reliable to be of much use. The difficulty lied in the uncertainty about the portions of the objects which would emerge and remain immersed, and the geometry of each. No model was available to simulate the behavior of drums lying on the seabed. To gain some insight on the fate of the TDI drums Townsend Thoresen had a dummy drum equiped with a transmitter and released where the real drums had been lost, but currents in the area vary much with weather and tide, and the tracking experiment produced little usable results.

4. ENVIRONMENTAL MONITORING OF THE CASUALTY

4.1. *Objectives and organization of the monitoring.* The environmental monitoring pursued two main objectives: to detect the presence of dangerous substances in the air or water inside the wrecked vessel in order to control the safety of personnel involved in the operation, and to assess the extent and impact of a spillage in the surrounding marine environment if one occurred. Safety inside the casualty was the concern of both the salvage company and the authorities, the latter having to conduct the judicial inquiry on board and to recover the remains of the victims. Air and water samples were therefore taken by both concerns and analyzed separately. The State oceanographic vessel "Belgica" provided a platform for on-site monitoring, and a container-laboratory was installed in Zeebrugge harbour. Gas measurements (CN detection and explosimetry) were made inside the casualty by personnel of the Air Pollution Section of the Institute of Hygiene and Epidemiology (IHE). The quantity of cyanide in the cargo was small, but because of the presence (actually prohibited) of unidentified acids on the same lorry, the formation of deadly cyanhydric acid could not be excluded. Cyanides in water were analyzed on board the Belgica by IHE personnel and by Navy personnel under MUMM's control. The hydroquinone was measured in the container by IHE personnel. Samples were dispatched every day to three laboratories of the Ministry of Agriculture and to the IHE's Water Pollution Section for confirmation of the field measurements and further analyses.

4.2. *Methods of analysis.* Cyanides in water were determined by colorimetry with pyridine-pyrazolone on a Technicon autoanalyzer using distillation. Toluene di-isocyanate and its hydrolysis product, toluene diamine, were determined by gas chromatography. Toluene, chosen as a typical toxic solvent likely to appear in paint and other organic products, was determined by gas chromatography coupled with mass spectrometry in order to eliminate the masking effect of oil. Hydroquinone was measured together with its degradation product - quinone - and total phenols by colorimetry with 4-amino-antipyrine in ammonia buffer on Technicon. Lead was determined by atomic absorption, after occasional separation of the organic fraction from the inorganic one by extraction.

4.3. *Sampling strategy.* Sampling was often made difficult and even dangerous by rough seastate and instability of the wreck. When boarding the casualty was impossible, seawater was sampled downstream at a safe distance. It was realized that any spilled substance would disperse so quickly that detection in the surrounding water would be impossible unless the spill was a massive one. This represented a major limitation for the monitoring strategy. To overcome this difficulty, marine organisms were monitored to see if a slow, unnoticeable leakage had caused bioaccumulation, and sediments were analyzed to find out if lead particles had escaped and spread around the vessel. Neither were found to be contaminated (Baeteman et al., 1987).

4.4. *Results of the water analyses.* The results of the water analyses are summarized in table II. All samples were below the detection limit for cyanides and toluene di-isocyanate. Total lead remained low, a maximum of 0.17 mg/l being recorded when suspended and oily matters in the water sample were included in the measurement. Solvents were detected occasionally, registering a maximum of 0.04 mg/l of toluene, well below the toxicity limit. Hydroquinone around the casualty remained below the detection limit at all times, but one single seawater sample taken on E deck on 11 April contained 1.8 mg/l of that substance. E deck had in fact been visited on 8 April at low tide while the ferry was held upright by the salvage vessels. Lorry C had been inspected on that occasion and found totally empty, no trace of the sacks of hydroquinone being visible. The April 11 sample indicated some residual contamination of the wreck. It is likely that the hydroquinone had dissolved away during stormy weather without being picked up in the scanty monitoring the seastate had permitted. Indeed, it can be seen from table III that the highest concentration predicted in case of spillage was below the detection limit (0.5 mg/l) of the only field method that could

322

practically be implemented. The detection limit also lied one order of magnitude higher than the toxicity threshold of the substance, which made the monitoring useful for warning purposes, but not for correct impact assessment.

5. PROTECTIVE MEASURES

A number of measures were decided in order to mitigate pollution effects. They included arrangements for oil containement and recovery, recovery of the packaged goods lost at sea and containment of the remaining cargo within the casualty, and attempts at locating and securing the dangerous goods on board. A keep-off zone for ships and aircraft was imposed by the State to protect the operation. The position of the authorities was that first-line protective measures were the responsibility of the shipowner and needed to satisfy the requirements laid down by the administration. The State insured a second line of defense by keeping vessels and equipment on stand-by, by supervising the operation, and by assisting in the search and recovery of lost cargo.

Absorbant material in 4 m boom sections was layed inside the ferry to retain leaking fuel. A net was fastened to the structure across the gaping opening of the open garage to prevent parcels and wrecks from escaping. A debris-retention net and an oil boom were deployed downstream of the casualty during the uprighting operation in order to intercept oil and floating objects. Very little oil escaped from the ship and in fact none was recovered at sea. No oil dispersants were used. Little could be done to remove the cargo from the ferry before it had been refloated and towed back into the harbour: only a few gas cylinders were located and removed while the vessel was at sea.

All vessels on scene participated in the recovery of floating objects. Navy mine hunters and salvage tugs did the search for sunken objects, but in spite of an intensive survey of the Zeebrugge ship channel and the Wielingen area only seven of the 25 TDI drums were recovered. One major difficulty was to sort out the various items among the recovered cargo and to recognize those containing harmful products. Labels peel off and drum markings quickly become obliterated in seawater. A full and precise description of the appearance of both the dangerous and the harmless packages would have been useful, but in many cases it was not available. It required a considerable amount of time and organizing to keep an accurate inventory of the recovered cargo and to verify the contents of suspicious containers. Yet it was essential to keep an updated inventory of the dangerous substances in order to correctly assess the residual danger at sea. A complete list of the dangerous goods picked up at sea or eventually removed from the casualty is given in table I. It can be seen that more than half of the goods known to have been shipped on board the HFE remained missing.

6. INTEGRATION OF THESE ACTIVITIES IN THE OPERATION

The integration of counterpollution activities in the salvage operation met at first with organizational and administrative difficulties that compounded the technical ones. They were due to the variety of simultaneous operations requiring co-ordination, to the conflicting interests of the State and the shipowner, and to a lack of correct understanding of the environmental implications of the situation by some of the parties involved.

For the owner of the HFE, the main concern was the salvage of the vessel in the shortest possible time and at the least cost. The salvage company's interest was to use simple, efficient refloating techniques with as little outside interference as possible. For the authorities the concerns were humanitary, judicial, nautical (freeing of the waterway) and environmental. Decisions were taken in a co-ordination committee where all the parties involved were represented. Thus the work of the salvage company was co-ordinated with the collecting of the bodily remains of the victims, the judicial inquiry on board (nothing could be disturbed before inspection by the judiciary) and ashore (by court order the recovered chemicals were held in custody), the search for, identification and recovery of the dangerous cargo, the environmental monitoring in and out of the casualty, and the protective measures against pollution. At each step of the operation an agreement had to be reached on priorities, communications, and ways of making these activities compatible.

As soon as the initial rescue operation was over and the presence of the dangerous cargo was known, the authorities announced their intention to set requirements in relation to counterpollution measures and to carry out a continuous survey of water quality. Immediately the shipowner went to court against the State, out of fear that the proposed measures could interfere with the salvage of the ship. The court appointed an independent expert and instructed him to reconcile points of view. All decisions relating to environmental protection were thereafter made under court supervision. This slowed down the procedure initially, the State having to use persuasion rather than authority in order to implement its policy, but it also presented the advantage for the State that any counterpollution measure agreed upon during the operation could not easily be contested by the shipowner afterwards.

Discussion

The Herald of Free Enterprise operation was long and difficult, but successful. Although more than half the dangerous cargo was lost in the accident, the ship was salvaged, and the environmental damage kept to a minimum. A number of lessons were learned by the authorities in charge of environmental protection, pointing to problems that deserve attention.

It is clear that a wide variety of packaged chemicals are transported by car ferries with completely inadequate protection in case they are lost at sea. Plastic and paper sacks have no chance of remaining intact in seawater. Partly filled drums sinking to the seabed do not resist the water pressure. It is also evident that such shipments are inadequately documented. Information is crucial in the management of chemical accidents (Jacques, 1985). The actual - not merely legal - requirements go well beyond the bill of lading to include specific data that, at this time, only direct contacts with shipping agencies and producers can supply: form under which the substance is containerized, concentration, nature of the solvent, degree of filling and weight/volume ratio of drums, recommended method of detection. It would be helpful if that information were required on shipping documents.

Many chemicals can only be detected with sophisticated instruments, and field methods applicable to seawater often lack the desired sensitivity. For those substances, harmful concentrations cannot readily be detected at sea, and only behavior/dispersion models can provide information on the level of risk, the fate of the spilled chemical, and its likely impact. There is a great need for data on the properties, behavior in seawater and marine toxicities of many dangerous substances, to ensure that mechanisms can be simulated and dangers assessed properly. A pressing need also exists for mathematical models capable of simulating the behavior of solid objects floating at the surface of the sea or drifting on the seabed.

324

From the organizational point of view, the operation revealed the absolute need to co-ordinate the recovery of drifting and beached wrecks, and to keep a precise inventory of these goods in order to permit a correct assessment of the residual risk. The need to formally incorporate environmental protection activities in general disaster management procedures and contingency plans was confirmed by the legal and organizational difficulties the administration met with in conducting the counterpollution program. One shoud recognize that pollutions do threaten the safety of man as well as nature, and it is imperative to consider the marine environment as a resource, to be treated on the same foot as other resources, and with all the more care that so many ecological mechanisms remain poorly understood.

Acknowledgements

I want to thank all MUMM, IHE and Belgian Navy personnel who participated in the counterpollution operation. A. Pollentier and G. Verreet assembled the information on cargo reproduced in table I. Table II was compiled from published information and internal reports kindly made available by the departments Environment (Water Pollution Section) and Toxicology of the IHE. S. Scory ran the computer simulations reported in table III.

References

Baeteman, M., D. Maertens, D. Declerck (I.W.O.N.L.), W. Vanhee, W. Vyncke, R. Gabriels, W. Van Keirsbulck, R. De Borger, Y. Van Elsen, P. Van Hoeyweghen, M. Termonia, A. De Meyer, J. Walravens & P. Dourck, 1987. 'Scheepsramp "Herald of Free Enterprise" DD.6.3.1987, Monitoringresultaten van het gemeenschappelijk urgentieprogramma "Visserij"'. *Mededelingen van het Rijksstation voor Zeevisserij (C.L.O.-Gent)*, Publ. Nr. 216: 11 pp.
IHE, 1987. 'Ongeval met de "Herald of Free Enterprise" (06/03/1987)'. (Internal report). Instituut voor Hygiëne en Epidemiologie. Brussels, 13 March: 19 pp.
Jacques, T.G., 1985. 'Scientific evaluations of an incident at sea involving a sunken ship carrying a dangerous cargo'. pp. 343-357 in: *Progress in Belgian Oceanographic Research* (R. Van Grieken & R. Wollast, eds.). Proc. of a Symposium held at the Palace of Academies, Brussels, 3-5 March. University of Antwerp Publ., Antwerp.
Sax, N.I., 1984. *Dangerous Properties of Industrial Materials*. Van Nostrand Reinhold Company, New York. 3124 pp.
UGMM, 1987. 'Herald of Free Enterprise Incident, 6 March 1987'. Report of the Ministry of Public Health and Environment, Belgium, to the 16th Meeting of the Bonn Agreement Working Group on Operational, Technical and Scientific Questions Concerning Counter Pollution Activities, Aalesund, 11-14 May. *BAWG OTSOPA* 16/4/7-E: 2 pp. + 4 annexes.
Vandenbussche, F., 1987. *'Berg de Herald', het gevecht na de ramp*. Aksis, Antwerp. 95 pp.
Verschueren, K., 1983. *Handbook of Environmental Data on Organic Chemicals*. Van Nostrand Reinhold Company, New York. 1310 pp.

The Dinoseb incident 1984

by Commander Preben S. Stamp
National Agency of Environmental Protection, Denmark
Strandgade 29
1401 Copenhagen K
DENMARK

Abstract

In the beginning of 1984, the Danish ship DANA OPTIMA lost 80 drums
of the chemical DINOSEB in the middle of the North Sea. Due to the
toxicity of the chemical the Danish Minister for the Environment
decided that the drums should be located and recovered . However,
inaccurate information on the exact position of the loss of the
drums and lack of experience in search techniques in the danish
contingency services resulted in a long - lasting search period.

The major part of the drums were found and recovered by various
means. This paper deals with the Danish search and recovery - operation, which gave a lot of experience, how to handle incidents
involving loss of packed chemicals from ships, a rather usual situation in the North Sea.

1. Introduction

On January 12, 1984 the Danish ship "M/S DANA OPTIMA" destined for
Esbjerg departed from North Shields in the United Kingdom.

The 1,599 GRT ship carried as deck cargo 42 containers and trailers, most of which were empty. However, one truck-trailer carried
80 drums of the chemical DINOSEB.

After departure from UK the ship ran into stormy weather which in
connection with the stop of the main and the auxiliary engines
caused that during Friday 13 of January 39 containers and trailers
from the deck cargo, including the cargo with the DINOSEB, fell
overboard (see Annex 1). During the night between the 13th and the
14th of January "M/S DANA OPTIMA" drifted among various Danish
off-shore installations but fortunately no collission took place.
During January 14 control over the ship was regained and the ship
could without assistance proceed to Esbjerg where it arrived
Sunday, January 15 at 0210. Information that the deck cargo fell
overboard was given to a Danish costal radio station, but no
information with respect to the loss of the DINOSEB was given.

P. Bockholts and I. Heidebrink (eds.), Chemical Spills and Emergency Management at Sea, 325—343.
© *1988 by Kluwer Academic Publishers.*

On Saturday, January 14 the National Agency of Environmental Protection's (NAEP) officer on duty was informed by the Flag Officer Denmark (FOD) that the missing deck cargo did not contain any substances harmful to the marine environment.

However, on Monday, January 16, the NAEP's officer on duty was informed by a private firm in Esbjerg that among the missing deck cargo was a truck-trailer with 80 drums of 200 litres DINOSEB each. This information was immediately after confirmed by the shipowner DFDS. The following day FOD issued a warning to mariners as well as to relevant authorities about the loss of the DINOSEB.

In the light of available information about DINOSEB's toxicity as regards man (see para 2), the Minister for the Environment requested the NAEP to investigate if it was technically feasible to find and recover the 80 drums. The NAEP confirmed that it was possible but that conditions as covering by sand or the movement of the drums might hamper the investigations.

The NAEP recommended that the search for the drums should be carried out from the NAEP's oil combatting vessel "Gunnar Seidenfaden" which should be equipped with a so-called Side-Scan Sonar (SSS), and special navigational equipment as well as with a so called Remote Operated Vehicle (ROV) equipped with TVcamera. Furthermore, divers should participate in the search.

2. Toxicity of Dinoseb

At that time the DINOSEB was used occupationally in Denmark by farmers and gardeners for weed control.

DINOSEB was one of the most acute poisonous sprays used in Denmark. After September 1984 import and sale of DINOSEB for use by farmers etc. is forbidden, because of its acute toxicity and its long term effect on health and environment.

The prohibition of the use of DINOSEB was a consequence of the re-evaluation of the pesticides which the NAEP had initiated under the provisions of the Act. on Chemicals etc.

Rat-tests had shown that consumption of approx. 40 mg DINOSEB per kilo body weight could be lethal. It was assumed that the toxicity for man was the same dimensions, which means that approx. 3 g of the substance could be lethal for adults. DINOSEB can be absorbed through the digestive system and the lungs, as well by skin contact.

The toxicity towards fish could be as low as 70 microgrammes per litre sea water (lethal limit). DINOSEB bio-accumulates in some organisms, but due to its high toxicity towards fish it is not expected that accumulation of a large quantity of DINOSEB is pos-

sible because the fish simply will die after a certain limited
accumulation. Due to the small amount of DINOSEB fish are able to
accumulate it is not reckoned that poisoned fish used for human
consumption will constitute any acute risk of poisoning. However,
when considering the toxicity of the substance for man consuming
fish with a DINOSEB-content the possible genetic effects as well as
other harmful effects should be evaluated.

Furthermore tests had shown that DINOSEB decomposes very slowly in
fresh water. The half-live period is approx. 1 year. The half-life
period in seawater is not known, but it is assumed to be the same
as for fresh water.

If a damaged DINOSEB drum or a lump of DINOSEB was fished up, the
greatest danger in respect of the crew would probably be that
DINOSEB , when in contact with the skin, could be absorbed in the
body. Furthermore, DINOSEB could - if spread over the deck of a
fishing boat or taken with cargo under deck - after evaporation be
absorbed through the lungs.

The risk of poisoning the fishermen who by chance would pick up
DINOSEB was estimated as considerably , as the fishermen would in
most cases be unprepared for handling the substance , coupled with
the fact that normally no immediate medical treatment would be
available.

In order to minimize the risk for the fishermen who had picked up
DINOSEB the NAEP sent out warnings about DINOSEB as well as in-
structions about personal protection when handling drums containing
DINOSEB.

3. Planning of the search operation

The search was divided in three phases.

Search phase 1 covered the period January 22-March 4, and included
the NAEP oil combatting vessel "Gunnar Seidenfaden", the Danish
net fishing vessel "Jesper Høj", and the Dutch mine hunters "Dru-
nen", "Haarlem", "Veere" og "Dordrecht".

Search phase 2 included recovery of 53 drums in the period March 31
- April 16, with "Gunnar Seidenfaden" and "Jesper Høj", and the
Dutch mine hunter "Haarlem".

During recovery of the 53 drums the following vessels took part:
naval defence vessel "Beskytteren" and "Vædderen", the oil combatting vessel of the NAEP "Gunnar Thorson", and the fishing vessel
"Jesper Høj".

Search phase 3 covered the period April 26 - May 7, and included the oil combatting vessel "Gunnar Thorson".

3.1 SEARCH PHASE 1

In order to be able to delimit a reasonable search area the NAEP
contacted the shipowner DFDS to get information on the exact location where the 80 drums were lost overboard from "Dana Optima".
DFDS informed that in accordance with the information from the crew
members, the vessel lost the deck cargo in the period between 0645
and 1500 hours, and the trailer with the 80 drums of DINOSEB was to
the best of their judgement lost at about 1500 hours.

The search area was delimited between the positions recorded between 0645 and 1600 hours. The course line between these positions
is indicated in Annex 2.

Based on the estimate given by DFDS as to the most likely time of
loss of the drums, "Gunnar Seidenfaden" made a preliminary sonar
search around the position of the ship at 1500 hours.

In the beginning of February DFDS informed the NAEP that a radar
location from 1500 which had been wrongly entered in the log book
had been corrected and - in the opinion of the shipowners - should
be investigated.

This location was then used as a starting point for the continued
search operations of "Gunnar Seidenfaden". The area is indicated in
Annex 3.

In the period February 11 - March 1 the three Dutch mine hunters
took part in the search, and searched an area of 1.25 nautical
miles on either side of the course line of 0645-1500 hours.

Having completed this search the Dutch vessels extended the search
area to an area north of the course line 1200-1600 hours, to take
account of the wind directions at the time of loss of the drums.

On February 17 the Minister for the Environment decided that the
fishing vessel "Jesper Høj" should also take part in the search.

Annex 4 shows the total searched area in search phase 1. "Gunnar
Seidenfaden" had searched a total area of approx. 60 nautical
miles^2, corresponding to 206 km^2, and "Jesper Høj" had searched
approx. 102 nautical miles^2 or 350 km^2, while the three Dutch mine
hunters had searched an area of approx. 166 nautical miles^2 or 570
km^2.

3.1.1 <u>Results of search phase 1</u>. In phase 1 a number of containers
and trailers were located. Of the 39 containers and trailers carried by "Dana Optima" 3 trailers and 3 containers were, thus, iden-

tified. None of the trailers were the Dinoseb trailer. Moreover, 6 containers from "Dana Optima" had been observed floating, or had been recovered from the water surface at varying distances from the search area.

On the sea bed was, moreover, found 1 trailer and 13 containers likely to originate from "Dana Optima". The one trailer was identified as not containing Dinoseb.

A number of other objects were also found, for instance container cars, steel frames, and wrecks of ships and aircraft.

A survey of the objects found is given in Annex 5.

It should be noted that the finds were concentrated in an area close to and on the course line of "Dana Optima" reported by DFDS between 0645 and 1200 hours, and that nothing was found in the neighbourhood of the 1500 hours position.

3.2 SEARCH PHASE 2

It appears from the above that at the end of search phase 1 none of the 80 poison drums had been found.

Plans were then made to suspend the organized search and to pass over to more indirect search operations, meaning that fishermen were to be asked to report to the NAEP if they were to catch one or several poison drums in their nets or trawls.

When studying the results of search phase 1 in more detail some doubt arose as to the verification of certain echo sound signals received in search phase 1. In view of this the Minister for the Environment decided that the search should be resumed as soon as possible.

The reason why it was decided to resume the search already in the beginning of April was among others that new technical information received in the NAEP showed that in unfavorable weather conditions the drums might start leaking already within 3 months, and not as previously estimated, after about one year. Moreover, fishing in the area in question intensifies in the spring and the summer, and the sonar conditions were better than later in the year.

Shortly before search phase 2, which covered part of the recovery phase, the first Dinoseb drums were found by Dutch fishermen. A total of 13 drums - 11 at one location - were recovered at positions only a few nautical miles from the later main find approx. 1 nautical mile SW of the position reported by "Dana Optima" at 0645 hours.

330

Search phase 2 was started on March 1, and included the Danish vessels "Gunnar Seidenfaden" and "Jesper Høj", and the Dutch mine hunter "Haarlem".

3.2.1 <u>Results of search phase 2</u>. On April 2 "Haarlem" found a large amount of drums (estimated number: 40) on the position 55° 36,0 N 03° 05, 5 E close to the 11 drums mentioned above.

Shortly afterwards "Gunnar Seidenfaden" identified the drums as Dinoseb drums, and the majority of the drums could be declared located at a position somewhat more than 1 nautical mile SW of the position reported by "Dana Optima" at 0654 hours. However, the exact number of drums found within an area of 50 x 100 m could not be stated.

In the days following this find, a further number of 18 drums were located close to the main find, and the trailer on which the drums had been carried was located somewhat over 2 nautical miles SE of the main find.

The search continued close to the main find, and at the end of search phase 2 the area indicated in Annex 6 (square 26 and 29) had been searched without further results.

3.3 SEARCH PHASE 3

As appears in para 5 "Gunnar Thorson" recovered a total of 53 drums. Taking into account the 13 drums recovered by the Dutch fishermen, a total number of 14 drums had not yet been located.

The Minister for the Environment therefore decided that - after the necessary time in port - "Gunnar Thorson" was to resume the search using a Side-Scan Sonar for up to a week, to find the remaining missing drums in the area surrounding the main find.

On April 27 "Gunnar Thorson" left for the search area. On that same date a Dutch fishing vessel had found one drum approx. 2,5 nautical miles west of the search area. The drum was considered to have been taken by a trawl and carried away from the main find. Then the number of missing drums was 13.

Search phase 3 was completed on may 7, 1984.

3.3.1 <u>Results of search phase 3</u>. No drums were found in search phase 3. The search took place with Side-Scan Sonar by "Gunnar Thorson", and was hampered by a large number of stones and other objects in the area. "Gunnar Thorson" therefore had to concentrate on verification of echoes in groups of 2-3 (10% of about 100).

On May 5 at 1600 hours the search was suspended, and "Gunnar Thorson" returned to base.

After search phase 3 was completed another 9 drums have been recovered by Dutch fishermen close to the main find.

The overall result of the search operations is then as follows:

Number of drums recovered by Dutch trawlers -   23
Number of drums recovered by "Gunnar Thorson" - 53
Number of drums lying on the sea bed -   4
Total number of drums -   80

Annex 7 shows the locations of the recovered drums.

4 Ships and equipment used in the search

4.1 SHIPS

4.1.1 "Gunnar Seidenfaden" proved to be a good platform for the activities, even in conditions as unsteady as the North Sea in the quarter of January.

Due to the size and stability of the vessel movements are not very violent, and only at wind speeds up to 15-18 m/second, work had to be suspended.

The bridge is large enough to make room for the additional equipment required in the search phase. Besides the vessel is designed so that the additional number of men (from 17 to 35 men) created no problems.

The large square quarterdeck - totally plane - proved very well fit for welding of containers holding special equipment, winches, cranes etc.

4.1.2 "Jesper Høj" is a steel vessel (net fishing) of 75 tons. It goes without saying that it was very suitable for navigation in the area in question, also in bad weather.

4.1.3 The Dutch mine hunters also served very well in the search operation.

4.2 EQUIPMENT

4.2.1 Klein Side-Scan Sonar

A Klein Side-Scan Sonar (100 KHZ) was installed temporarily in "Gunnar Seidenfaden" and was one of the main elements of the search equipment.

The sonar is normally used in connection with mapping of raw material in Danish waters, and for searching of wrecks and archological finds. The Side-Scan-Sonar, which looks like a small torpedo dragged after the vessel, emits sound waves at right angles to the course line. The sound waves reflected from the sea bed or from objects on the sea bed are recorded in the side-scan sonar hydrophone (submarine microphone) and transferred to a printer which draws a shadow picture of the sea bed. To interpret the picture the operator identifies the reflections and shadows from the normal topography of the sea bed (surface shapes) and different material compositions. Having done so he can see whether stones or other large objects lie on the sea bed (containers or poison drums). The intensity and the character of the echo received from an object on the sea bed depends, first, on the size of the objects, and, also, on the character of the sea bed and the reflecting properties of the object. To interpret the picture fully the operator also has to calculate the size and position of the objects.

As the knowledge of the sea bottom conditions increased the possibilities of identifying echoes improved. Due to the solution efficiency of the Side-Scan Sonar, and resulting narrow search width (150 m between the search lines when searching for trailers, an 50 m when searching for drums) the speed had to be rather low (maximum 3.2 knots), which explains the size of the area searched by "Gunnar Seidenfaden" compared to the area covered by other vessels taking part in the operation.

However, the interaction between the Side-Scan Sonar and the sonar operated by "Jesper Høj" proved very satisfactory, the fishing vessel being capable of operating at a larger speed and search widths, but having difficulties interpreting the type of objects found. On the other hand, the Side-Scan Sonar gave a more detailed picture of the sea bed and the objects found. Generally speaking the search techniques used were considered as succesful.

4.2.2 <u>Simrad Sonar</u>. The fishing vessel "Jesper Høj" permanently operates a SIMRAD SR2 NAVSONAR (49 KH) with printer. The search width is max. 1500 m, and the distance between search lines was therefore fixed at approx. 500 m. The search speed was 5-6 knots, and searching operations could be carried out up to wind speeds of 10-12 m/s.

"Jesper Høj" was thus able to cover a much larger area than "Gunnar Seidenfaden" per unit of time. The equipment however, proved more suitable for locating large objects like containers and trailers than individual drums.

During the search operations the vessel was fitted with additional navigation equipment.

4.2.3 <u>Dutch Sonar Eqipment</u>. The NAEP has no detailed information on the sonar efficiency of the Dutch mine hunters. However, the equipment was generally considered to be well suited for the search operations.

4.2.4 <u>Navigation equipment</u>. In view of the fact that ordinary navigation equipment is not sufficiently accurate for searches where navigation lines are down to 50 m from each other, the search vessels had to be fitted with special precision navigaion equipment.

During the first part of the search operations carried out by "Gunnar Seidenfaden" the vessel operated a navigation system "ARGO" supplemented with the system "SYLEDIS".

As was expected the "ARGO" system was rather sensitive to atmospheric conditions, causing functional problems at sun rise and sun set.

"Jesper Høj", "Gunnar Thorson", and in the last part of the search phase, also "Gunnar Seidenfaden", used the navigation equipment "Pulse/8".

This system is less sensitive to atmospheric disturbances, especially in the area where the drums were found. The system is less precise than "ARGO" (20 m as opposed to 6 m), but sufficiently precise to allow exchange of positions between the two systems without problems.

Even if "ARGO" presented many advantages during this search operation it is the general opinion that "Pulse/8" should be preferred by searches in this part of the North Sea.

4.2.5 <u>Submarine video camera</u>. The video cameras used (ROV = Remote Operated Vehicle) during the search phase were of the types SCORPIO and SCORPI, both borrowed from and operated by a Danish salvage company.

a. SCORPIO is fitted with the strongest submarine engines. Beside TV camera and search sonar SCORPIO is fitted with remote control grab.

b. SCORPI is also fitted with a TV camera and search sonar, and was used by "Gunnar Seidenfaden" at a later period of the search phase. The equipment consists of a Rov parked in a garage which will be lowered almost to the sea bed. Then the ROV can move around 360° and at distances up to 150 m from the garage, i.e. searching an area three times as large as the area covered by SCORPIO.

If the swell is too high the ROV can be difficult to steer back into the garage, an this limited the work to some extent.

When comparing the SCORPIO and the SCORPI, experience shows that SCORPIO was probably most suitable because of the stronger engines, even if the anchoring requirements of the mother ship are more stringent than for SCORPI.

However, SCORPI was considered more suitable in internal Danish waters with less high seas, as the equipment requires less acurate anchoring and covers a search area three times as large from the< same anchoring place.

Finally, experience showed that a ROV must be fitted with a search sonar.

4.2.6 <u>Dutch Equipment</u>. Also the Dutch mine hunters were equipped with submarine video cameras, but of another type than those operated by the Danish vessels.
The NAEP has no detailed information on the Dutch equipment. The NAEP is of the opinion, however, that the equipment functioned satisfactory during the search operaton.

5. Salvage operation

Salvage of the Dinoseb drums located during the search was carried out under the command and responsibility of the Royal Danish Navy. The ships operated by the navy were not satisfactory platforms for the diving operations in the North Sea. It was therefore decided that the special oil combatting vessel of the NAEP "Gunnar Thorson", and the fishing vessel "Jesper Høj" were to assist the navy.

Mobile equipment for detection of possible Dinoseb pollution to be expected in connection with salvaging of leaking drums was not available. The NAEP Marine Pollution Laboratory therefore developed a method which could be used on board "Gunnar Thorson".

Two recovery methods were planned: recovery by divers, and recovery by submarine robot, ROV. When the recovery operation was planned and prepared, account had to be taken primarily of the lack of knowledge of the toxicity of the poison in contact with salt water, and the resistance of the drums to the pressure at the depth in question. It was decided to use a navy fishery inspection vessel in addition to "Gunnar Thorson", mainly in order to maintain a dedicated helicopter contingency in the area, and thus give fast evacuation possibilities in case of personal damage. This proved to be a wise decision during an operation involving divers.

The drums were located in one main area, with single drums spread around the main find at distances up to approx. 1700 m. Water depth was 40-50 m. To begin with it was planned to recover the drums by means of divers in the day time, and by ROV at night. Later this procedure had to be changed, and the main drum concentration was recovered by divers, and the individually located drums by ROV. At the end of the salvage operation, a total of 34 drums had been recovered by divers and 19 drums by ROV.

6 Final Disposal

The recovered drums - almost all leaking - were kept on board "Gunnar Thorson" in special steel containers. They were carried to Esbjerg, and by the order of the civil defence authorities transported to the consignee "Kemisk Værk Køge" for final disposal.

7. Costs of the Operation

The total Danish costs in connection with the search and recovery of the Dinoseb drums were about 13 million dkr. (or 1,9 million US$).

An action for damages has been brought against shipowners, but is still not decided.

8. International Co-operation

As will appear from the description of the search operation, it took place in close co-operation between the Danish and Dutch authorities. The Netherlands assisted with several mine hunters from the Dutch navy. Coordination of the operation of the vessels took place in close co-operation between the NAEP, Rijswaterstaat, The Dutch navy, Flag Officer Denmark, and the participating Dutch and Danish ships. The co-operation was very satisfactory, in mutual and complete understanding of the special problems which arose from time to time in the two countries in connection with the operation.

It should be noted that the Dutch assistance, carried out without costs to the Danish state, had great influence on the final result of the search.

9. Lessons learnt

The search and recovery of the Dinoseb drums were the first operation of its kind in the history of the Danish contingency services.

Much and varied experience was gained during the operation; only a few lessons learnt will be mentioned here.

a. The reason why the search phase was so long was that Dana Optima

336

reported a position of loss of drums approx. 28 nautical miles away from the actual position. This, however, contributed much to the experience gained in relation to search strategy.
b. Co-operation with fishermen with local knowledge and with suitable sonar equipment is very important.
c. The Danish procedure, with fishing vessel sonar for searching of large areas, supplemented with Side-Scan Sonar for detailed scanning/identifcation proved very useful.
d. The ability to distinguish the searched objects from other sonar reflecters increased significantly during the operation. This indicates a need for regular training and exercises in connection with searches for drums.
e. The value derived from the use of ROV also increased considerably during the operation. This also points to the need for regular training in the use of this identification/recovery technique.

337

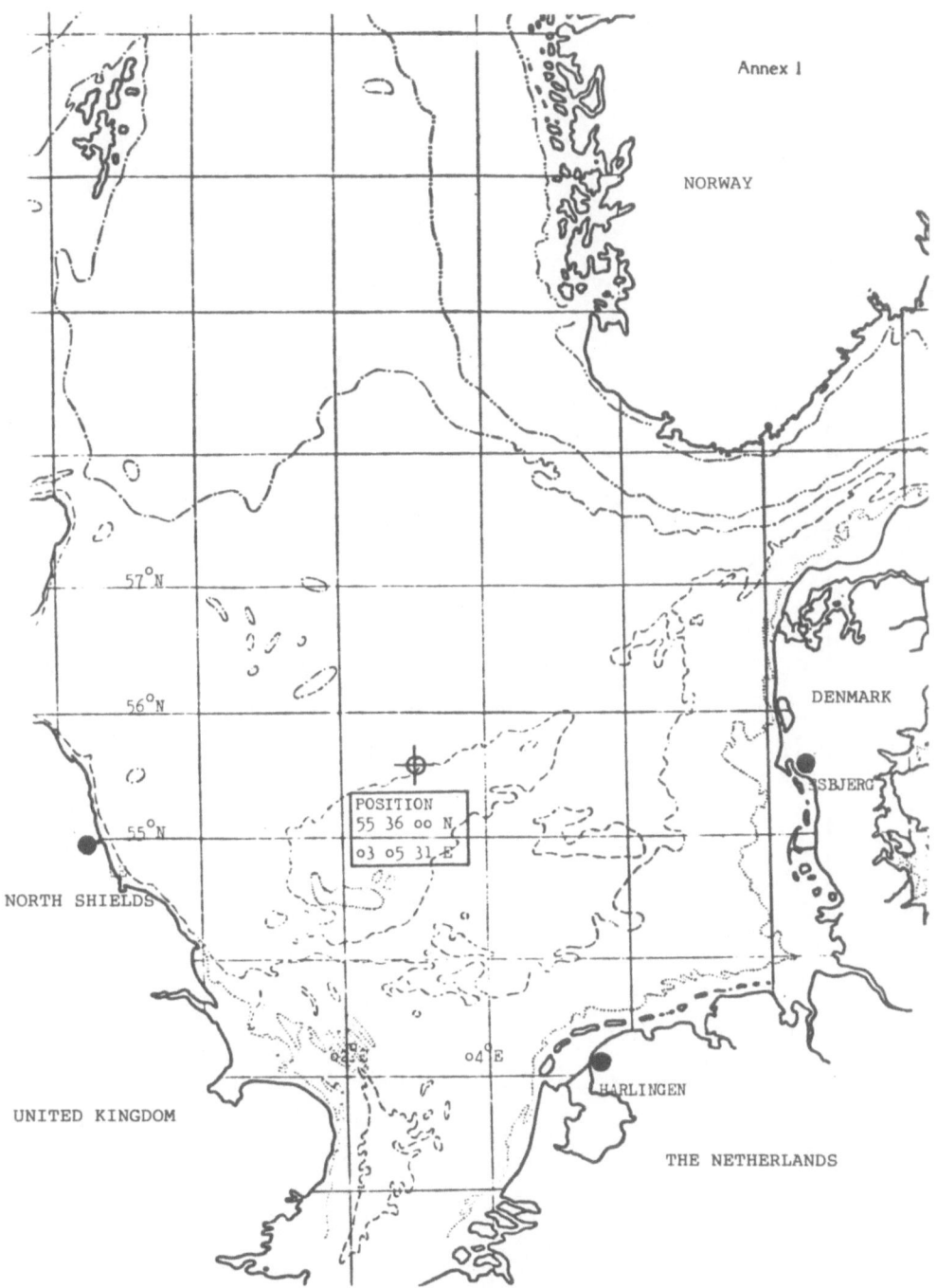

Annex 1

NORWAY

57°N

56°N

DENMARK

ESBJERG

POSITION
55 36 00 N
03 05 31 E

55°N

NORTH SHIELDS

04°E

HARLINGEN

UNITED KINGDOM

THE NETHERLANDS

Annex 2

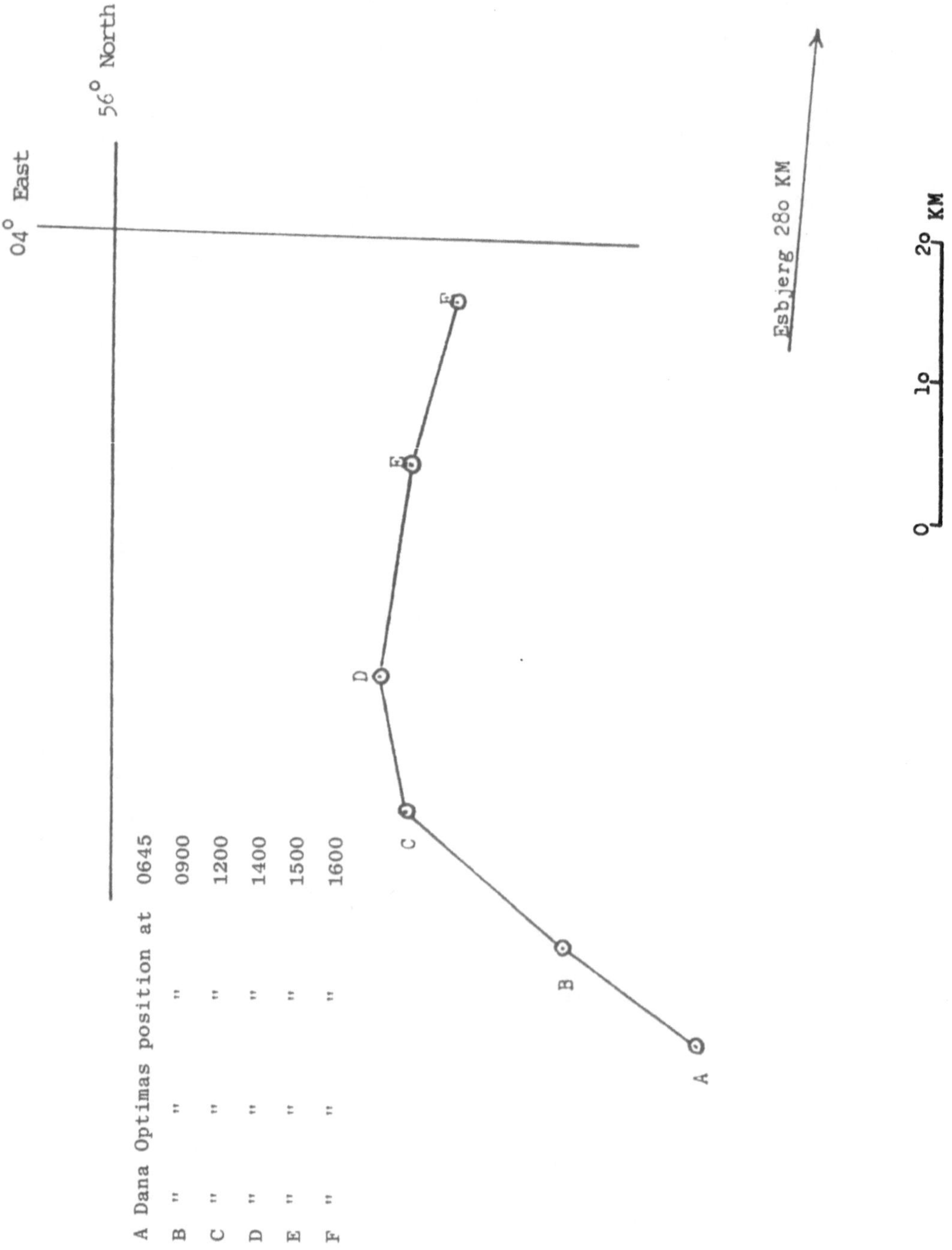
04° East
56° North
Esbjerg 28o KM
0
1o
2o KM
A
B
C
D
E
F
A Dana Optimas position at 0645
B " " " 0900
C " " " 1200
D " " " 1400
E " " " 1500
F " " " 1600

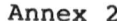

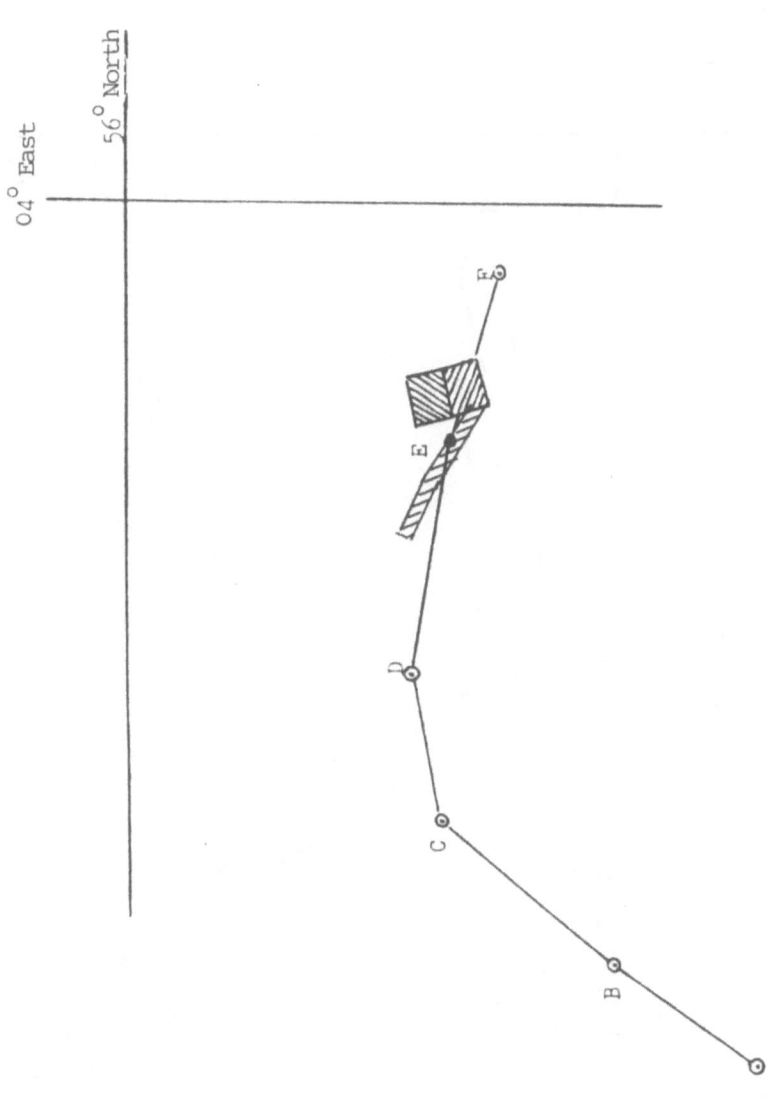

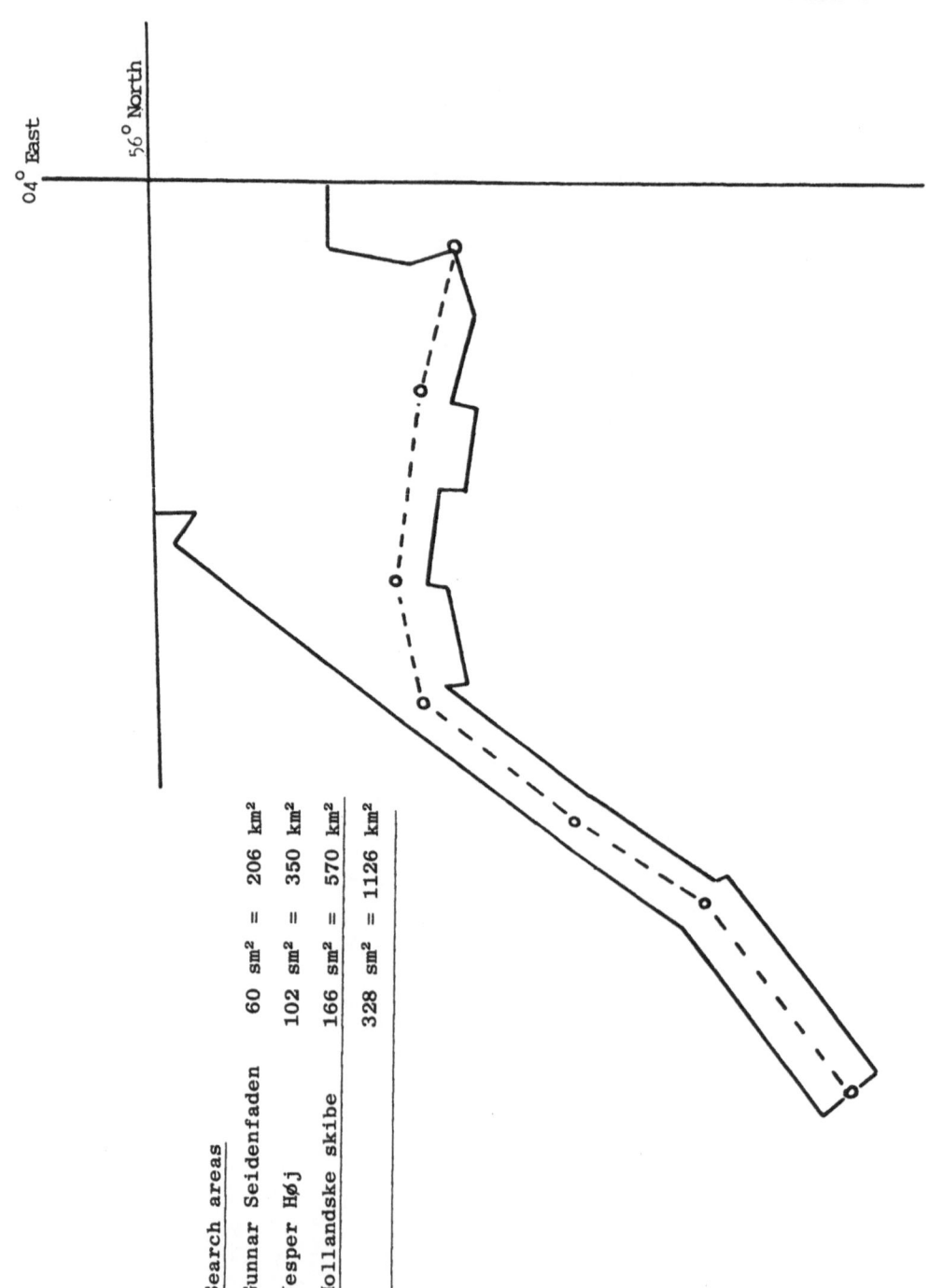

Search areas

Gunnar Seidenfaden   60 sm² = 206 km²

Jesper Høj   102 sm² = 350 km²

Hollandske skibe   166 sm² = 570 km²

                     328 sm² = 1126 km²

56° North

04° East

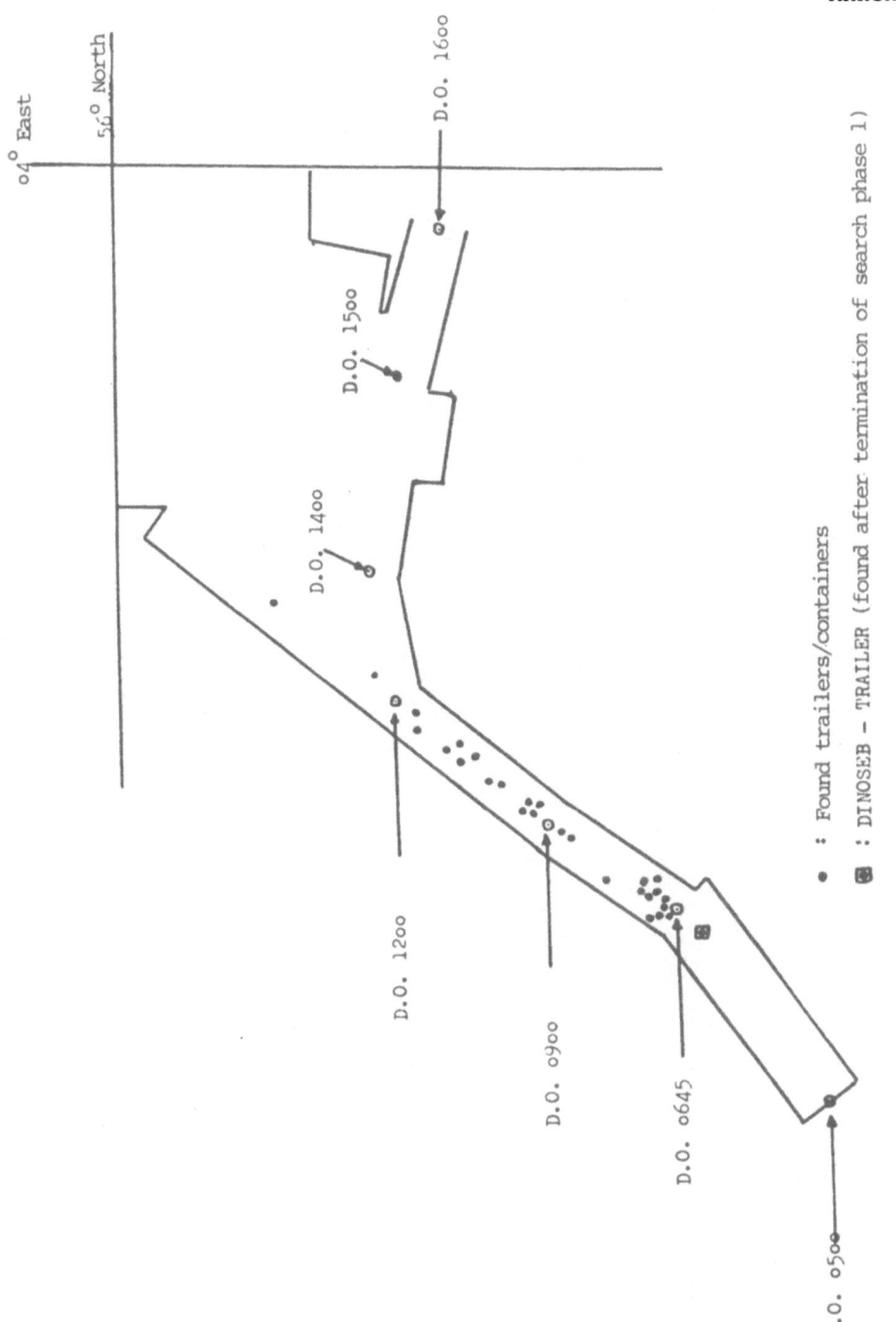

• : Found trailers/containers

▣ : DINOSEB – TRAILER (found after termination of search phase 1)

342

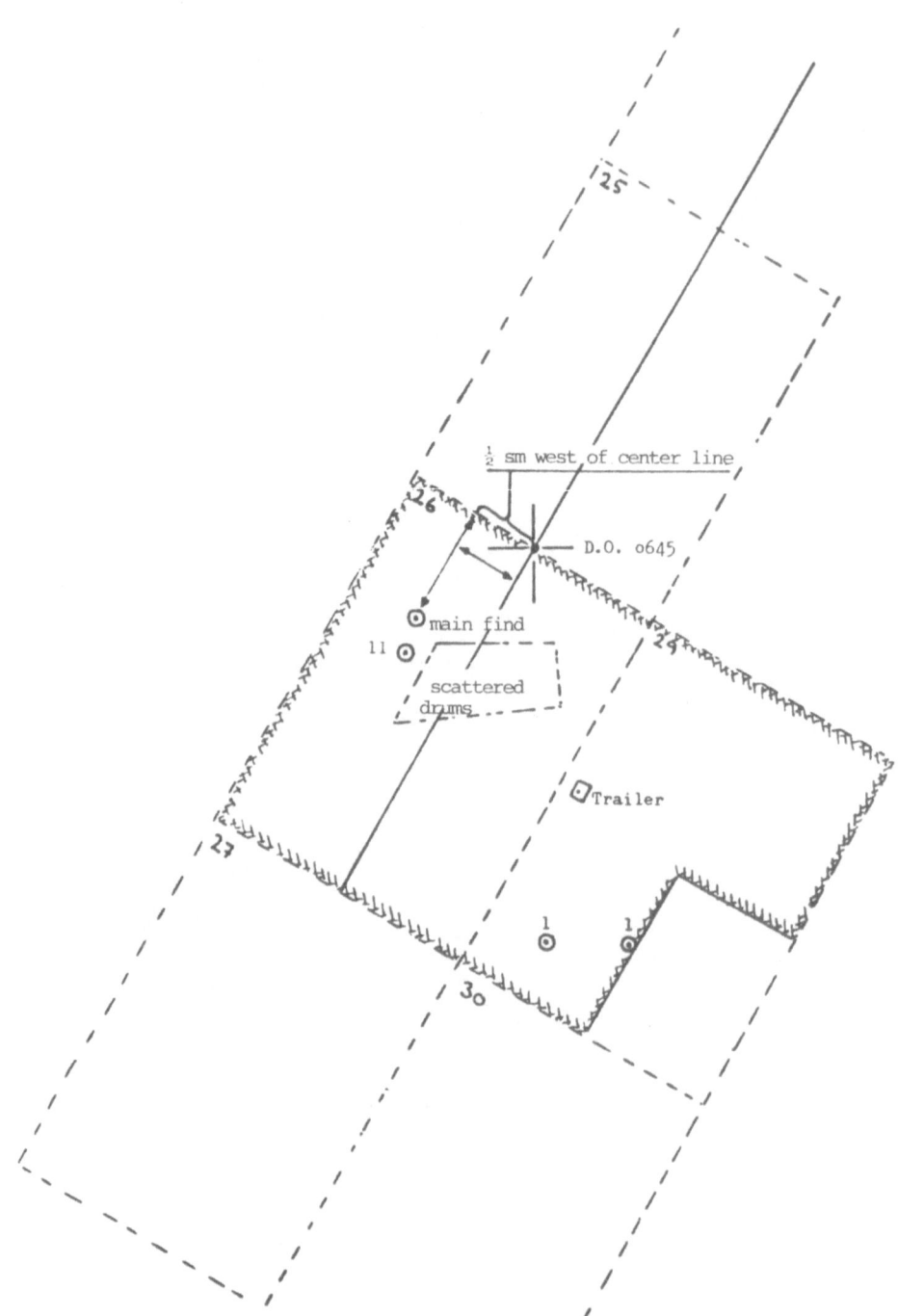

$\frac{1}{2}$ sm west of center line

D.O. o645

main find

11

scattered drums

Trailer

25

26

24

27

30

1

1

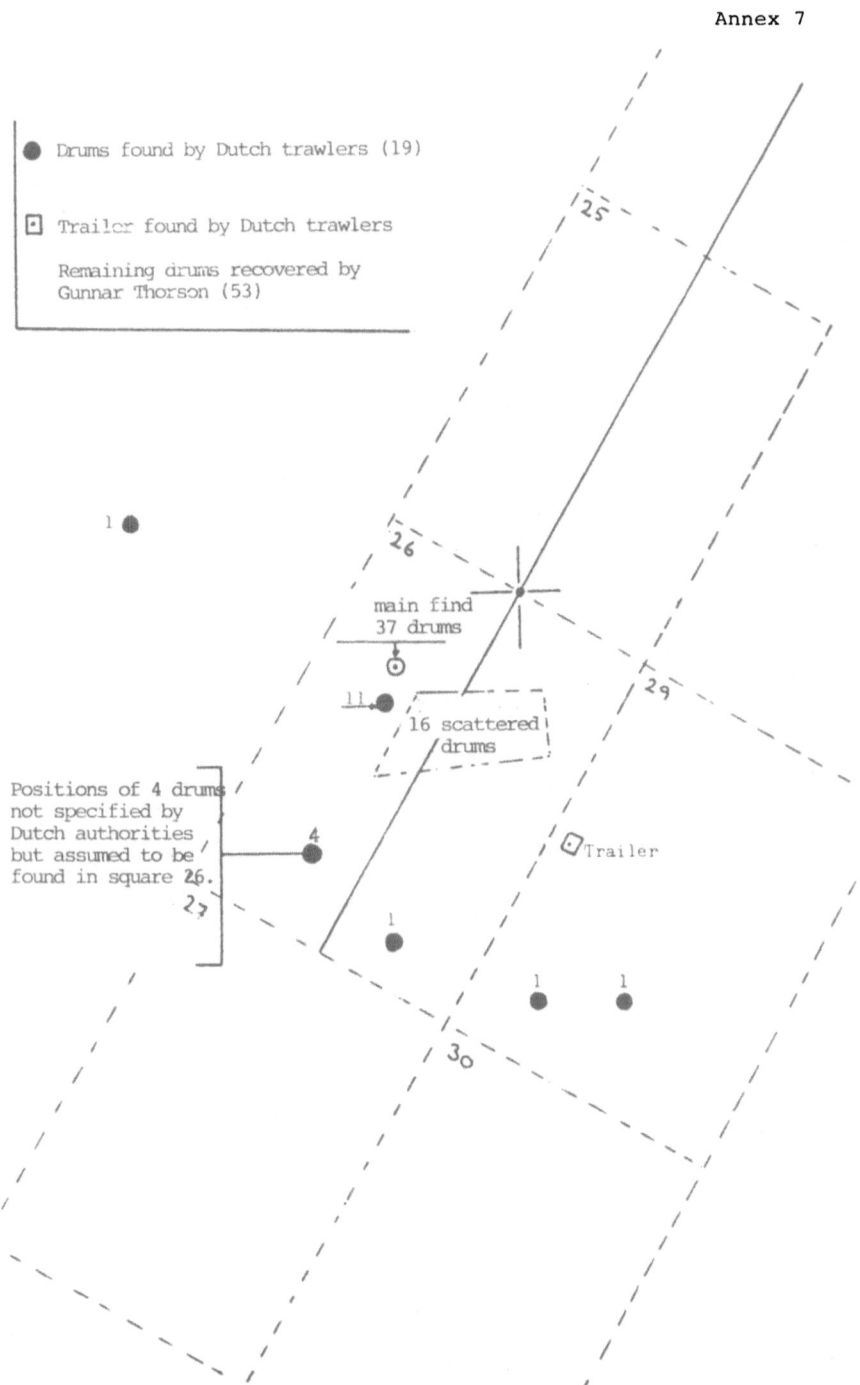

● Drums found by Dutch trawlers (19)

⊡ Trailer found by Dutch trawlers

Remaining drums recovered by
Gunnar Thorson (53)

25

1 ●

26

main find
37 drums
⊙

11 ●

16 scattered
drums

29

Positions of 4 drums
not specified by
Dutch authorities
but assumed to be
found in square 26.

4 ●

⊡ Trailer

27

1 ●

1 ●   1 ●

30

LESSONS TO BE LEARNT FROM PAST INCIDENTS
SUGGESTIONS FROM ENVIRONMENTAL ORGANIZATIONS

Gerard Peet (1)
Stichting S.E.A.
Oosteinde 167
2611 VD DELFT
The Netherlands

ABSTRACT. The past has been witness to many incidents involving
(potential) chemical spills. Many of these could have been prevented,
others happened in spite of a large number of measures which had
already been taken to prevent such incidents. Environmental
organizations have always stressed the importance of preventing spills
and emergencies. One of the lessons that has to be learned from the
past is, however, that prevention has to include measures with a view
to being able to combat those spills and emergencies which will
inevitably happen in spite of preventive measures. As an observer at
the International Maritime Organization the environmental organization
Friends of the Earth International has made several suggestions for
such measures of secondary prevention.

1. INTRODUCTION

In December 1979, the Iraqi vessel Sinbad lost 51 cylinders of chlorine
gas during a heavy storm off the Dutch coast. Efforts to recover these
cylinders were undertaken after the incident but only 12 cylinders
could be recovered shortly after the incident. In 1984, some of the
cylinders were caught incidentally by fishermen. The damaged state of
the cylinders thus recovered prompted the Dutch Government to undertake
the search for the remaining missing cylinders, most of which were
found and destroyed. The recovery of these cylinders has been difficult
due to inaccurate information about the course followed by the vessel
concerned (2).

In January 1984, the Danish merchant ship Dana Optima en route from
North Shields in the United Kingdom to Esbjerg in Denmark lost 39
containers and trailers from its deck cargo during stormy weather and a
stop of its main and auxiliary engines. The first information given was
that no substances harmful to the marine environment were contained in
the lost deck cargo. Three days later, however, new information
surfaced that among the missing deck cargo there was one truck-trailer
with 80 drums of Dinoseb. In the light of available information about

345

P. Bockholts and I. Heidebrink (eds.), Chemical Spills and Emergency Management at Sea, 345—354.
© *1988 by Kluwer Academic Publishers.*

Dinoseb's toxicity efforts were made to recover the 80 drums. After some time most of the drums were recovered. A considerable number of these drums were leaking probably because these drums could not resist the 40 meters water pressure; others may have been damaged by bottom trawls (3).

In June 1984, a barge (OK Menga) which was being towed from Port Moresby to the Fly River (both in Papua New Guinea) capsized in rough seas 24 kilometers off the east coast of the mouth of the Fly River. From this barge 15 containers with a total of 2,700 drums containing sodium cyanide, a toxic substance, were lost overboard. Salvage operations were hired to ensure the fastest possible clean-up. It was reported that preparation for salvage activities started about a week after the incident. Some 94 drums washed ashore and were recovered. A further small number of drums was recovered at sea. The majority of the drums has not been recovered (4).

In August 1984, the French vessel Mont Louis sank off the Belgian coast after a collision. It took several days before information regarding the presence of uranium hexa-fluoride in its cargo was available and adequate action could be undertaken. Salvage operations resulted in the recovery of the uranium hexa-fluoride (5).

In November 1984, the Brigitta Montanari sank some 15 km off the coast of Yugoslavia carrying a cargo of 1,300 tonnes of vinylchloride.
In view of the risks for human health and the marine environment it was decided that ship and cargo had to be salvaged. Salvage operations started only in 1987; no information was available whether salvage operations were completed (6).

In July 1986, the Olav sank in the North Sea off the Dutch coast, carrying a cargo which contained high levels of heavy metals. It took considerable time before information was given regarding the nature of the cargo of this vessel. When this information was received it was decided that, inn view of the risks for the marine environment, ship and cargo had to be salvaged. Salvage operations were completed successfully. As regards the costs of the salvage operation it is unclear whether the Dutch Government will be able to recover these costs (6).

In June 1987, the Junior sank in the North Sea off the Dutch coast, carrying a cargo which contained high levels of cyanide and heavy metals. In view of the risks for the marine environment it was decided that ship and cargo had to be salvaged. Salvage operations were completed successfully. As regards the costs of the salvage operation it is unclear whether the Dutch Government will be able to recover these costs (6).

In May 1988, the Dutch chemicals tanker Anna Broere sank in the North Sea off the Netherlands coast, carrying a cargo of acrylonitril.' In view of the nature of its cargo and the risks the cargo posed for the environment and human health it was decided to salvage the ship and its cargo. Salvage operations were completed after some difficulty; a considerable portion of the cargo was lost. As regards the costs of the salvage operation it is unclear whether the Dutch Government will be able to recover these costs (6).

From this list it is clear that incidents involving chemicals or other hazardous substances other than oil are not rare. The list is not complete.

2. PREVENTION AND SECONDARY PREVENTION

Prevention has always been the cornerstone of the environmental policy advocated by environmental organizations. Some of the incidents listed above may have been preventable, others may not have been. The fact that incidents do happen calls for some considerations regarding the scope of the concept of 'prevention'.
It is possible to distinguish different levels of prevention (7):
-Retrospective prevention could be defined as an effort to prevent problems from recurring (eventually in combination with efforts to cure the symptoms of the perceived problem). This includes efforts to develop new regulations undertaken after incidents have taken place, as was, for instance, the case after the oil spills of the Torrey Canyon and Amoco Cadiz.
-Genuine prevention could be defined as an effort to prevent environmental problems from occurring. Whenever activities are known to result in serious environmental damage the option for a genuine preventive policy is clear: no such activity should take place. Certainty as regards such risks is hard to find, however. Reasonable certainty will have to be the basis for preventive activities and the chance that something will, nevertheless, go wrong will have to be taken into account as well. This then, leads to the third level of prevention.
-Secondary prevention could, taking this into account, be defined as those measures which are taken to prevent or minimize environmental effects of incidents. There are several examples of secondary prevention, such as contingency plans for oil and other spills and salvage operations as described above.

This Conference focusses its attention on this secondary prevention, and more specifically secondary prevention regarding chemicals and other hazardous substances.
The fact that environmental organizations have primarily focussed their attention on the prevention of incidents might be taken to indicate that they have forgotten about this secondary prevention.
They have not.
In the maritime field, for instance, the environmental organization Friends of the Earth International (8) has called for several measures to be taken by the International Maritime Organization which can be considered to be secondary prevention.

3. LESSONS TO BE LEARNT FROM PAST INCIDENTS

Setting aside those lessons which should be learnt as regards how the
incidents mentioned above (and others) could have been prevented
lessons can (and should) also be learnt regarding secondary prevention,
lessons regarding chemical spills and emergency management at sea.

From the examples given a number of important problems can be
identified in relation to minimizing or preventing further damage after
these incidents had taken place:
-In several cases (Dana Optima, OK Menga, Sinbad) when efforts were
made to recover dangerous packaged substances lost by ships it has been
extremely difficult to find these back.
-In a number of cases (such as the Sinbad and the Dana Optima) the
packages in which the hazardous substances were transported could not
withstand the environment (pressure or other causes of damage) without
spilling their contents.
-In a number of cases (specifically the Dana Optima and the Mont Louis)
it took considerable time before information was available that cargoes
were involved which could be harmful for the marine environment.
-In at least one of these cases (the Sinbad) inaccurate information
regarding the course of the ships involved caused considerable
difficulty for efforts to recover the lost cargo.
-In at least one case (OK Menga) it took considerable time before
salvage operations could start.
-In most cases there appears to be considerable difficulty to recover
costs involved in salvage or other operations undertaken to prevent
(further) damage to the marine environment.

4. THE RECOVERY OF DANGEROUS CARGOES LOST BY SHIPS

Practice proves that it is difficult to find back packaged chemicals
which have been lost overboard by ships. In some cases (Dana Optima)
incidental catches of such packages in fishing nets help, in other
cases (OK Menga) most packages could not be found back.
A second problem which surfaced from practice was that packages
suffered more damage when under water than was expected.

Both issues have been addressed by Friends of the Earth International
(FOEI) at meetings in the framework of the International Maritime
Organization.

During the discussions regarding further improvements to Annex III of
the MARPOL Convention, FOEI presented several papers (9, 10, 11, 12)
to the Marine Environment Protection Committee (MEPC) of the
International Maritime Organization (IMO) to address the difficulties
encountered in the efforts to recover lost packages lost by the Sinbad,
the Dana Optima and the OK Menga.

FOEI took these incidents as clear signals that more adequate regulations had to be developed regarding the transport of dangerous goods in packaged forms: new regulations were to be developed to facilitate the preventive recovery of hazardous packages lost overboard.
Measures suggested by FOEI included regulations to:
-ensure that hazardous packages are shipped in a manner which would prevent packages from being dispersed over a large area (e.g. concentrated shipment in minimum size packages), and
-ensure that packages lost overboard and gone adrift or sunk can be successfully traced (e.g. by means of radiobeacons or other homing devices for certain categories of substances).

Neither of these suggestions has found approval within the IMO.

FOEI has also pressed for stricter regulations regarding immersion testing of packages (13) especially as regards their durability under 'deep sea' conditions. The state in which the cylinders of the Sinbad and the drums of the Dana Optima had been recovered was the basis of these efforts. Several delegations of IMO-members supported the view taken by FOEI in this respect.

5. THE AVAILABILITY OF NECESSARY INFORMATION

A major problem in some the incidents mentioned earlier in this paper has been the lack of adequate information at the time of, or even after the incidents occurred.
Information was lacking as regards the question whether environmentally hazardous substances were onboard these ships; in the case of the Sinbad information was lacking regarding the exact course of the ship at the time it lost part of its cargo.

Important efforts are being made as regards the availability of information about chemicals which are transported by ships (e.g. the SEABEL hazard identification and decision support system with regard to chemical spillages at sea).
The major problem, however, might not be the availability regarding the various substances involved, but might well be the availability of information as regards which ships carries which substances (or which substances are being carried by a specific ship which is involved in an incident).

An important issue discussed within the maritime community in this respect is that of Vessel Traffic Services (VTS). Friends of the Earth International considers Vessel Traffic Services as a major instrument to solve problems regarding the availability of the necessary information in cases of incidents.
In one of its submissions (14) to the IMO, FOEI has given its considerations why VTS would also be extremely important for the protection of the marine environment. According to FOEI, the knowledge

a VTS centre should have of the shipping in its area could very well
put such a centre in a unique position for contributing to the
limitation of environmental damage after an incident has occurred.
To do so a VTS should have the following power and information:
–information on the dangerous goods or other pollutants carried by
ships operating in the VTS area;
–information on the identity of each ship entering the VTS area;
–information about the destination of the ship with an estimated time
of arrival;
–in the case of ships carrying significant quantities of potential
pollutants, a passage plan should be available, with a requirement that
the VTS centre should be notified in the case of any significant
deviation from that plan;
–information on the equipment available for limiting the escape of
polluting substances and for immediate salvage work, including both
ship-board and shore-based equipment;
–close liaison facilities with local, national and international
pollution control services, search and rescue services; in some cases
it may be appropriate to have these functions integrated with a VTS
centre;
–power, at least in emergency situations, to give instructions
concerning the general movement of shipping in an area where an
incident has occurred or where there is an imminent possibility of
danger for life or of environmental pollution;
–power, to authorize additional or alternative rescue or salvage work
where existing efforts appear to be inadequate to protect life or the
environment.

A VTS centre thus equipped would be a powerful instrument for emergency
management in general; it would ensure the availability of the
necessary information ashore.

Depending on the (radar) facilities of a VTS centre it is also
conceivable that such centres could help in backtracking the exact
position of ships which have lost dangerous cargoes; this, of course,
would greatly facilitate the recovery of such cargoes.

The FOEI proposals regarding VTS have not resulted in any decisions by
the IMO in accordance with these proposals. FOEI's proposals still seem
to be considered as going to far.
In the wake of incidents in years after the beforementioned proposals
were submitted to the IMO, other environmental organizations have also
advocated VTS systems both for the prevention of incidents and for
better emergency management once incidents have taken place.

6. SALVAGE OPERATIONS

Salvage operations are increasingly important for the protection of the
environment. No longer are such operations aimed at salvaging property
only. In the North Sea regions salvage operations after the incidents

of the Olav, the Junior and the Anna Broere have been exclusively aimed at the protection of the marine environment.

Environmental organizations have generally welcomed and applauded these efforts as well the policy which made these efforts possible.
The problems encountered during the salvage operations of the Anna Broere have nevertheless prompted some critical questions from the environmental organization Werkgroep Noordzee (15).
Some doubts were raised whether in the present circumstances salvage companies were still in a position to undertake salvage operations optimally. These doubts were raised after statements regarding the (insufficient) level of payments for such activities.

In this respect, Friends of the Earth International has been following the discussions regarding a new Salvage Convention in which environmental considerations will be given more weight intensively. In 1989 this new Salvage Convention will be discussed and (hopefully) concluded at an International Conference.
FOEI intends to take an active part in that Conference with a view to ensuring that environmental interests are adequately covered in this new convention.

7. LIABILITY AND COMPENSATION

A major issue to come out of several of the cases mentioned in the first paragraph of this paper has been the question of liability and compensation. Governments involved in efforts to prevent environmental damage after incidents (Denmark and the Netherlands after the Dana Optima, Belgium after the Mont Louis, the Netherlands after the Olav, the Junior and the Anna Broere).

The basic problem is that there is no adequate international regime regarding liability and compensation for damages caused by hazardous and noxious substances carried by sea.
The lack of such a regime might in future prove to be a major obstacle to obstruct the necessary action to prevent environmental (and other) damage after incidents have taken place.
The incidents mentioned before have been costly to the respective Governments and it is more than doubtful whether it will be possible to recover these costs under the present circumstances. This might discourage future efforts.

Friends of the Earth International has been very active in the discussions regarding the necessary liability and compensation regime. Aimed at the 1984 International Conference on Liability and Compensation for Damage in Connexion with the Carriage of certain Substances by Sea FOEI defined its position in a document submitted to this Conference (16).

FOEI has advocated both a realistic and a cost effective approach as

regards liability and compensation. FOEI welcomed a new Convention on these issues and was therefore disappointed when the 1984 Conference did not produce the results it had aimed at.
FOEI still considers the development of an adequate liability and compensation regime for damages by hazardous and noxious substances to be of the highest priority; a regime which should cover both substances carried in bulk and those carried in packaged form.

8. CONCLUSION

There is still considerable need for new regulations and arrangements regarding chemical spills and emergency control.
Friends of the Earth International has, in the past, contributed several ideas and suggestions for such measures, the most important of which might well be the need for VTS systems in certain areas (such as the southern North Sea) and proper salvage and liability and compensation regimes.

REFERENCES

(1) Gerard Peet is a consultant to the environmental organizations Friends of the Earth International and Werkgroep Noordzee and represents Friends of the Earth International as observer at the International Maritime Organization

(2) R. C. Schriel (directie Noordzee); Draaiboek Opsporing en Berging Chloorgascylinders ex m.s. Sinbad (NZ;R-84.10.24; Rijswijk, 24 October 1984

(3) Information taken from Document MEPC 21/INF.2, submitted by Denmark to the 21st session of the Marine Environment Protection Committee of the International Maritime Organization; Loss and Salvage of Drums containing DINOSEB; 30 November 1984

(4) Based on information and newspaper articles received from the environmental organization Friends of the Earth Papua New Guinea

(5) Information taken from Document MEPC 21/10/5, submitted by Friends of the Earth International to the 21st session of the Marine Environment Protection Committee of the International Maritime Organization; Implementation of Annex III of MARPOL 73/78; 22 March 1985

(6) Information taken from Document MEPC 26/INF.26, submitted by Friends of the Earth International to the 26th session of the Marine Environment Protection Committee of the International Maritime Organization; Enforcement of pollution conventions: casualty investigation in relation to marine pollution; 5 August 1988

(7)   Gerard Peet; The anticipation principle as a basis for policy?; published in 'Environmental Protection of the North Sea, edited by P. J. Newman and A. R. Agg; Oxford, 1988

(8)   Friends of the Earth International is an umbrella organization for national Friends of the Earth groups in some 30 countries in all continents. Major groups are in the Netherlands, the United Kingdom, the United States and Malaysia. Other countries include Nicaragua, Poland, Ireland, Argentina, Ecuador, France, Spain, etc.. Friends of the Earth International has observer's status at several international organizations such as the International Maritime Organization, London Dumping Convention and International Oil Pollution Compensation Fund. The Dutch environmental organization Werkgroep Noordzee assists Friends of the Earth International in its work at these organizations.

(9)   Document MEPC 20/9, submitted by Friends of the Earth International to the 20th session of the Marine Environment Protection Committee of the International Maritime Organization; Review of the implementation of Annexes III and V of MARPOL 73/78; 27 July 1984

(10) Document MEPC 21/10/5, submitted by Friends of the Earth International to the 21st session of the Marine Environment Protection Committee of the International Maritime Organization; Implementation of Annex III of MARPOL 73/78; 22 March 1985

(11) Document MEPC 23/8/3, submitted by Friends of the Earth International to the 23rd session of the Marine Environment Protection Committee of the International Maritime Organization; Implementation of Annex III of MARPOL 73/78 and amendments to the IMDG code to cover pollution aspects; 6 June 1986

(12) Document MEPC 24/9/7, submitted by Friends of the Earth International to the 24th session of the Marine Environment Protection Committee of the International Maritime Organization; Implementation of Annex III of MARPOL 73/78 and amendments to the IMDG code to cover pollution aspects; 16 January 1987

(13) During the 24th session of the Marine Environment Protection Committee of the International Maritime Organization in February 1987

(14) Document NAV 30/7/2, submitted by Friends of the Earth International to the 30th session of the Sub-Committee on Safety of Navigation of the International Maritime Organization; Ship Reporting Systems, guidelines for Vessel Traffic Services; 13 November 1984

(15) See 'Springtij 88-2': Drie keer scheepsrecht? Lessen van de Anna Broere; published by Werkgroep Noordzee Amsterdam; August 1988

(16) Document LEG/CONF.6/INF.4, submitted by Friends of the Earth International to the International Conference on Liability and Compensation for Damage in Connexion with the Carriage of certain

354

Substances by Sea; <u>Consideration of the draft HNS Convention and draft Protocols to revise the 1969 CLC and the 1971 Fund Conventions</u>; 13 April 1984

DECISION MAKING

Decision Making for Chemical Accidents at Sea

J.R. Taylor
ITSA
Jernbanegade 52A
DK-4000 Roskilde, Denmark

When a ship carrying hazardous chemicals runs aground, is holed, or sinks, the time available for decision making may be minutes, hours, or months. Quickly, a large number os questions must be answered. Will the accident lead to a release? How quickly? What will the consequences be? Can the release be stopped? Can the toxic substances be recovered? Can the operation be carried out safely? Will the recovery be reliable? Will it be complete? One thing is certain. For effective and safe spill recovery, it is necessary to be prepared.

Risk analysis provides a planning background for accidents of involving toxic releases. Properely used, it gives not only a quantitative view of the extent of any accident. It will also give scenarios describing the sequence of events in a wide range of potential accidents. The scenario descriptions give not only the possibility of presenting the results of theoretical calculations. They also make it possible to incorporate practical experience from actual accidents.

Planning for marine chemical spills

The results of a risk analysis tell what kinds of accident can arise, indicate their extent and frequency, and tell something of the effects on fishing, tourism, coastal population etc. Long term effects are more difficult to evaluate, and decision making here will often be politically determined. It will only rarely be the case that resources can be spared to cover the full range of potential accidents. Some criteria will therefore be needed to determine what degree of preparedness is necessary, what accidents will be treated, and what will be regared as beyond the scope of planning.

P. Bockholts and I. Heidebrink (eds.), Chemical Spills and Emergency Management at Sea, 357–362.
© 1988 by Kluwer Academic Publishers.

358

Ideally, planning could be made on the basis of average frequency and cost of chemical accidents. The average yearly cost of accidents is balanced against the cost of acquiring and operating emergency facilities. In order to do this, the spectrum of accidents needs to be listed and the equipment type and placement alternatives outlined. The effect on overall risk then involves checking that each new decision results in a decrease in cost.

Such studies have been made in this idealised fashion, primarily for third world countries in the process of building up their emergency preparedness. The much more usual situation is that a facilities are already available, and the question is then whether they should be supplemented or reorganised.

Practical decision making under these circumstances is much more complex. It involves evaluating the possibilities of improvisation, the possibilities of using the facilities of salvage companies, in fact, of thinking through each possible accident, and determining how the effects of accidents could be reduced, and the overall costs of treating the accidents as they arise.

Traditional risk analysis methods are little suited to this kind of investigation. Few of them take emergency preparedness into account at all, and if they do so, the number of alternatives is limited. A new approach combines decision making training with risk analysis. Experienced emergency managers are presented with accident situations, and are requested to suggest their response. The reliability of each response, mobilisation times, and effort are retrieved if known, and the overall effect is recorded. If these values are unknown, then it is necessary to estimate or calculate them. Comparison of the resulting scenarios provides a background for evaluation of risk.

Decision making under accident conditions

When the accident has actually occurred, the decision making situation is in some ways much clearer than during planning. The weather conditions, tides and location are generally known, and the resources which are immediately available are clear. On the other hand the substances which threaten the environment are often only poorly known, and there is a potential for unpleasant surprises. This contrasts with the risk analysis situation, where the range of substances is known or is assumed.

There are several different time scales for decision making in this situation. Of these, advice to the captain of the daamged or threatened ship is the most pressing. Although the captain will most often have a description of his cargo, with safety procedures and emergency cards, this will not always be the case. He may need advice on choice of sailing route, suitable areas for emergency grounding etc. and will often need advice on how to treat any spill.

A rapid access to chemical properties, and assistance in property identification will be of enormous benefit in this situation. Several chemical identification data base services are available on an international basis (refs 1,2,3) but do not generally include information concerning marine risks. "Paper" reference information is even more valuable, since it can in many cases be more rapidly available. In either case it is necessary to train in the use of such tools, otherwise much time is lost in communicating and the tools become an additinal problem rather than an aid.

Determining the potential size and extent of consequences is perhaps the next most important aspect of accident decision making. This information is needed in order to be able to select the type, scope, and size of the response. For accidents involving liquid release and pollution, there exist several consequence calculation packages, (refs 4,5,6) of which at least two are specially adapted to dealing with marine accidents.

In an actual emergency there is often considerable uncertainty concerning the situation, with factors such as release sizes, the poosibility of ship break up, unexpected chemical cargoes, the success of any cargo transfer, and future weather conditions as factors. For these reasons, it is not possible to make a simple prediction of consequences, and decide on action on that basis. Rather, there will be a range of possible alternatives.

In practice, most emergency managers make decisions based on the most pessimistic interpretation of known facts. All but the most experienced emergency managers have a tendency to overlook the fact that surprises are normal when making decisions about emergency response approaches. The range of potential problems considered is often limited. There are good reasons for this, in particular the short time available for decision making. By contrast, in actually carrying out an emergency plan, commanders often exercise considerable caution when committing personnel. For example proper use of protective equipment is standard even when the immediate need is not obvious.

An ideal decision making approach would involve investigating a range of alternative future developments in the emergency, and investigating the consequences and responses to each. In practice this will be impossible in all but the slowest emergencies, for example ships which have sunk with holds or containers unbreached.

All of this points to one approach to decision support, that is, prior training in decision making for using risk analysis and

360

consequence calculation tools. Ideally such training should be
related to actual emergency cases. Ideally, too, the tools used
should be such that they can if time permits be used in actual
incident conditions.

More complete decision support and emergency management facilities
can be provide by modern equipment. A list of possible facilities
is as follows:

- Initial priority communication of alarm and warning messages
  to emergency personnel.
- References to emergency and support services, shipping
  offices etc.
- Chemical materials properties and identification guide.
- Safety precautions information.
- Guidance on appropriate use of emegency equipment.
- Inventory of available emergency equipment, and communication
  channels for obtaining support.
- Consequence calculation models.
- Training facilities.

Example

Fig 1 shows the overall design for an experimental emergency
training system. It is intended to allow users the ability to play
through a large number of scenarios, and to prepare scenarios for
larger training exercises, or to practice decision making.

In use, the accident situation is first described in terms of the
ship size, type, and cargo, a map of the accident area, and the
weather and current conditions. On the basis of this information,
the user can work through the various stages in the accident,
using check lists to identify the possibilities for release, for
fire or explosions, for new failures, and for changes in weather
conditions. This basic structure is represented as an event tree,
with appropriate timing information. The tree can contain just one
simple sequence of events, or many parallel sequences, and many
alternatives.

Onto the basic structure of the event tree, possible emergency
actions can be attached. These can either be fixed as part of the
scenario, or can be listed as alternatives to be selected by
trainees. For each combination of events, weather, and emergency
response, the corresponding consequences can be calculated. The
results can then be displayed, both in mapped and pictorial form.

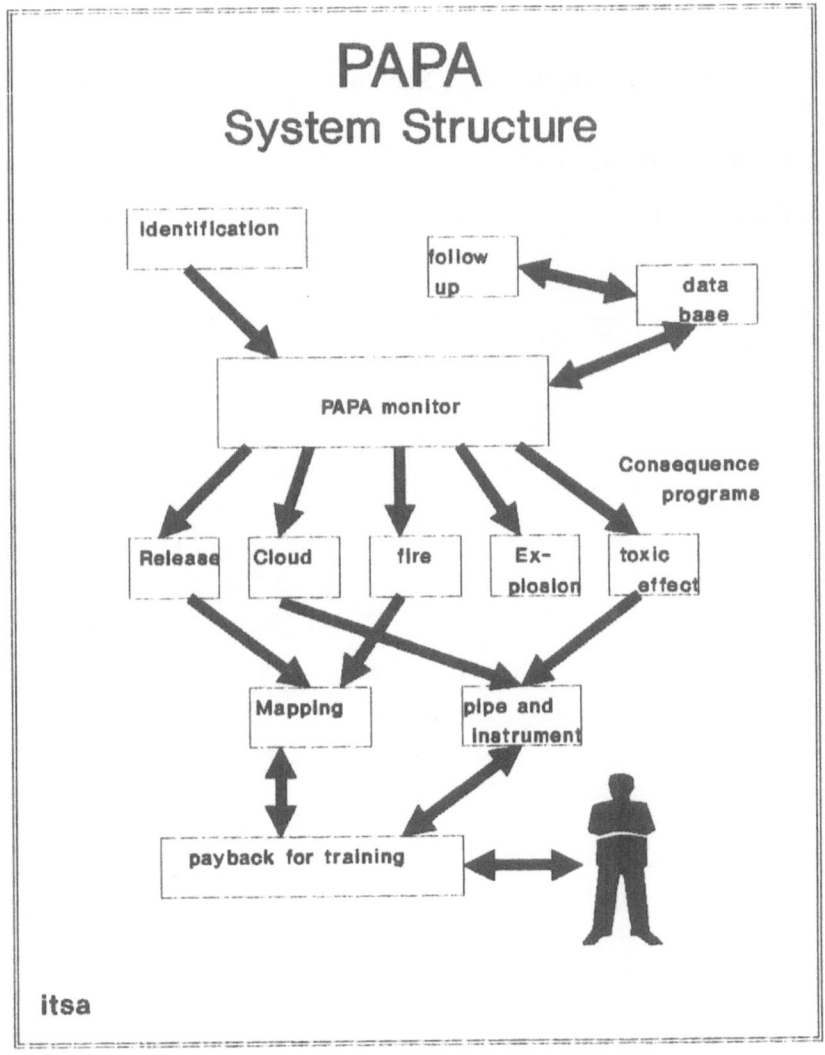

PAPA
System Structure

Identification

follow up

data base

PAPA monitor

Consequence programs

Release

Cloud

fire

Ex-plosion

toxic effect

Mapping

pipe and instrument

payback for training

itsa

References

1. HACS   Described in Melguen and Kantin, Nato CCMS workshop, Brest 1987

2. SEABEL TNO, Holland 1987

3. PAPA ITSA, 1988

DECISION MAKING FOR EMERGENCY RESPONSE

Laurens M. Schrijnen
Netherlands Organization for Applied Scientific Research
Division of Technology for Society
Department of Industrial Safety
P.O. Box 342, 7300 AH Apeldoorn, The Netherlands

Abstract
Recent years have shown a great increase in the transportation and
storage of hazardous materials which may present a danger to both people
and environment if not handled in a proper way. Public concern over
hazardous materials has justifiably grown and both governments and
industry have taken steps to prevent and respond to hazardous materials
incidents.

These steps concern the protection of people, special packaging procedures, the availability of emergency response equipment etc.
However the quality of these precautionary measures is too often in
violent contrast with the attention governments and industry give to the
actual decision making in emergency response circumstances.

This paper will explain why emergency response decision makers should
give more attention to their decision making. An example of decision
support for emergency response decision making is sketched.

1. CRISIS DECISION MAKING

Decision making is an activity people do every day. Simple situations
will lead to simple decision processes. Complex decisions however need
formal decision making strategies. Good decision making strategies
follow this procedure:

    Step 1: define the parameters affecting the decision,
            define relations between those parameters,
            assign values to these parameters and
            measure the importance of each parameter.

    Step 2: define all possible decisions,
            focus on probable applicable decisions.

363

P. Bockholts and I. Heidebrink (eds.), Chemical Spills and Emergency Management at Sea, 363–369.
© 1988 by Kluwer Academic Publishers.

Step 3: define outcomes measures.

Step 4: quantify the possible consequences of each decision.

Step 5: order the applicable decisions to their optimal score on
all relevant aspects.

Step 6: carry the best decisions into effect.

Emergency response circumstances are often crisis situations. Crises are
defined by the following aspects: there are large uncertainties,
surprises occur, important values are threatened, and the time available
for making decisions is limited [1,3]. Important new alternatives,
events, or outcomes in a crisis are not always known in advance; even
those that are known may occur in an unexpected order.
A crisis often exists because such unanticipated and haphazard variables
make standard decision and operating procedures difficult to apply.
Overlooking these surprise variables may cause flawed decision making.
On the other hand, scrutinizing every variable may waste time on parts
of the problem that have little effect on the decision. Analysis of only
those variables that may change the decision will make best use of the
limited time available.

The danger in a crisis situation comes from a threat to important
values, such as human life or the eco-system. Values considered in the
midst of a crisis can easily be biased. For example, long range values
can seem relatively unimportant when immediate concerns are threatened.
Thus, assessment of the decision makers' preferences among different
outcomes could best be done in advance of a crisis, as long as flexibility is maintained to include the influences of surprise variables added
to the crisis model. During a crisis, quick and efficient techniques are
needed to assess the probable outcome ranges for each alternative as the
model evolves.

The time pressure of a crisis can intensify either as the time period is
shortened for a specific number of decisions, or as the number of decisions is increased in a specified period of time. For instance, decision
makers may want to know the best decision immediately, before the
weather changes. Or, they may want to analyze as many separate decisions
as possible in a limited period of time, perhaps to solve several
smaller problems before attacking a larger one. A crisis that involves
both types of time limitations is particularly challenging.

2. THE DECISION MAKERS

Traditional predictions of the psychological effects on decision makers
include emotional, physical and cognitive stresses [1].
Emotional stress produces anxiety that, if not controlled by adaptive
coping behaviors, can lead to fear, frustration, hostility and other
maladaptive behaviors. Physical fatigue results as continual tensions

breaks down the body's defenses. Cognitive effects are evident in dis-
torted perceptions of time, perceptions of fewer elements in the
environment, increased perception of risk and an inability to differen-
tiate personal and institutional threat.
Misperceptions such as these can result from a dogmatic approach to
decision making that includes greater conceptual rigidity, biases
against new information, less tolerance of ambiguity and an inability to
handle complex cognitive processes.

Decision makers can be affected by these psychological stresses.
They may choose alternatives prematurely and on a basis of insufficient
information. They may discriminate less between alternatives and be less
able to predict possible consequences. They may consider fewer alternatives and be more likely to choose a less desirable one. The longer the
duration of the crisis, the more likely they are to change their goals,
often in the direction of rationalizing what already has been done.

Rosenthal [2], in his book on crisis decision making in the Netherlands,
gives fourteen propositions in this field based upon empirical evidence.
Most of these statements are so widely applicable they should not be
wanting in this paper on emergency response decision making. Some of
these propositions and a number of recommendations for each proposition
are the following:

1. Decision makers become obsessed by the time aspect. They concentrate
    on very short term matters.

    Some recommendations for this proposition are:
    Decision makers should not translate all incoming information in
    necessary decision time. A special data processing team can unburden
    the decision makers. The decision team should be aware of all
    consequences of its decisions.

2. Crisis situations can lead to group thinking. There will be little
    room for doubts and objections.

    Recommendations:
    The members of the decision team should be challenged for other
    solutions. The team has to keep in touch with its grass-roots:
    inform them and listen to them. Inviting experts and outsiders can
    be very useful. Very critic decisions should be followed by so
    called "second-chance" meetings.

3. In crisis situations decision makers tend to underestimate their own
    capabilities and to overestimate other peoples possibilities.

    Recommendations:
    The decision team should be aware of pessimism and realize specific
    crisis interactions like: misleading, bluff and overasking.

4. In crisis situations decision making usually occurs in a small team. Formal regulations are less strictly observed. Crisis decision making can be characterized by improvisation.

    Recommendations:
    The decision team will have to ensure that informal processes are defendable retrospectively and that all decisions are recorded. The decision team must prevent that involved experts become decision makers.

5. In crisis situations information and communication-processes increase.

    Recommendations:
    The team should be aware of information from others than 'trusted, liked sources' and bad news messengers should not be excluded. Information media can be useful, are sometimes unavoidable, but usually cause jams in telephone exchanges. Precautionary measures are relatively simple.

3. CRISIS DECISION ANALYSIS

Crisis decision analysis can be helpful for both contingency planning and crisis management [1]. Contingency planning encompasses any "what-if" thinking done prior to a crisis; whereas, crisis management refers to the decision-making process used during a crisis. A well structured contingency plan can be the basis for effective crisis management, or perhaps even avoidance of the crisis entirely; however, a contingency plan so thoroughly structured that no allowance can be made for surprise events is inflexible and, thus, undesirable. An ideal compromise is to design contingency plans as small, pre-structured parts of a larger decision model of an anticipated crisis. Then, as new information becomes known before the crisis occurs, these parts of the model can be updated and the decision strategies reanalyzed.

Crisis decision making in any organization could be improved by using a logical procedure to model decisions made under stress; moreover, the effectiveness of such a procedure would be enhanced by reducing the stressful factors in the organizational environment. Prescriptions for such procedures can prevent some of the typical crisis decision-making pathologies, such as premature consensus, information distortion, group judgement errors, rigidities in programming, unpreparedness and imple- mentation failures.

An interactive computer program that asks the appropriate questions and rapidly performs the required calculations would be indispensable for decision making in an actual crisis, one with the characteristic com- plexity and time pressure of the real world. Acceptance of a normative modeling procedure by decision makers "on the firing line" would clearly depend on how well the computer program processes and displays the information it receives. An analyst, acting as an intermediary between the decision maker and the computer program, may be vital to the successful use of the modeling procedure.

The modeling procedure, adapted from standard decision analysis methods, that guides the structuring, assessment and analysis of crisis decisions will be able to:

1. Identify and structure ALL those, and ONLY those, variables that may change the decision.

2. Assess probable outcomes QUICKLY and EFFICIENTLY.

3. Indicate, AT ANY MOMENT, the best decision and the VALUE OF FURTHER MODELING.

4. PRACTICAL IMPLEMENTATION OF DECISION SUPPORT FOR EMERGENCY RESPONSE

The steps in the decision process as sketched in section 2 do not all have to be taken the moment a decision has to made. Some steps can be performed in advance. Decision support and information systems are shown very valuable in this field.

Good decision support systems for emergency response decision making should include as many decisive parameters as possible and contain as many decisions as possible. (Step 1 and 2). If step 3: "define outcomes measures" is not implemented in these systems or in the procedures of the decision team, probably an enormous amount of decision time will be lost by discussions on this subject.

Simulation models of physical processes can, if inputted with actual data, be very helpful in establishing possible consequences of each decision. (Step 4). These simulation systems predict outcomes of measures and physical behaviour of substances. The precise values of outcomes is usually not very important. A quickly estimated best and worst value can give a lot of information. The limiting conditions of each model must be very clear to the decision team.

A clear, unambiguous presentation of preferable decisions and accessible background information on the pros and cons of the preferable and also dropped choices, are necessary conditions to a good crisis decision making process.

A fine example of such a decision support system is SEABEL: The hazard identification and decision support system for emergency response of chemical spills at sea. The structure of one of the modules of SEABEL, the emergency response decision module, is explained here.

SEABEL's process to come to a balanced decision on the optimal response at a certain accident is divided in two consecutive activities:

    the selection of measures and
    the evaluation of measures.

Measure SELECTION is done by searching the total set of available measures for those that are possibly appropriate. That means these measures are worth thinking about. This specific definition of measure selection prevents the loss of high value decision time on evaluating absolutely-non-applicable measures, gives the opportunity of a list of reminders to the decision makers and leaves the factual decision to the decision makers.

The total set of available measures is a database containing all by the emergency response organization usable response methods.

The selection procedure can be based on the hazardous materials and other aspects of the specific accident. These aspects discriminate response measures. Important aspects of the accident scenario are properties of involved chemical substances, characterizations of the accident, its threats and finally some meteo-data.

The name of the chemical substance is not mentioned. However if the name is available, or classification numbers like the UN-, Kemmler-, or IATA-number and the decision-team has databases available with such a classification, many aspects of the mentioned substances can be retrieved from these databases.

The measure EVALUATION is done in two steps. Before hand a preference can be established based on predictable aspects such as effects of measures on people and environment, possible nuisance in surrounding areas and costs.

Usually the second step can only be made with case-dependent information. The decision maker can add ad-hoc aspects as decision criteria. This implies a presentation of the best selected measures instead of one and only best measure. Is there ever such a measure?
Again: the factual choice of a measure or a set of measures is to the decision maker.

5. CONCLUSIONS

Emergency response decision making can be improved a lot by preparation of the members of the decision team and a structured decision procedure.

Part of the preparation can be the development of a decision support system. Such a system is not just applicable during a crisis but can also be used as a decision team training device.

6. ACKNOWLEDGMENT AND REFERENCES

Jacques F.J. van Steen (TNO, department of Industrial Safety) deserves
credit for his helpful suggestions and comments.

[1] Burke E. Robinson. 1979.
    Crisis Decision Analysis.
    Menlo Park: SRI International.

[2] U. Rosenthal. 1984.
    Rampen rellen en gijzelingen: crisisbesluitvorming in Nederland.
    (in dutch).
    (Disasters, riots and hostage-taking: crisis decisionmaking in
    the Netherlands)
    Amsterdam: De Bataafsche Leeuw.

[3] Charles F. Hermann (editor). 1972.
    International Crises: Insights From Behavioral Research.
    New York: Free Press

[4] Jacques van Steen. 1987.
    A methodology for aiding for hazardous materials transportation
    decisions. In: European Journal of Operations Research 32.
    Amsterdam: North-Holland.

LIABILITY/ASSURANCE

LIABILITY AND COMPENSATION FOR DAMAGE BY HAZARDOUS AND NOXIOUS
SUBSTANCES IN A MARINE ENVIRONMENT

A. van Eden, Capt. R.NL.N.(Retd)
Deputy Director North Sea Directorate. RWS. (Retd)
Environmental Consultant to the Commission of the
European Communities.

Abstract
This paper addresses certain aspects arising from shipping accidents
which may result in or threaten to cause damage to the marine
environment by hazardous, noxious substances carried. In particular
attention is given to the desirability of arriving at an international
regime of strict-liability and compensation for those matters which
demand preventive measures and clean-up operations and if factual
evidence is available the restoration of the marine ecology. Reference
is made to the present liability and compensation regimes for oil-
pollution and nuclear damages of the marine environment. Work undertaken by the IMO, Legal Committee in this field is discussed and
reasons are given for the failure of arriving at an HNS Convention by
the Diplomatic Conference on HNS held in London in 1984. Finally
transport of chemicals in various modes is looked into whilst criteria
are being mentioned regarding the dangerous and toxic characteristics
of chemical substances. Conclusions are drawn.

1. GENERAL BACKGROUND

a. The question raised in the wake of the "Torrey Canion" disaster,
   which resulted in the adoption of the 1969 International Convention
   on Civil liability for Oil Pollution Damage was: Should there be a
   special liability and compensation scheme in respect of maritime
   transport of hazardous and noxious substances? The Governments then
   present urged the IMO (then IMCO) to concentrate the activities in
   this field on "all" aspects of pollution by substances other than
   oil which could be hazardous and noxious to the marine environment.
   The IMO referred this concern to the Legal Committee which at its
   seventh session in Januari 1970 recognised the problem of pollution
   by H.N.S. "as a matter of urgency". It was easier said than done.
   The Legal Committee did not have the technical information and
   knowhow on the various types of dangerous substances and the impact
   thereof on the marine environment. Moreover transport of chemicals
   in the sixties was, in comparison with the present volume, carried
   out on a moderate scale. Incidents with ships carrying chemicals

373

P. Bockholts and I. Heidebrink (eds.), Chemical Spills and Emergency Management at Sea, 373—380.

were sporadic and far between. The chemical industry and the
shipowners of chemical tankers realising the inherent dangers of
such transports started to improve the safety precautions in the
design, construction and equipment of these tankers long before the
IMO started to make regulatory efforts. So there was a general lack
of experience with maritime accidents involving chemicals. Aspects
of civil liability and compensation with regard to damage by H.N.S.
were not considered to require immediate regulatory measures by the
Legal Committee and it was decided to await the 1973 MARPOL
Convention. It is well known however that there is a strict
liability regime under the 1960 Paris or 1963 Vienna Convention
dealing with nuclear damage, and third party liability in the field
of nuclear energy. To be precise the Convention relating to Civil
Liability in the field of Maritime Carriage of Nuclear Materials
1971 was adopted in Brussels, which exonerates the maritime carrier
of nuclear material from liability if the operator of a nuclear
installation is liable for nuclear damage under the Paris 29 July
1960 or Vienna 21 May 1963 Convention or comparable national law.
Next to these provisions in the nuclear field the International
Maritime Community adopted, as already mentioned, in 1969, the
inter-governmental regime for strict liability and compensation for
damage caused by oilspills from tankers. This regime was based on
two international conventions:
- The 1969 International Convention on Civil liability of Oil
  Pollution Damage (C.L.C. '69) which entered into force in 1975 and
- The 1971 International Convention on the Estabishment of an International Fund for Compensation for Oil Pollution Damage (F.C.71)
  which entered into force in 1978.
The Civil Liability Convention lays down the principle of strict
liability for the shipowner and provides a system of compulsory
insurance to secure adequate compensation for the damage caused by
oil pollution; i.e. damage inflicted on the environment, the costs
involved for cleaning up oil spillages and when factual evidence
exist, the restoration of the marine ecology.
The Fund Convention is supplementary to the Civil Liability Convention and provides compensation to victims when the compensation
under the C.L.C. is inadequate. The International Oil Pollution
Compensation Fund (I.O.P.C. Fund) was established on 16 October 1978
to administer the regime of compensation under the Fund Convention.
The I.O.P.C. Fund is financed by persons who receive crude and heavy
fuel oil in States party to the Conventions. There were 38 nations
member to the I.O.P.C. Fund on the 1st January 1988. The C.L.C. 69,
registers 60 States as member to this International regime.
It should be mentioned that prior to these international conventions
two voluntary schemes were conceived by the industry, the tanker
owners and the owners of the oil cargoes, at the time when the
corresponding international conventions were being negotiated. Today
there are 3200 members to TOVALOP covering some 6.000 tankers
including OBO's, barges etc., representing over 95% of the world's
tanker tonnage. CRISTAL was devised to provide supplementary
compensation to that available from tanker owners under TOVALOP.

Today it is estimated that the total crude and fuel oil receipts of the parties to CRISTAL exceed 80% of all receipts. Both agreements are intended to be interim solutions and to remain in operation only until the international conventions have world-wide application. Although the International Community welcomed the initiative by the shipping and oil industries to provide for a voluntary system of compensation, in the long run it is preferable to have a legally binding regime, provided for at an inter-governmental level through international agreements concluded at diplomatic conferences.

b. At the moment, November 1988, no special international regime exists relating civil liability and compensation for damage arising from the carriage by sea of hazardous and noxious substances other than persistent oils and nuclear material. In fact there is a gap between the 1969 C.L.C. and its consort the 1971 Fund Convention, on the one hand, and the 1960 Paris Convention on third party liability in the field of nuclear energy etc., on the other hand. Moreover in the draft Convention on Salvage which will be the subject of a Diplomatic Conference in 1989 special compensation will be awarded to a salvage operation if a salvor has prevented or minimised damage to the environment.

Therefore there should be constructed an international instrument which should cover the following damage possibilities to third parties and for which a system of strict liability (with no proof of fault) and a very limited number of exonerations should meet the following requirements: Damage by;
- HNS in bulk;
- HNS in packaged form;
- explosion and fire by gas and other substances on board ships;
- bunkerfuel from merchant ships.

Furthermore a new system should meet the requirements of:
- a liability that should rest with an easily identifiable party;
- as far as practicable liability should be strict;
- any limit of liability should be sufficiently high to provide adequate compensation for the damages mentioned above of a reasonable magnitude;
- where damage exceeds the limit of liability an obligation could be placed on an international fund.

c. Claims for damage to third parties by H.N.S. may fall under the Convention on Limitation of Liability for Maritime Claims, 1976, or the Convention on limitation of Liability of Owners of Seagoing Ships, 1957, if applicable, and have to proven on fault. In recent years the principle of "the polluter pays" has been generally accepted. However the inherent harmful characteristics of the substances carried, the substandard shipping, and the enormous increase of liability risk, the "shared" responsibility between shipowner and shipper, consigner or consignee was raised as a new form of modern thinking. The International Chemical Industries (CEFIC and HMAC) are strongly against any form of responsibility for any damage by HNS when transported by sea.

The Council of European Chemical Manufacturers considers the principle of "the polluter pays" somewhat ambiguous. CEFIC maintains

376

that individual liability for damage should remain the rule. At the
same time it stresses the necessity of introducing reasonable and
equitable provisions on liability that may benefit insurance cover
under economically acceptable conditions. One of the possibilities
to create a voluntary or legally binding Fund, for example by the
Chemical Industries for the coverage of excess damage will be a
rather difficult proposition to sell. The draft Convention for
inland transport prepared by UNIDROIT and presently under discussion
within the E.C.E. puts, in short, the liability for damage by
substances carried with the carrier irrespective what kind of
carrier that may be. This is in the line of thinking of CEFIC. The
disadvantage of this system is that it would only apply to the
transportation of goods in the European area and not world wide.

2. WORK UNDERTAKEN BY THE IMO

a. As already mentioned, the Legal Committee of the IMO recognized the
   problem of pollution by HNS as a matter of urgency in January 1970.
   There was however in the beginning of the seventies insufficient
   expertise and a general lacking of experience in the field of
   judicial development concerning liability and compensation of damage
   by HNS to the marine environment. Before arriving at a satisfactory
   regime a lot of questions and information gathered from governments
   and interested organisations had to be fully considered. One of the
   options under discussion was a limited extension of the 1969 C.L.C.
   and the elaboration of a new convention for damage by pollutants
   other than oil. Some delegations of the Legal Committee were of the
   opinion that not enough data were available concerning all the
   possible types of substances capable of causing pollution damage of
   a specific and serious nature. In mid-1975 there were even
   delegations to the Legal Committee who failed to see a real need for
   the creation of a new liability and compensation regime because:
   - the nature of most types of marime pollution by HNS was such that
     clean up costs were unlikely to be high;
   - the higher limitation amounts most likely to be adopted by the
     1976 International Conference on Limitation of Liability for
     Maritime Claims, which was to be convened to revise the 1957
     Convention Relating to the Limitation of Liability of Owners of
     Sea-going Ships should solve most civil liability problems that
     could arise in connection with damage other than pollution damage
     resulting from the transportation of dangerous goods by sea.
   Prior to the limitation of Liability Conference in 1976 the Legal
   Committee showed however the intention to deal with the whole
   subject of civil liability and compensation for pollution damage
   arising from the carriage of hazardous and noxious substances by sea
   through a step-by-step approach. Whether that meant a protocol to
   the CLC-69 convention for HNS or a separate comprehensive convention
   covering all hazards other than persistent oil would encompass in
   fact more or less the same amount of work. It was, and it still is,
   a policy decision which had nothing to do with the scientific and
   technical data to be supplied by the operational sectors. The

dangerous cargoes carried by ships were primarily controlled by the
International Maritime Dangerous Goods Code. The regulation of which
were in the first place intended for the safety of the ship and its
crew. Concern for the environment slowly started to carry its weight
in the seventies and was embodied in Marpol 73/78, the Annexes II,
III, IV and V.

b. Nevertheless several Governments were afraid that the world would
suddenly be faced by a major catastrophe involving hazardous noxious
substances. The Amoco Cadiz disaster underlined such a possibility.
Thus they insisted on the development of an international legal
regime covering strict-liability for damage to third parties and
adequate compensation. In 1978 an informal working group was set up
under the Legal Committee which arrived at three alternatives for an
HNS regime.

Alternative I   : A system of party liability, including joint and
                  several liability of the shipper and shipowner.
Alternative II : A two-tier system of liability providing for a
                  primary of the shipowner and an excess (residual)
                  liability of the shipper.
Alternative III: An exclusieve liability of the shipper.

Next to these three openings, thoughts were given to another
alternative text allocating liability to the shipowner alone
(Alternative IV), whereas also the possibility was being considered
in which liability should be borne exclusively by the cargo
interests, a product liability, which entailed either a cargo levy
fund or a system of cargo insurance. The idea of a shipper's
liability was received as a rather new development in Maritime Law.
The final draft text of the new HNS Convention, which was forwarded
to the 1984 Diplomatic Conference on Liability and Compensation in
London, incorporated a "mixed" two-tier system of strict liability,
Alternative II.

c. The International Diplomatic Conference of the IMO in May 1984 was
to consider the adoption of the Draft HNS Convention but also
protocols to the 1969 C.L.C. and the 1971 Fund Convention. Serious
discussions were held regarding the various aspects of a new HNS
Convention. However the seventy participating States did not arrive
at a consensus of opinion. The draft on an H.N.S.-regime was
referred back to the IMO for further study, amendments and clarifi-
cations by the Governments delegations in the Legal Committee. The
Diplomatic Conference of 1984 failed to reach agreement on a number
of issues:
- The definition of the shipper.
- The definition of dangerous substances and the minima of
  quantities still dangerous.
- The geographical application.
- The limitation of strict liability of the shipowner either in
  accordance with the 1976 L.L.M.C. Convention or linked to the
  tonnage of the ship concerned.
- The liability of the shipper on top of the shipowners liability.
- The limitation of the liability of the shipper; alternative I -
  fixed amount; alternative II - a tonnage linked liability.

- The restriction of the draft to bulk cargo.
Especially the last item was carrying weight.
d. Since then various representatives of the Legal Committee have
intensified their work on a new draft convention. It should be
mentioned that even during the diplomatic conference a redraft of
the H.N.S. text was graciously welcomed but in view of the time
available and lack of official instructions no results were
obtained. The failure of adopting the draft Convention on H.N.S. was
a matter of very serious concern to the Council of the IMO. But
during the fifty-eighth session of the Legal Committee at the IMO
Headquarters in October 1987, among other items, a new paper was
tabled: "Considerations of the question of liability for damage
caused by the maritime carriage of Hazardous and Noxious
Substances". Professor Cleton, Chairman of the Legal Committee, who
will present the next paper, will enlarge on this new approach.

3. THE TRANSPORT OF CHEMICALS AND ITS DIVERSITY

a. The products of the chemical industry come in a great variety and
their diversity increases every day. Chemical products are
indispensable for our society. Everyone is aware of the world wide
ramifications of the oil industry. It is taken for granted that our
industries and society needs the transport of crude oil and/or heavy
fuel oil by sea on international voyages. The chemical industry
however depends similarly on transport by sea. Most of the chemicals
are carried in bulk, only 10% is shipped in drums, containers,
portable tanks or in rail or road wagons. The substances, mostly
liquid can be quantities of raw chemical materials, half or finished
products. These substances originate from the fossil fuels (coal,
natural gas and petroleum) water, air, salt, limestone, sulphur,
phosphate and fluorspar. Basic feedstocks, which are converted by
the chemical industries into primary, secondary and tertiary
products, or even further, the last being closest to the consumer.
It is clear that not every chemical product is hazardous to the
environment and mankind and of course the chemical industry is its
own best customer because its products mostly need to be processed
further in order to be made suitable for passing it to the person
who ultimately pays the bill, the consumer.
b. It is estimated that roughly 30 million tons of chemical are
transported world wide every year. Twenty percent of that amount is
or can be toxic to the environment. In the worldtrade there are at
the moment (Jan. 88) about 160 bulk chemical tankers belonging to
about ten major companies. Their sizes vary between 6,000 and 36,000
tons. It is not expected that the size will exceed the 40,000 dwt.
because these ships are constraint by draught and port-facilities.
The parcel-tanker, the most sophisticated technically speaking, can
carry as many as 56 different bulk liquid cargoes in separate tanks.
Tank capacity ranges from 200 tons to a few thousand tons each. The
safety of bulk chemical tankers has been of matter of national and
international concern for many years. The basic philosophy of IMO's
Code for the construction and equipment of ships carrying dangerous

chemicals and liquefied gases in bulk, the BCH and I.B.C.-Code and
the G.C. and I.G.C.-Code, is that of relating ship types to the
hazards of the chemicals carried. Modern chemical tankers are in
fact floating factories, quite safe and accidents seldom happen. The
work of IMO in developing international regulations on the carriage
of hazardous chemicals and liquefied gases in bulk is largely
complete. The need remains however to update the Codes to include
new chemicals and liquefied gases for bulk shipment, take
technological advancements into account and clarify requirements as
necessary.
The transport of chemicals in packaged form is quite a different
story. The packaged chemicals are mostly the more dangerous
substances contained in drums, cilinders, barrels, containers,
tanks, road and railwagons, etc.
These parcels are conveyed by ordinary merchant ships, most of the
time as deck cargo or in containers on board of containerships.The
Solas Convention puts down regulations for the safety of life at sea
without taking care of possible marine pollution. In stowing
packaged dangerous substances on board of merchantmen it is always
necessary to find an optimum between the safety of the ship and its
crew and proper concern for the environment. Packaged chemicals
carried by ships on deck are lost at sea quite regularly. They may
be lost to the sea through adverse weather, collisions, groundings
or accidents on board. They may be jettisoned in emergency
circumstances to secure the safety of the ship and crew. They may
remain on board a sunken ship or start drifting away from a sinking
or sunken ship. Lost packages may float on the sea surface, float in
submerged condition, be swept on to the beach or sink to the seabed.
In these ways lost packages may be accidentally encountered by
seafarers, by fishermen in particular or by the general public. Such
encounters can be lethal. The loss at sea of chemicals in packaged
form has dominated the list of chemical incidents. To search and
recover sunken packages of chemicals can be cumbersome, difficult,
time consuming and sometimes technologically not feasible, apart
from the finances involved. To salvage sunken ships with bulk
chemical cargoes can be even more demanding. Their recovery is a
very costly affair which can be a heavy burden on coastal states.
The following extravagant costly salvage operations can be
mentioned:
1984: Recovery of chloride gas containers, off IJmuiden.
1984: Recovery of hexafluoride from Mont Louis, off the Belgium
      Coast.
1985: Recovery of anti-knock dopes, off Mogadishu, from "Ariadne".
1986: Salvage of the coaster Olaf with a cargo of fly-ash containing
      substantial quantities of heavy metals, off Den Helder.
1987: Salvage of the Cason, off Cape Finisterre.
1988: Salvage of the Anna Broere with a bulk cargo of acrylo-nitril,
      off Den Helder.
c. The question arises as to what kind of hazard of a specific chemical
has to be taken into consideration and of course combinations
thereof. How do we approach these spilled chemicals when estimating

the various aspects as: dangerous to mankind, the environment, the
community, port installations, ship and crew, the marine ecology as
a whole or specific layers in the specimen etc. How can we deal with
such hazards to neutralise their effect. The identification and
evaluation of the various kinds of hazards is of vital importance.
In itself it is a complex problem stemming from the combination of
several factors like: flammability, toxicity, the amount spilled,
the sensitivity of the area, prevailing currents, temperature,
salinity etc. whether it is a bulk spill or in packaged form. By
dividing the chemicals in different catagories it is possible to
develop more appropriate response methods and get a better
understanding of the danger aspects to mankind and the aquatic
environment. The following categories or classifications can be used
in the field of response to hazardous material spills in a marine
environment;
- classification by transport,
- classification by physical hazard,
- classification as marine pollutant,
- classification by behaviour,
- classification by reactions.
At the IMO, the allocation of substances to an international
classification system for recorvery is still under discussion.
Making use of two fundamental categorisations, i.e. the Gesamp
profiles and the I.M.D.G.-Code, could provide an International
Recovery Response Category System which should be considered as a
lead-in when evaluating the liability and compensation of hazardous,
noxious substances spilled in a marine environment.
The scientists and technical experts are in a position to define the
degree of hazardness and toxicity. IMO should establish an
international classification system for recovery.

4. CONCLUSIONS

There is at the moment in the maritime world a dire need for an international instrument, an internationally adopted regime, which should
cover the costs for damage or potential damage by:
-   Hazardous and noxious substances spilled in bulk.
-   Hazardous and noxious substances spilled in packaged form.
-   Explosion and fire by ships.
-   Bunkerfuel spilled by merchant ships.
The new system should meet the requirements of:
-   liability that should rest with an easily identifiable party;
-   as far as practicable liability should be strict;
-   any limit of liability should be sufficienly high to provide
    adequate compensation;
-   An International Fund should be available for excessive damage.
Professor Cleton will discuss the pro's and con's of such a regime and
its feasability.
Next to that there should be established by the I.M.O. an International
Classification System for the recovery of spilled hazardous, noxious
substances in a marine environment.

Compensation for damage caused
during transport of hazardous
and noxious substances by sea

Robert Cleton
Ministry of Justice, The Hague
Professor of Transport Law,
Erasmus University Rotterdam

*In this paper the main outlines for a compensation
scheme relating to carriage by sea of hazardous and
noxious substances (HNS) are discussed. First some
basic requirements for a new compensation scheme are
considered. Attention is also given to compulsory
insurance and channelling of liability. The most
controversial issue is the question on whom the
liability should be imposed and very divergent views
have been expressed.*
  *Several alternative solutions for a compensation
scheme are discussed. They include some form of
exclusive shipowner liability, exclusive shipper
liability, some form of shared liability or a possible
international compensation fund.*

1. Introduction

In his contribution to this International Conference Mr. van
Eden has already covered a number of issues relating to the
preparatory work sofar done with respect to an international
compensation scheme for damage caused to third parties
during the transport of hazardous and noxious substances
(HNS) by sea. In view of the limited time available I will
restrict my intervention to the main issues which have
arisen during the discussions in recent years on this
subject and to some of the alternative solutions which have
been suggested. After the failure of the IMO Diplomatic
Conference of 1984 to agree on a new international
convention, the matter has been referred back to the Legal
Committee of the IMO which has made a fresh start by
discussing the basic requirements for a new scheme and the
general outlines for possible alternative solutions. The
Legal Committee has not yet reached the stage for taking
final conclusions and at this moment it is difficult to
forecast in which direction the solutions will be found.

P. Bockholts and I. Heidebrink (eds.), Chemical Spills and Emergency Management at Sea, 381—391.
© *1988 by Kluwer Academic Publishers.*

2. Basic requirements for any new compensation scheme

There is, however, a certain consensus within the Legal
Committee about the basic requirements which any new
compensation scheme should meet. These requirements are the
following:
  (a) not only HNS carried in bulk but also in packaged form
      should be covered by the new instrument;
  (b) liability for compensation of damage should rest with
      an easily identifiable party;
  (c) the liability should as far as practicable be strict;
  (d) any limit of liability should be sufficiently high
      to provide adequate compensation for HNS damage.

2.a. HNS IN PACKAGED FORM

Mr. van Eden has already gone into the question whether or
not HNS which are carried by sea in packages should be
covered by the new compensation scheme and I can be brief
about this issue. Many governments accept that HNS carried
in packaged form, for example in drums, can pose a serious
threat to the environment and that they have to be recovered
after an accident in order to prevent later pollution
damage. These operations can be very expensive for the
authorities involved. During the preparation of the draft
HNS Convention which was submitted to the 1984 Conference it
became apparent that the inclusion of this kind of transport
of HNS will cause some difficult legal problems, in
particular with respect to limitation of liability and
compulsory insurance. When the scheme would only cover bulk
transport it would be applicable to a limited number of
specialised ships, but HNS in packaged form may be carried
by any cargo ship. The scope of the scheme would be
broadened considerably by the inclusion of HNS in packaged
form.

2.b. THE PARTY LIABLE

One of the most controversial issues is the question who
should be liable to pay compensation under the new scheme.
It is not surprising that none of the industries involved,
shipowners and P&I Clubs on the one hand and the chemical
industry on the other hand, are eager, to put it mildly, to
accept this new financial burden.

Shipowners and P&I Clubs reject their liability under a new scheme on the ground that the special risks posed by HNS are not caused by the movement of the goods but by the inherent dangerous qualities of the substances. They are prepared to accept liability for any negligence on the part of shipowners or their servants or agents, but not strict liability. The chemical industry points out that shipowners are well aware of the nature of the cargo and know or should know which precautions to take and that the cargo once on board is out of control of the shipper and cargo-owner.

Governments, however, have also to look at the interests of the persons who have suffered damage caused by HNS. The reason to establish a new scheme is to provide the victim with a new instrument for adequate and speedy compensation of his damage. This requirement is not fulfilled when there is uncertaintity about the identity of the party liable. Therefore the liability should rest with an easily identifiable party. In most cases the identity of the shipowner can be fairly easily established by consultation of the ships' registry. The identity of the cargo-owner or shipper is much more difficult to discover for the victim who is an outsider with respect to the contract of carriage and the sales contract. Nobody will be willing to assume liability after an accident has occurred.

Nevertheless there is some sympathy for the arguments advanced by the shipping industry. It must be admitted that it is the nature of the substances which causes the enhanced risks compared to the transportation of other cargo rather than their transportation. One may argue that the HNS risk lies more within the sphere of the responsability of the cargo interests than within that of the shipowner. Therefore some governments consider it both necessary and equitable that some part of the liability should be borne by the cargo interests. It was this consideration that led the Legal Committee of IMO to introduce in the draft HNS Convention of 1984 a dual liability system: the shipowner was made liable up to a certain limit and the shipper of the HNS had to assume liability for damage exceeding the shipowner's limit. However, this system has not been accepted by the Diplomatic Conference.

There are other aspects which have to be taken into consideration such as the question which insurance market could best cover the new liability risk created by the compensation scheme. Without adequate insurance cover there will be no guarantee for the victim that he will be able to recover his damage from the party liable.

2.c. STRICT LIABILITY

There is, I think, among governments a general consensus

that any new system for compensation for HNS damage should
be based on strict liability and that the party liable
should only be able to rely on a limited number of defences.
Such a system of strict liability is already operative in
the field of nuclear damage and oil pollution damage. An
alternative system would be a system based on the presump-
tion of fault enabling the party liable to exonerate himself
from liability if he can proof that the damage caused could
not be attributed to negligence on his part or on the part
of his servants or agents. But such a system would not meet
the requirements for an acceptable compensation scheme.

2.d. LIMITATION OF LIABILITY

The fourth basic requirement formulated by the Legal
Committee is that any limit of liability should be
sufficiently high to provide adequate compensation for HNS
damage. The issue of limitation of liability is an important
one and it has dominated to a large extent the debates.
  Limitation of liability is of course a well known feature
in maritime law but this is not the only reason to introduce
it into a new HNS compensation scheme. The party liable
should be able to protect himself by insurance against the
financial consequences of his liability, in particular when
this liability is strict. One cannot expect from the
insurance industry that they will provide unlimited cover.
The wording of this requirement is, however, very vague.
Accidents with ships carrying HNS in considerable quantities
can be of a catastrophical nature and from the point of view
of victims the only adequate level of liability would be
unlimited liability. But this is not a practicable approach
to the problem and it will not induce to take out adequate
insurance cover. Some compromise must be found.
  The Legal Committee will have to rely heavily on information to be provided by the insurance industry about
available capacity and costs. During the 59th session of the
Committee, held in April of this year, the limits of the
1984 Protocol to the Convention on liability for oil pollution damage (CLC) have been mentioned as the minimum
starting level of compensation under the new scheme. The CLC
Protocol provides for the limitation of shipowners liability
a minimum level of 3 million SDR up to 5000 tons, 420 SDR
for each ton in excess thereof and a maximum of about 60
million SDR. The insurance industry has been asked to
prepare a document giving information on the insurance
implications of a new compensation system for HNS damage.
This will probably not be an easy job to do, taking into
account that so many different options for a compensation
scheme are still afloat.

2e COMPULSORY INSURANCE

In order to guarantee the claimant that the person liable under the new compensation scheme will be able to meet his financial obligations it is envisaged that such person shall be required to carry insurance to cover his liability. The amount of this compulsory insurance will be fixed by applying the limits of liability prescribed in the Scheme. The authorities of the States Party will have to check the insurance cover and will issue or certify the insurance certificates for the ships and/or shipments falling under their competence. The authorities will also have to inspect ships carrying HNS cargo entering their territories and see to it that the regulations of the new Scheme have been complied with. Also in this respect any new compensation scheme must be tested on its practicability.

A compulsory insurance scheme for liability of the shipowner is already in operation with respect to liability for oil pollution damage under the 1969 Brussels Civil Liability Convention for Oil Pollution Damage. The liability for HNS damage could in principle be arranged in the same way through the P&I Clubs. To introduce a compulsory shipper's liability insurance for carriage of HNS by sea would be a novelty and it is not certain whether the insurance market would be prepared to arrange such new type of insurance.

2f. CHANNELLING OF LIABILITY

Under a new system of strict liability combined with compulsory insurance the question may arise whether the victim should maintain possible recourse actions against other persons than the person liable or against the person liable for unlimited liability. Those actions are based on general rules of contractual or extra-contractual liability. Following the example of the Conventions in the field of nuclear liability and the 1984 Protocol to the 1969 Civil Liability Convention on Oil Pollution Damage one could envisage a channelling of liability by excluding claims for compensation of HNS damage against certain persons who are closely connected with the transportation and are exposed to recourse actions. In this connexion one could think of the crew members of the ship and other servants and agents, salvors and persons taking preventive measures after an accident and possibly other persons. The main reason to relieve those persons from liability for damage are, dependant on the group to which they belong, avoidance of double insurance, social protection of employees and to encourage salvage operations and the taking of preventive measures with respect to HNS accidents. Such channelling of liability should, however, not exclude all possible recourse actions.

3. The different options

3.1 INTRODUCTION

When discussing the issue which party should bear the liability under the new compensation scheme I have pointed out that very divergent views have been expressed and that attempts have been made to find a compromise solution. I would like to come back to this issue by explaining the various alternative solutions which have been discussed. These can be divided into three categories and some of these categories can be subdivided into sub-options:

  I. channelling of all liability to the shipowner;
  II. channelling of all liability to one of the parties who has a direct economic interest in the cargo;
  III. shared liability of the shipowner and cargo;
  IV. an international compensation fund.
These options will be discussed below.

3.2 FIRST OPTION: EXCLUSIVE LIABILITY OF SHIPOWNER

Under this option the liability under the new compensation scheme would rest exclusively with the shipowner. This is for obvious reasons not the option favoured by shipowners and their P&I Clubs. This option raises the question how the limitation of shipowner's liability should be regulated. The 1976 London Convention on the limitation of liability for maritime claims governs the global limitation of shipowner's liability and it covers limitation of liability for HNS damage as well. However, the level of compensation provided for under the 1976 Convention is regarded as insufficient for the purpose of adequate compensation for HNS damage. Therefore some additional provsions have to be made to raise the level of compensation. To this end several possibilities are under discussion.

  a.*Exclusive shipowner liability combined with a general increase of the 1976 limitation amounts*
The increase of the limitation amounts would be general and no specific reference would be made to HNS damage. This option would require a Protocol to amend the 1976 Convention and a separate HNS Convention introducing strict liability and compulsory insurance. The new HNS Convention would be

closely connected with the amended 1976 Convention, a State Party to the HNS Convention would have to become also a Party to the amended 1976 Convention.

One could perhaps question the feasibility of this option from an insurance point of view because all limits of liability would have to be raised to a level which would be adequate for the special HNS risk but too high for other liability risks. This might result in an unacceptable general rise of insurance premiums or even be impossible to cover in view of the capacity of the insurance market. But this is a question to be dealt with by the insurance industry.

b. *Exclusive shipowner liability combined with a*
   supplementary layer above the 1976 limits for HNS damage
In order to meet the possible objection raised above one could introduce a special layer for HNS damage into the 1976 Convention or even in the HNS Convention.The general limits of the 1976 would also be available to compensate HNS damage, but in case such damage would exceed his global limitation the shipowner would have to constitue a supplementary limitation fund to compensate any HNS damage which has not been paid out of the general fund.

c. *Exclusive shipowner liability with separate*
   limitation amounts for HNS damage
This option would be in conformity with the 1969 Brussels Convention on Civil Liability for Oil Pollution Damage. The HNS Convention would contain its own limitation regime independent from the 1976 Convention. Unless the latter Convention would be amended for the purpose to exclude HNS claims, this option would create a legal conflict between the two Conventions. Such a situation existed under the predecessor of the 1976 Convention, the 1957 Brussels Limitation Convention, after the 1969 Convention had come into force, but this has been remedied in the 1976 convention by excluding oil pollution claims.

d. *No special liability regime for HNS damage but only a*
   general adjustment of the 1976 limits
Some governments have expressed the opinion that the problem of compensation for HNS damage is not caused in the first place by the liability regime but rather by the low level of limitation under the 1976 Convention. In their view it would be sufficient to adjust the limitation amounts without creating a new HNS Convention. This rather simple looking option has the same possible objection as mentioned under option *a* but it has the additional disadvantage that no uniform strict liability would be applicable and no international compulsory insurance scheme would be introduced. Each State might create its own national regime.

3.3 SECOND OPTION: EXCLUSIVE LIABILITY OF SHIPPER

In this option the liability would be exclusively imposed on one of the parties who has an direct interest in the shipment of HNS. In view of the requirement that the party liable must be easily identifiable, the producer and the owner of the shipment should not be chosen. Even if the identity of the producer could be known, he is often a person who has brought his product already on the market long before the goods are loaded on board the vessel. He has no connection with the transportation by sea, unless he is also the shipper of the cargo. Moreover products liability is a liability for *defective* products and the new HNS compensation scheme it is irrelevant whether the damage has been caused by defective or non-defective products. Products liability should not be mixed with HNS liability for carriage by sea. The identity of the owner of a cargo is often difficult to discover for the claimant since the goods are often sold when sailing and he is a complete outsider as to the underlying contracts of sale. The shipper is the person by whom or on behalf of whom the cargo has been handed over to the maritime carrier for transportation and who is mentioned in this quality on the b/l or sea waybill. Even then it may not always be possible to find out the true identity of the shipper of a particular consignment. This objection can be overcome by requiring that every shipment will be accompanied by a valid insurance certificate stating the identity of the insurer against whom a direct recourse action may be instituted. A shipowner who would allow any consignment of HNS on board his ship without a valid insurance certificate would become liable as if he were the shipper.

This option has been rejected by the IMO Legal Committee when it was discussed during the preparation of the 1984 Diplomatic Conference.

3.4 THIRD OPTION: SHARED LIABILITY

Under the third category of options the liability is shared by the shipowner and the shipper. Both have to take out insurance to cover their liability. Three options have been discussed:

a. Joint and several liability of shipowner and shipper
The claimant could action against both parties who would be strictly liable, each up to his own limit. After settlement of the claims with the claimants the final bill would have to be settled by way of recourse actions between the two parties liable. The main disadvantage of this sub-option is that it requires full insurance cover for both parties and

would involve double insurance for the same risk which is probably very expensive. This option was also rejected by the Legal Committee.

b. First layer of liability imposed on the shipowner and a supplementary liability of the shipper

This sub-option contains a strict liability of the shipowner up to a certain limit. One could opt for the limit of the 1976 Limitation Convention or for a seperate limit. If the damage cannot be fully compensated out of the limitation fund of the shipowner, this option requires the shipper to constitute a supplementary limitation fund. The supplementary liability of the shipper operates in a similar way as the International Compensation Fund for Oil Pollution Damage, supplementary to the 1969 Civil Liability Convention, but based on an individual liability combined with a seperate insurance contract. This system has been submitted by IMO to the 1984 Diplomatic Conference in the form of a draft HNS Convention but it has not been accepted by that Conference.

From a theoretical point of view this option is attractive, but it raises a number of problems which are also connected to other forms of shipper's liability and shipper's compulsory insurance:

1. the liability system is complex because two seperate persons are liable;

2. it is difficult to indicate precisely the liable person representing the cargo interest (definition of "shipper");

3. the introduction of a new form of liability insurance;

4. in case of multiple shipments of HNS each shipper has to take out full insurance protection, even for small parcels of packaged HNS.

c. Shipowner liability supplemented by compulsory shipper insurance

This option was presented during the 1984 Diplomatic Conference as a possible alternative for the preceding option mentioned under *b*. There was no time available to discuss it. Under this option the responsability should rest primarily with the shipowner on the basis of strict liability up to his limit under the 1976 Convention. However, the supplementary liability of the shipper would be replaced by a direct supplementary insurance cover for the benefit of the victims to be taken out by the shipper. Each shipment of HNS should be delivered to the carrier with a valid insurance certificate to ensure that insurance is actually available. The shipowner would be held responsible if the shipment had been accepted without such certificate. The victim would have no direct action for compensation against the shipper but only against his insurer.

Several objections raised under option *b* are also applicable to this option. Moreover I have the impression that the insurance industry does not favour to provide an

insurance cover for this type of risk without any liability
resting with the shipper who is responsible for taking out
the insurance.

3.5 FOURTH OPTION: INTERNATIONAL COMPENSATION FUND

The idea to create an international compensation fund stems
from the actually operative International Compensation Fund
for Oil Pollution Damage which was established by the 1971
Brussels Fund Convention, supplementary to the 1969 Civil
Liability Convention. This IOPC Fund is financed by
contributions made by companies who receive crude oil and or
fuel oil in the territories of Member States of the Fund.
This scheme operates on the basis of contributions to be
made for two types of oil only and is applicable to a
restricted number of contributors so that the administration
of the Fund is relatively simple.
The situation with respect to the transport of chemicals is
completely different, a large number of substances and
persons trading in those products are involved. The administration of an international HNS Fund may be much more
complex and expensive, because the contributions will have
to levied from a large number of contributors in the member
States. One has to prevent that a large percentage of the
financial burdens to be imposed by a new compensation scheme
will be caused by the costs for the administration of the
scheme.
The United Kingdom delegation in the IMO Legal Committee
has suggested a compensation fund which would supplement the
shipowner liability up to a fixed limit per incident. This
fund would be financed by a levy on HNS shipments. The fund
would compensate the victims by a collective insurance to be
taken out in the international insurance market and the
premium of such insurance would be paid out of the levies to
be collected from the contributors. The United Kingdom
delegation has been invited to prepare a note containing
more precise outlines for such a scheme and a feasibility
test. An absolute condition for the realisation of such a
scheme is the availability of adequate cover for such a
collective insurance which is at this moment uncertain.
The United States delegation has suggested an IOPC-type or
other fund arrangement (with a practical financing scheme
supplementing the shipowner liability. Also this delegation
has been invited to specify its ideas in more detail.

5. Conclusions

In this contribution I have only outlined the major issues
regarding the establishment of a new compensation scheme for
HNS damage. There are many other important questions which I

have not dealt with. The discussions started already after
the 1969 Conference on oil pollution damage and have lasted
with temporary interruptions until 1988 without the
expectation that a solution is very near. The discussions
have proved that we have to deal with a very complex subject
involving a number of difficult policy decisions. However,
time is running out and decisions have to be taken and
choices made. So far there has been no great enthousiasm
from the part of the industries involved to contribute to
the establishment of a compensation scheme. I hope that they
will realise that it is important that an internationally
agreed scheme will come into force, because otherwise there
will be a risk that governments will take action on a
national level which will not promote the international
transport and trade of chemicals. The problem of HNS risks
can no longer be denied.

RECOVERY OF COSTS - A LIABILITY INSURER'S POINT OF VIEW

by Captain Erik Stein

Assuranceforeningen SKULD, Copenhagen

1. HISTORY

Since World War II the carriage of hazardous and noxious substances
(HNS) has grown substantially owing to the enormously increased use
of many of these commodities. Including products carried in bulk such
as chemicals, LPG/NPG as well as all the products of the oil refinery
industry HNS substances are to-day estimated to reach about 2/3 of
all cargoes carried at sea which can be classed as dangerous.
Excluding the commodities carried in special designed vessels such as
tankers it is believed that 15 to 20 per cent of the commodities carried in either dry cargo container or similar vessels should also be
classed as dangerous transport.

Seen in a historical perspective the carriage of commodities which
for any reason were likely to endanger the lives of either passengers
or crew or the safety of the vessel was in principle forbidden in
ships up to the time of the SOLAS Conference held in 1948. Suddenly
the notable expansion in the traffic of HNS substances was acknowledged which immediately led to a radical review of the maritime mode
of transport. The Conference also recommended a further study with
the object of drafting international regulations.

Since then measures had been taken in most maritime countries all
over the world to regulate by legislation or by recommendation the
carriage of dangerous goods in ships. The various codes and practices
however differed in their framework and in the labelling of the
goods. The terminology was different and the provisions for packing
and stowage also varied from country to country, and this somewhat
complicated situation created difficulties for all who were in any
way directly or indirectly concerned with the tranport of HNS substances.

Meanwhile, in 1956, the United Nations Committee of Experts on the
Transport of Dangerous Goods, which had been actively considering all
the international aspects of the carriage of dangerous goods by all

P. Bockholts and I. Heidebrink (eds.), Chemical Spills and Emergency Management at Sea, 393—401.
© *1988 by Kluwer Academic Publishers.*

forms of transport, completed a report dealing with the classification, listing and labelling of dangerous goods, and with the suggested transport documents required for such goods (Orange Book). This report with the subsequent modifications offered a general framework to which existing regulations could be adapted. The goal was of course to develop the ultimate aim of world-wide uniformity for maritime transport of dangerous goods.

As a further step towards meeting the need for international rules governing the carriage of dangerous goods in ships the International Conference on Safety of Life at Sea was held in 1960 (SOLAS 1960).

In addition to laying down a general framework of provisions in chapter VII of the convention IMCO, to-day IMO, which means International Maritime Organization, was invited to undertake a study with the view to establishing a unified international code for the carriage of dangerous goods by sea, pursued in co-operation with the U.N. Committee of Experts, taking into account existing maritime practices and procedures.

To-day the International Maritime Dangerous Goods Code (IMDG Code) adopted by the organization by Resolution A.81 (IV) together with the relevant chapters as well as matching parts in Enclosure B in the Code of Safe Practice for Solid Bulk Cargoes (BC Code) adopted by the organization by Resolution A.434 (XI) are incorporated in the legislation of most maritime countries.

An interesting point is that transport by sea of dangerous cargoes has been regulated over the past 60 years or more in order reasonably to prevent injury to persons or damage to the ship, and it is only in the latest amendments to the IMDG Code 24-86 where remarks have been inserted if a commodity is considered to be a **MARINE POLLUTANT**.

2. THIRD PARTY LIABILITY

Approximately 96 per cent of the world tonnage have their Third Party Liability insured with one of the 18 P & I clubs available to the market to-day. 14 of the clubs covering approximately 91 per cent of the world tonnage and among them the 3 Scandinavian clubs are closely related within the International Group of P & I clubs in London. The remaining 4 clubs standing outside the group cover approximately 5 per cent of the world tonnage.

P & I clubs (Protection and Indemnity) are mutual associations with the shipowners and their ships as members. The scope of cover pertaining to Third Party Liability is almost identical within all the clubs mainly because of the Inter-Club reassurance system.

Third Party Liability could be damage to or loss of cargo, wreck removal, oil pollution, other types of pollution such as air or sea plus a long row of other types of losses for which a shipowner can be held liable towards a third party. The scope of Pandi cover is only related to fault or negligence on the part of the vessel, her master and crew and not to liability arising out of negligence in the shipowner's office if not directly related to the operation of the vessel entered.

According to the association's rules it is a condition precedent of insurance cover that the entered vessel is also entered with and classed by a classification society approved by the association and that the vessel remains in class complying within the stipulated time limits with all regulations, recommendations, requirements and demands made by the classification society. Further it is a condition precedent of the insurance cover that the member has not allowed the entered vessel to be used in the performance of any unlawful, illegal or prohibited acts.

The P & I insurance covers the member's liability and costs arising out of oil pollution or other pollution or threat of pollution. However, coming to Nuclear risk, the standard insurance shall not cover any liability, cost or expense arising out of or in consequence of the emission of ionizing radiation from all the toxic, explosive or other hazardous properties of Nuclear fuel or radioactive products or waste carried as cargo with the exception of radioisotopes of radioactive material processed in reactors and adjoining chemical installations and for use in industrial commercial agricultural medical or scientific purposes.

3. STRICT LIABILITY v. CULPABLE NEGLIGENCE

The basic legal principle in most maritime countries throughout the world is that a shipowner will be held liable for damage caused by fault or neglect during service by the master, crew, pilot or any other person performing work in the service of the ship. The shipowner may under normal circumstances limit his liability according to the legislation applicable in the country which can establish proper jurisdiction, and the same right is due to a holder of title to a ship, who is not a shipowner, operator, charterer, manager and any other person who performs service in direct connection with salvage work. Also persons for whom the shipowner or others mentioned above are responsible may limit their liability according to the law applicable. Finally and most important for P & I clubs is that any person who insures against liability for claims subject to limitation of liability has the same right to limit the liability as the insured himself.

Another basic legal principle is of course that a responsible party cannot limit his liability if it is proved that he himself has caused the loss or damage intentionally or grossly negligently and with the understanding that such damage would probably be caused. Such "PRIVITY" shall not be covered under the owner's P & I insurance.

396

The right to limitation of liability exists irrespective of the basis of liability, cf. however what I just said about PRIVITY, with respect to claim made as result of:

1) personal injury or damage to property if the damage or injury occurs on board the ship or in direct connection with the operation of the ship or with salvage.

2) loss as a result of violation of a non-contractual right if the damage occurred in direct connection with the operation of the ship or with salvage.

3) raising, removal, destruction or demolition of a ship which is sunk, stranded, abandoned or wrecked and of what is or has been on board such ship.

4) removal, destruction or demolition of the ship's cargo.

5) measures which are taken with a view to preventing or limiting losses which are or would be subject to limitation of liability, and losses which are due to such measures.

The right to limitation of liability does however not include:

1) claims for salvage money, general average contributions or considerations according to contract for measures as mentioned above, Nos. 3, 4 and 5.

2) claims as a result of damage or costs in connection with liability for oil pollution.

3) claims subject to international convention or national law which regulates or prohibits limitation of liability for nuclear damage.

4) claims as a result of nuclear damage caused by a nuclear powered ship.

5) claims for interest and costs of proceedings.

The amounts of limitation are regulated by 3 international conventions, the Limitation of Liability of Owners of Sea-going Vessels, 25th August, 1924 - the Limitation of the Liability of Owners of Sea-going Ships, 10th October, 1957 or the Convention on Limitation of Liability for Maritime Claims (LLMC 1976), whichever is ratified or incorporated in the national law of the country of jurisdiction.

It has just been mentioned that the right to limit liability did not include liability for oil pollution as such liability is regulated by a different convention, namely the International Convention on Civil Liability for Oil Pollution Damage, 1969, which imposes strict liability upon a shipowner. Thus an owner of a ship carrying oil in bulk as cargo shall irrespective of the question of fault be liable for all damage occasioned outside the ship by reason of pollution caused by the escape or discharge of oil including bunker oil from the ship. The ship shall moreover be liable for any expense or damage resulting from reasonable measures taken <u>after</u> the accident in question has occurred with a view to preventing or minimizing such damage.

Because of the strict liability other limitation amounts are applicable and further a number of funds and/or compensation schemes are available to claimants who have suffered loss from oil pollution, funds or compensation schemes which are not available to victims or claimants in connection with all other types of maritime claims such as those we are discussing during this scenario, but is is important to note that there is a sharp distinction between strict liability as applicable in connection with oil pollution damage and culpable negligence which normally always gives the shipowner a right to limit his liability in accordance with the liability conventions, whichever is applicable in the country of jurisdiction.

4. THE SCENARIO

In the early hours of 15th November, 1988, the Dutch chemical tanker "TNO" of Apeldoorn collides with the Danish dry cargo vessel "ORD" of Copenhagen. The collision takes place in the southern part of the North Sea in international water about 20 nautical miles off the Frisian island nearest to the Dutch Schiermonnikoog and the German island Borkum.

Immediately after the collision the Dutch tanker catches fire in one of the chemical products loaded, namely Benzene, and because of the imminent danger of an overall explosion on board the master decides to abandon vessel. The "TNO" starts thereafter drifting towards the coastline. As a bunker oil tank is ruptured in connection with the collision, fuel oil escapes into the sea. Within few hours a number of oil slicks are drifting towards the coastlines.

The master on the Danish dry cargo vessel "ORD" decides, although the vessel being heavily damaged, to seek for nearest port of refuge but with an increasing list to the starboard side he decides to abandon ship after a few hours whereafter the vessel starts sinking. Ultimately the "ORD" sinks in Dutch territorial waters.

As to the "TNO" several attempts are made by salvors in order to get the vessel under tow, however unsuccessfully because of the extreme heat generated from the fierce and total blaze on board, and totally burnt out the vessel sinks in German territorial waters.

All the crew members from the two ships are saved by German rescuers and taken into a hospital in Borkum for treatment and examinations.

Immediately after the collision is known to the owners in Copenhagen the hull underwriters and P & I club are notified and an early morning meeting is established. At the same time technical experts are heading for the site to make their investigations into the casualties. Local representatives representing the shipowners and their insurers are notified and instructed to proceed on scene with the clear instruction to try to avoid any information being released from the crew to anybody, especially the media as in order to protect the shipowners' interests we intend to do our utmost to avoid undue speculation in the incident.

The reason for this is that a master on board a Danish ship must - in accordance with the Danish Merchant Shipping Act - appear in a Danish court or before a Danish foreign service representation in order to provide to the greatest possible extent complete information about the actual circumstances of and the reason for the occurrence.

In case of collision the maritime declaration shall - where possible - be heard for both ships at the same time and for this purpose the Minister of Industry in Denmark may grant a postponement. Should such postponement not be possible the master having the right to demand declaration given behind closed doors i.e. a court hearing without access from the public.

Simultaneously with the legal formalities being dealt with and the statements from persons involved being properly secured, the casualty investigators from the Hull and P & I Underwriters are carefully considering which line of action to be taken in the coming few hours. Most probably the Hull Underwriters will declare constructive total loss and leave the scene paying out the insured amount to the shipowner in accordance with the hull policy.

Thereafter the investigator appointed on behalf of the shipowner and his P & I Underwriter will consider the situation and draw up a preliminary advice to his principals with an estimate of all claims which he deems will be forthcoming in due course and which claim in his opinion may fall upon the shipowner and consequently the P & I Underwriters to compensate.

Following claims forthcoming can immediately be seen:

1) Wreck removal

2) Oil pollution or preventative measures from same during removal or demolition of the wreck

3) Removal of cargo which endangers the environment

4) Loss to or damage of cargo under the Bill of Lading contracts

5) Other cargo claims in tort

6) Repatriation of crew

7) Other social secured claims from the crew

8) Excess collision - division of liability

Assuming that the Danish "ORD" is a 500 gross tons dry cargo vessel the 1976 Liability Convention which is applicable in Denmark and adhered to under Danish law, the maximum liability amount available for all claims arising out of the same incident will be approximately USD 230.000.

If we further assume that the Dutch flag "TNO" is a tanker of 1.600 gross tons it will be the limitation amount as stipulated in the 1957 Liability Convention which is applicable for all claims arising out of the same incident equal to approximately USD 116.000, and if both vessels are to blame in connection with the collision the sum of the two amounts, totalling approximately USD 346.000 will then be available provided of course that the claimants can establish jurisdiction and liability based on culpable negligence.

If we change flag on the "ORD" from Denmark to Vanuatu, thus suddenly the 1957 Liability Convention becoming applicable, the limitation amount is dropping from approximately USD 230.000 to USD 38.000 and if "TNO" is the same size and under the same flag as mentioned before the sum available if both to blame will be less than half of what we saw before, now approximately USD 154.000 to be shared by proportion for all claims arising out of the same incident.

Finally I would like to show what happens if the "ORD" is a product carrier or a container liner of 43.000 gross tons under Danish flag. Then approximately USD 9.302.000 will be the limitation amount for the Danish vessel and assuming now that the "TNO" being a 500 gross tons vessel under Dutch flag, then the amount of limitation will be approximately USD 38.000, totalling approximately USD 9.340.000 provided of course that both vessels are to blame in the collision.

Such alternating of the possibilities between flags and sizes clearly illustrates the difficulties the claimants may find seeking for recovery against the shipowners. Looking at the last example the easiest solution will be, provided the bulk of the claims does not exceed the amount of limitation, to establish jurisdiction and seek recovery by the big vessel's owner, leaving it to him to seek recourse under the division of liability with the other vessel's owner.

Realistically seen, however, I believe that the first or second example is most likely to be the case and as one may have a certain feeling that the bulk of the claims will exceed the two vessels' amounts of limitation several times, the claimants seeking recovery may find themselves in a difficult position where they may have to establish jurisdiction for both vessels in two different countries with all the uncertainties related thereto, especially where vessels are registered under flag of convenience. Considering the costs of litigation it may not be worth while to run for USD 38.000 in Vanuatu.

As far as Danish and Dutch law concerns there appears very little doubt that the State is entitled to recover the costs incurred in connection with removal of a wreck situated close to its territorial waters or to the approach to one of its waterways from the owners of the responsible vessel in so far as the navigability or safety of the sailing route justifies the incurring of those costs. A judgment confirming that has been rendered in Holland by Hooge Raad der Nederlanden 26th May, 1978, M/S "GAASTERDIJK" v. M/S "ZUIDPOOL" and in Denmark by a Supreme Court judgment in 1940 where a sunken ship was found to endanger the safe navigation into Danish waters outside the territorial waters. The State was entitled to remove the obstruction and recover costs from the shipowners by doing so even if the shipwreck had taken place in international water.

As a consequence of the demands in connection with the implementation of the 1973 MARPOL Convention and the 1978 MARPOL Protocol new legislation tool place in Denmark called Act No. 130 of April 9th, 1980 on the Protection of the Marine Environment.

The purpose of this act is to prevent and combat pollution of the marine environment from ships by oil and other substances or materials which may result in hazards to human health, harm to living resources and marine life, hindrance to legitimate uses of the sea and reduction of amenities.

Further the purpose of this Act is to ensure that a contingency system is maintained to combat pollution at sea, on coasts and in ports and harbours.

The Act applies to Danish ships as well as other ships operating in Danish territorial waters, also foreign ships operating outside Danish territorial waters without prejudice to international law.

According to this law expenses for all reasonable measures pertaining to contingency or combating measures shall be borne by the owners of the ship. This law, however, does not give the shipowner the right to limit his liability which I have previously discussed at length.

A few years ago a Danish flag vessel lost a container with 80 drums of pesticides over board en route from U.K. to Denmark. The Danish authorities, although the incident took place in international water, decided to send an expedition to the area in search for the sunken drums which were scattered over a large area of the seabed. Most of the drums were recovered and the Danish State claimed all costs in connection with this operation against the shipowner who counterargued that if at all liable he was entitled to limit his liability in accordance with the Danish Merchant Shipping Act.

This argument was denied by the State and the dispute has now been heard at The Maritime and Commercial Court of Copenhagen which this summer decided that the shipowner was entitled to limit his liability in accordance with the Danish Merchant Shipping Act. At this moment of writing it is not yet known whether the judgment will be appealed to the Supreme Court of Denmark. However, considering the limitation amount of approximately DKK 800.000 against the total claim of approximately DKK 10.000.000 one would believe that the Danish State will indeed try its luck at the Supreme Court.

For those who now feel that what has been said necessitates a rapid change in international law in order to establish the same resources to pay for damage after an extensive chemical spill by the same means which are available in connection with oil pollution, it is my opinion that one ought to consider the situation carefully. There appears no need for drastic changes in the shipowners' liability such as strict liability, compulsory extra insurance or changes in the governing rules on limitation of liability. Any regulation or legislation deemed necessary should be so framed as not to impede unnecessarily the movements of HNS substances.

Strict rules will undoubtedly lead to a concentration of HNS substances in fewer ships with a far greater risk exposure as a result. The only reasonable steps to be taken are by extensive training courses, loss prevention and sufficient education to those who are involved in the shipping of dangerous substances. It is of paramount importance among all parties concerned to realize that if one link in the chain fails, this may cause severe danger and hazard to the following.

A heavy responsibility lies also on the manufacturer of HNS substances, the shipper of such goods, the charterer, the shipowner and the stevedores. They should all be properly trained and educated in all aspects of handling HNS substances and last but not least, in detail they should know how to use The International Maritime Dangerous Goods Code as a tool, not only in order to carefully load, handle, stow, carry, keep, care for and discharge the goods, but primarily to protect vessels, crews and the environment exposure from a potential danger.

ACCIDENTAL POLLUTION OF INLAND WATERWAYS

MODELLING OF ACCIDENTAL SPILLS AS A TOOL FOR RIVER MANAGEMENT

J.A.G. van Gils
Delft Hydraulics
P.O. Box 177
2600 MH Delft
The Netherlands

SUMMARY. The Sandoz accident induced a rapid development in the mathe-matical modelling of accidental spills. River Basin Authorities are now using new models as support for determining the possible measures to be taken in case of spills. This article discusses the requirements of such models. A model application used in the Dutch management of the Rhine river is described. Examples of its use, derived from the Sandoz accident are presented.

1. INTRODUCTION

On November, 1, 1986 a fire broke out in a storage area of the Sandoz chemical firm near Basel. The water used to extinguish the fire streamed into the river Rhine, taking a significant amount of chemicals with it. The spill consisted mainly of herbicides and insecticides and severely damaged the ecosystem of the river Rhine. The accident made newspaper headlines for several weeks and caused a lot of concern in Germany, France and the Netherlands.

The Netherlands is a small densely populated country. The water supplied by the rivers Rhine, Meuse and Scheldt is very important to its inhabitants. The water is used as raw water source for drinking water, for agricultural and industrial purposes and for recreation. Furthermore it provides important nature areas with fresh water. Before reaching the Netherlands the river water passes large industrial areas in Switzerland, Western Germany, France and Belgium. Therefore, a con-siderable risk exists that polluting substances are released into the rivers.

When an accidental spill occurs, there is an immediate demand for information about the effects of the spill. River management authori-ties need to decide what actions should be taken. It may be necessary e.g. to stop the intake of drinking water from the river. It may also be necessary to stop the distribution of river water to polders and lakes. On top of that, radio, television and newspapers need to be in-formed.

This article deals with the role mathematical models can play in supporting management decisions in case of an accidental spill.

P. Bockholts and I. Heidebrink (eds.), Chemical Spills and Emergency Management at Sea, 405—414.
© *1988 by Kluwer Academic Publishers.*

The specific requirements a mathematical accidental spill model should meet will be described. After that the basic concept of a model application will be explained. A few mathematical aspects are also considered. Finally, an example of its use will show the possibilities of the model, its accuracy and some problems to be dealt with in the future.

The model described in this article is being used by the national water authority in the Netherlands, Rijkswaterstaat. It has been developed by Rijkswaterstaat and Delft Hydraulics during the past year.

2. REQUIREMENTS WITH RESPECT TO MODEL DESIGN

In general, there are a number of requirements a model has to meet, in order to be useful to water management authorities. Very important, of course, is that the model should be able to yield results easily and quickly in case of an emergency. Apart from that, the maintenance of the model also needs attention during model design. Both viewpoints will be explained in this section.

2.1. User Requirements

Accidents do not always happen on working days between 9 a.m. and 5 p.m. Therefore, it is not convenient when an accidental spill model is only available in a scientific research department, that probably will be closed on Sundays. The model should be present at the location where a spill alarm is received. This ensures that communication lines are as short as possible and enables responsible authorities to react in an efficient way.

Usually, modelling experts will not be on hand immediately after the report of an accident. For this reason everyone having some knowledge about the characteristics of the modelled area should be able to run the model. Feeding the model with input data has to be easy and the interpretation of the model output straightforward.

In case of a spill, often only limited information is available about the nature of the accident. It is not clear what substance or how much of it was released. The physical, chemical and biological characteristics of the substance(s) are not very well known. There is no report on the duration of the spill. The mathematical spill model however should yield useful results with a limited amount of input data.

The spreading of pollutant in the river system depends on the actual hydrological situation. Relevant information about this situation is necessary to enable calculations. In order to guarantee quick results the model should use only directly accessible data and should understand the available information directly. For example, a measured water level can be converted to an estimated river discharge using tables. This procedure is time consuming, gives rise to errors and should therefore be avoided. The conversion from water level to river discharge can be carried out by the model instead!

When very specific river management measures are possible in the modelled area, the various alternatives need to be included in the model. In the Dutch Rhine-Meuse estuary there are two branches flowing into the sea. In one of them, the Haringvliet, large sluices control the local discharge. The sluices enable river authorities to influence the distribution of water from the Rhine and Meuse in the estuary. Accidental spill models of the estuary are able to simulate different control options for the sluices. A prediction of the effects helps the authorities to select the best alternative.

Table I summarizes the basic user requirements.

TABLE I. User requirements

Availability 'on the spot' Independence of the presence of modelling experts Possibility to yield results with few input data available Direct use of accessible, actual hydrological information Possibility to simulate the effect of alternative measures

2.2. Maintenance Requirements

After the design, the construction and the installation of a model the model will be subject to small changes. Modifications in a model can only be realised easily when the model is designed and constructed in a modular way. This means that the overall function of the model has been split up into several partial functions, carried out by separate modules. Each module performs its task independently of the others. Modifications in the model can be made by removing the module in question, adapting it and then returning the modified module.

It may be desirable to expand the modelled area or to apply the model to another area. For this reason it is advantageous to design a model that is independent of the area that will be modelled, as much as possible. Parts of the model typical for the modelled area should be included in a few separate modules. Applications to other areas involve only the exchange of these modules, while the bigger part of the model remains unchanged.

In most cases, accidental spill models need to produce answers within a few minutes. Therefore the calculation of the hydrodynamic behaviour of the river system is impossible. Empirical relations are used to estimate river discharges and velocities. Due to the slowly varying characteristics of river systems, empirical relations must be updated from time to time. The model should offer the possibility to adapt all empirical data involved, in a simple way.

Table II summarizes the requirements concerning maintainability.

TABLE II. Maintenance requirements

Modular construction of the model Inclusion of all aspects typical for the modelled   area in a few separate modules Easy modification of all empirical data in the model

3. DESCRIPTION OF AN ACCIDENTAL SPILL MODEL

The Sandoz accident caused a demand for adequate mathematical models, supporting measures to restrict the effects of an accidental spill. Rijkswaterstaat, the authority responsible for river management in the Netherlands at a national level, has started the development of such models. One of them will be described here. It will be called the 'Rhine model', after its most important application.

The model has been developed in agreement with the requirements mentioned in the previous section. It is meant to give a first estimate of the effects of a spill. The model is implemented on a Personal Computer in order to be 'portable'. Also, a PC offers the possibility to make computer models userfriendly. Communication can be interactive and may be simplified by using computer graphics.

The model can be applied to systems of rivers and canals including bifurcations and tributaries. Since February 1988, applications to the Rhine, the Meuse and the Dutch Rhine-Meuse estuary are operational. Different applications were realised using interchangeable hydrodynamic modules and input files containing a schematic representation of the modelled area.

The input of model data is done by editing several 'forms' on the computer screen. While the user is doing this, additional comment appears on the screen to assist him. If possible, the model supplies default values. The model uses only directly accessible input data. In the information center of Rijkswaterstaat in Lelystad, an automatical link has been made with a hydrodynamic monitoring system. All relevant input from the database is accessed without intervention of the user.

The amount of matter spilt can be defined in three different ways. An instantaneous spill is totally characterized by the time of the accident and the mass spilt. If information is available, also the duration of the spill can be taken into account. As an alternative the spill may be defined by a measured concentration curve on a certain location. This option can also be used to feed the model with output data from other models.

The Rhine model calculates the travel time of pollutant matter from the place of the accident to an observation point. Also the expected concentration curve at that location is calculated. For this purpose the model uses an analytical solution to the advection diffusion equation (Delft Hydraulics, 1988). This equation represents the transport of a dissolved or suspended substance in a flowing medium. An extra term has been added to account for decay processes. Empirical relations are used to describe the hydrodynamic characteristics of the river system. Due to this approach the calculation takes less than one minute on a low cost IBM/XT compatible Personal Computer.

The model output consists of figures and tables on the screen that may be manipulated by the user. Hardcopies of the computer screen make up the final output of the model.

4. THEORETICAL ASPECTS

4.1. Behaviour of Polluting Matter

The quantification of the amount of released matter in case of an accident forms a special problem. Furthermore, the exact behaviour of the substance spilt is often not very well-known. Therefore the behaviour of pollutants is modelled in accordance with the primary goal of the model: it should yield a quick estimate of travel times and concentration levels, based on concise information about the spill.

The model assumes the spilt matter to behave as a dissolved or suspended substance immediately after the spill. In reality released matter can volatilize, float, dissolve, go into suspension or sink after the accident. Combinations are also possible.

The volatilizing and sinking parts spread only in the direct surroundings of the spill. Therefore they normally are excluded from model calculations.

Usually floating substances will be isolated and cleared away as quickly as possible, so that they most likely will not travel great distances. For this reason the floating fraction of substances in the model is not distinguished separately.

The floating and sinking fractions of the spill can be important from another point of view. The material on the bottom or on the surface of the river may enter the water column slowly and reach the dissolved or suspended state. This effect can be modelled as an extra, time dependent source of dissolved or suspended pollutants.

The behaviour of the released substance during transport in the river system is characterized by a reaction rate coefficient. This coefficient represents the sum of physical processes like volatilization and sedimentation, chemical reactions that consume the modelled substance and bacterial decay. The reaction rate in the model is expressed as a half-time value and is assumed to be constant. The reaction rate of a certain substance can be estimated from literature values. If information on this point is missing a calculation without decay can be carried out. In this way a 'worst case' answer is obtained.

4.2. Transport of Pollutants

The Rhine model calculates the transport of dissolved or suspended matter by solving the one dimensional advection diffusion equation in an analytical way. This choice was made in order to make calculations as quick as possible. The approximations in this solution were considered acceptable bearing in mind the expected accuracy of input data in case of an emergency.

The transport of spilt matter in the model is determined by advective and dispersive transport. Advective transport is the movement of substances with the main flow. In a one dimensional model the cross sectionally averaged concentration is transported with the averaged velocity. If the discharge of a river is constant (no dilution) advective transport shifts an existing spatial concentration pattern along the river.

Dispersive transport is the spreading of substances due to the small scale velocity field and cross sectional concentration differences. It represents all transport phenomena that are not explicitly modelled. The effect of dispersive transport is the spreading of a cloud of pollutants. The center of the cloud is not moved by dispersive transport.

Dispersive transport in the model is proportional to the local longitudinal concentration gradient. The constant of proportionality turns out to be an empiric parameter. It can vary in space and time and should be calibrated using tracer experiments.

A possible modification in the future is the modelling of cross sectional mixing in the region downstream of the location of the spill. Also attention may be paid to the effects of salt induced stratification in estuaries or the effect of dead zones due to irregular river banks (Griffioen, 1987).

5. AN EXAMPLE: THE SANDOZ ACCIDENT

The Rhine model will be demonstrated on the Sandoz accident. This choice was made because a set of measured concentrations in the Rhine is available for the Sandoz case. A calculation will be carried out for the German part of the Rhine to show the accuracy that can be obtained. Afterwards different measures to minimize the effects of the spill in the Netherlands will be simulated and evaluated.

5.1. Travel Times and Concentrations in Germany

In Table III the input data for the basic calculation are collected.

TABLE III. Input Data

Location	: Sandoz AG, Schweizerhalle
Substance	: disulfoton, an insecticide
Released mass	: 6800 kg (estimate)
Duration of the spill	: 6 hours (estimate)
Dispersion coefficient	: 220 m2/s
Time for 50% decay	: 2.888 days (estimate)
Hydrodynamical data	: water levels measured on November 1, 1986 (BfG,1987)

The amount of disulfoton that was released and the reaction rate were estimated using measured total mass fluxes at different locations (DKRR, 1986).

In Table IV the calculated travel times at a number of locations in Western Germany and the Netherlands are shown. Measured travel times are also given (BfG, 1987 and DKRR, 1986).

TABLE IV. Travel times at different locations

Location	km	Travel time (calc.)	Travel time (meas.)
Maxau	362	4 d 4 h	3 d 16 h
Ludwigshafen	428	4 18	4 11
Mainz	498	5 12	5 5
Koblenz	591	6 8	6 3
Bad Honnef	640	6 18	6 16
Dusseldorf	744	7 15	7 8
Lobith	862	8 21	8 9

The differences between calculated and measured values can have different causes. For example, the variation of water levels during the travel of the cloud from Basel to the Netherlands might have been important. However, a calculation using the measured water levels at the passage of the cloud gave no significant improvement.

Another cause of errors may be the insufficient calibration of the empirical hydrodynamical relations in the model. Data from the Sandoz accident and supplementary tracer experiments will be used to increase the accuracy of the empirical relations as much as possible.

In Table V the calculated concentration levels at a number of locations in Western Germany and the Netherlands are shown and compared with measured values (DKRR, 1986).

TABLE V. Maximum concentrations at different locations

Location	km	Cmax (ug/l, calc.)	Cmax (ug/l, meas.)
Maxau	362	37.3	24.3
Mainz	498	16.9	18.3
Bad Honnef	640	9.1	9.0
Lobith	862	4.8	5.3

The results in Table V were obtained using the input data from Table III. A constant dispersion coefficient of 220 m2/s was used. The results show a reasonable accuracy. A further calibration of the dispersion coefficient, using data from tracer experiments, should be carried out in the future.

5.2. Measures taken in the Netherlands

The large scale river system in the Netherlands, dominated by Rhine and Meuse, is shown in Figures 1 and 2.

412

Figure 1. The rivers Rhine and Meuse in the Netherlands.

Figure 2. The Rhine Meuse estuary.

After crossing the Dutch border near Lobith the river Rhine splits into three branches, the Lower Rhine/Lek, the Waal and the IJssel. The Lower Rhine/Lek and the Waal make up a complex, branched estuary, together with the river Meuse. The estuary discharges into the North Sea through two branches, the Nieuwe Waterweg near Rotterdam and the Haringvliet about 20 kilometers to the South. The former is a narrow river having an open connection to the sea, the latter is a very wide river with large sluices to prevent seawater from entering the river. The IJssel discharges into Lake IJssel and eventually into the Wadden Sea.

There are two vulnerable areas that authorities wanted to protect from the Sandoz cloud, Lake IJssel and the Southern branch of the Rhine-Meuse estuary.

The weirs in the Lower Rhine/Lek can be used to protect Lake IJssel. Normally, when the Upper Rhine has a low discharge, the weirs are closed to guarantee high water levels in the IJssel and the Lower Rhine for shipping. During the passage of the Sandoz cloud the weirs were open, in order to limit the amount of spilt matter flowing towards Lake IJssel. The model predicted the effect of this measure to be small, due to the Rhine discharge of about 2200 m3/s. A 4% decrease of the load on Lake IJssel was calculated.

The Southern branch of the Rhine-Meuse estuary was protected by closing the Haringvliet sluices (Figure 2). In this way the polluted river water was forced to flow into the North Sea through the Nieuwe Waterweg. The effect of this measure was also calculated with the Rhine model (Figure 3). The predicted decrease of the disulfoton load due to the closed sluices was about 80%, the decrease of the maximum concentration about 75%.

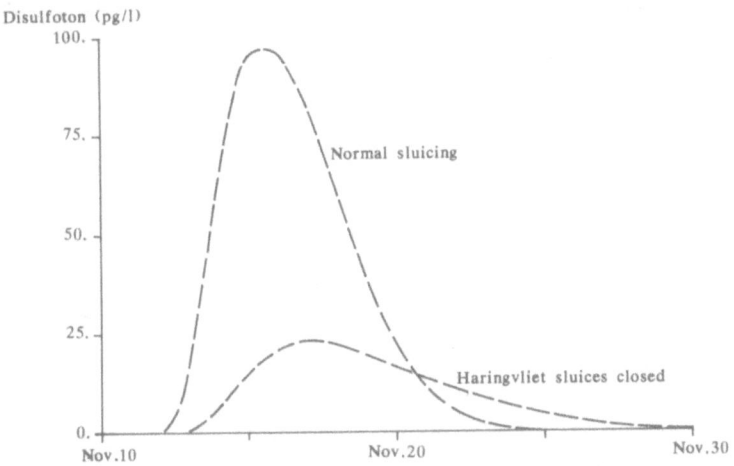

Figure 3. Simulated concentration curves in the Haringvliet, under different circumstances.

6. CONCLUSION

An accidental spill model, satisfying requirements discussed in this article, is presently being used by Dutch river authorities. Up to now, the model appears to work satisfactory. The model in question has been described and examples of its use have been given. The results of a calculation of the Sandoz spill show that additional tracer experiments are necessary to calibrate the transport of released matter in the model.

The continuous development in computer hardware and software will induce subsequent improvements of accidental spill models. Better interactive software will enable modellers to make their products more and more userfriendly. The increasing speed of microprocessors also creates new possibilities. Spill models that calculate the hydrodynamics as well as the transport and decay of pollutants already exist for river systems. Such models are expected to be applied to coastal waters and lakes in a few years.

7. REFERENCES

BfG, 1987.
'Fliesszeiten im Rhein aus Flugelmessungen.', Bundesanstalt fur Gewasserkunde, Koblenz, 1987.
'Kalamiteitenmodellering Rijn en Maas, modelformulering.', concept rapport onderzoek T380, Delft Hydraulics, Delft, 1988.
Griffioen, 1987.
'The difference between travel time and transport time: the effect of stagnant zones.', Griffioen, P.S., to appear.
DKRR, 1986.
'Deutscher Bericht zum Sandoz Unfall mit Messprogramm.', Arbeitsausschuss Messmethoden der Deutschen Kommission zur Reinhaltung des Rheins, 1986.

EMERGENCY MANAGEMENT RELATED TO THE INTERNATIONAL RHINE COMMITTEE

J.H. Jansen
Rijkswaterstaat
Directorate Zuiderzee Werken
P.O. Box 17
8200 AA Lelystad, The Netherlands

INTRODUCTION

The river Rhine basin is one of the most populated areas in the world.
A large number of industries and municipalities are discharging their waste water into the river Rhine.
Rhine water is an important source for the production of drinkingwater. Appoximately 20 million people are depending on a good Rhine water quality for their drinking water.
The Rhine is also one of the most intensively used rivers for inland navigation.
The increasing pollution of the river basin has lead to the setting up of an international commission and an international treaty for the protection of the Rhine against chemical pollution. The treaty is signed by Switzerland, France, The Federal Republic of Germany, Luxemburg, The Netherlands and the European Community.
According to this treaty the International Rhine Commission (IRC) is playing an important role in preparing measures for the protection of the river Rhine basin.
Furthermore the Rhine water quality is continuously monitored. Figure 1 indicates the international measuring locations in the Rhine basin.

P. Bockholts and I. Heidebrink (eds.), Chemical Spills and Emergency Management at Sea, 415—420.
© 1988 by Kluwer Academic Publishers.

figure 1
measuring points in the international Rhine basin.

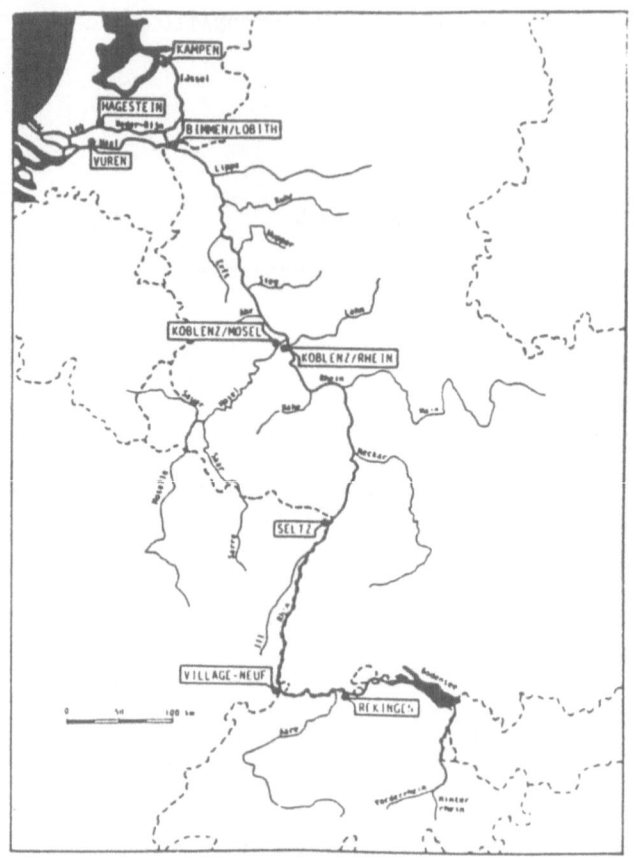

Some of the measuring locations are located at the international
borders, knowing:

Village Neuf	:	Swiss French border
Selz	:	French German border
Bimmen	:	German Dutch border (German measuring point)
Lobith	:	Dutch German border (Dutch pontoon)

Considering water pollution one can distinguish between two kinds
of pollution, knowing the regular pollution (point source
/diffuse source) and the pollution caused by spills (e.g. caused
by accidents or irregularities in production processes).
The contribution of spills represents a substantial part of the
total pollution of inland waters.

A synopsis of spills for the Dutch inland waters is given below.
A distinction is made between the Rotterdam harbor area
(transshipment of more than 100 million metric tons of crude and
mineral oil products a year) and the remaining part of the
Netherlands.

figure 2

spilled products	registered number of spills for the Dutch inland waters (excl. Rotterdam)	
	1985	1986
.mineral oil products	596	711
.edible oils and fats	11	16
.chemicals	36	46
.garbage, manure etc.	38	131
.unknown products	23	13
total	704	917
caused by shipping	127 (of 704)	166 (of 917)
cause unknown	316 (of 704)	478 (of 917)

An estimation of the spilled quantity for all Dutch inland Waters
is ranging up to approximately 1500 tons of product every year.

figure 3

spilled products	registered number of spills for the Rotterdam harbor area	
	1985	1986
.mineral oil products	379	394
.edible oils and fats	24	15
.chemicals	66	38
.garbage, manure etc.	11	15
.unknown products		
total	480	462
caused by shipping	≈85 (of 480))*	≈106 (of 462))*
cause unknown	72 (of 400)	50 (of 462)

)* best guess

418

The impact of an accidental spill on the Rhine water quality was
emphasized by the fire at the Sandoz chemical works at Basel
Switzerland in november 1986.
It lead to a special action programm being adopted for speeding
up the clean)up of the river Rhine. Additional measuring devices
are installed at the measuring points, including bio-alarming

systems. Furthermore, additional safety measures have to be taken
by those dischargers that pose an extensive ris_ for the river
Rhine basin.

Now the question is, what will happen when an accident, resulting
in a waterborne discharge, takes place.

First of all it is worth while mentioning that the river Rhine
basin is devided in a number of regions. In every region a main
alarm agency is located, knowing:

1. Rhine area in Switzerland   : Basel city agency   (R1)
2. Rhine area in France   : Strasbourg agency   (R2)
3. Mosel area in France   : Metz agency   (M1)
4. Mosel area in Luxembourg   : Luxembourg agency   (M2)
5. Rhine area 1 in Germany   : Mannheim agency   (R3)
   (Baden Würtenberg)
6. Rhine·area 2 in Germany   : Koblenz agency   (R4)
   (Bavaria, Hessen,
    Rheinland-Pfalz)
7. Rhine area 3 in Germany   : Düsseldorf agency   (R5)
   (Nordrhein-Westfalen)
8. Rhine area in the Netherlands: Arnheim agency   (R6)

A report of a spill will be given to the agency in whose area the
spill takes place. It is also possible that the effect of a spill
will be measured by the measuring point itself.
According to the international warning and alarming system Rhine
(IWASR), a standard procedure is given in order to inform up and
downstream agencies about the nature and extend of the spill, so
that (counter)measures can be taken.
The procedure of informing is given in the next scheme:

figure 4
dispatch model

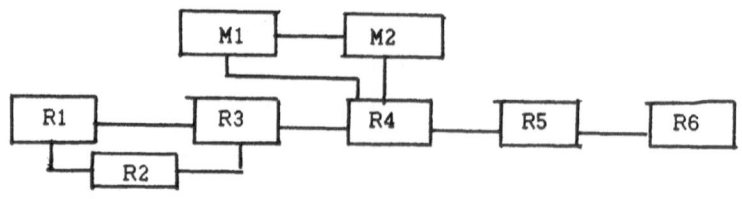

Figure 5 indicates the span of control of every alarm station.

figure 5
geografic display of alarming stations

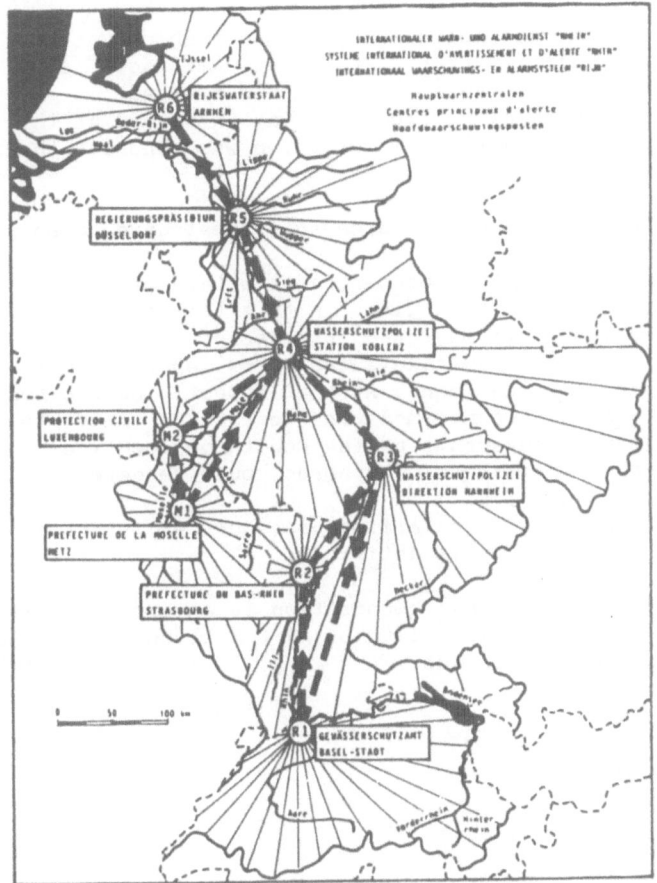

A case history of a spill detected by one of the measuring points
(also a spill not reported by the causer of the spill) is given
below.

As mentioned before, the Rhine water quality is continously
monitored. Among other substances the content of organic
micro-pollutants is measured by the Dutch measuring pontoon at
Lobith (German Dutch border; station R6 in figure 4).
At april 14th 1988 at the Lobith pontoon an increase was measured
of organic mircro's. A further check revealed the presence of
Isophoron in the Rhine water.
Isophoron is a solvent used in spray lacquers, pesticides and the
production of nitrocellulose resins.

420

Immediately after the measurement of this increase the Dutch
authorities contacted (according to the dispatch model) station
R5 (Düsseldorf agency). R5 reported the increase to R4 etc.
Using computer models an estimation was made of the possibly
spilled quantity.
The drinking water production companies, using Rhine water as a
source for the production of drinking water, were notified.
Intake of Rhine water by these companies was stopped.
In the meantime, the sampling frequency was increased at the
Lobith pontoon. All newly found information was reported through
the IWASR communication system.
The press was notified.
Soon it turned out that the Isophoron spill must have taken place
between stations R5 and R6.
Analysis of samples from the Emscher, a tributary of the Rhine
indicated, that the spill took place in this tributary.
Three days after the first measured increase of Isophoron in
Rhine water at Lobith, the German authorities reported that the
responsible discharger was found. At that moment 3 metric tons of
Isophoron were spilled. Countermeasures were taken immediately.
At april 20th 1988 measurements at the Lobith pontoon indicated
that no significant increase of Isophoron could be detected
(concentration $< 0,1 \mu g/l$).
An official international de-alarming procedure was started in
order to close Isophoron spill activities.

It is clear that international cooperation is an important issue
when dealing with accidental pollution of transboundary inland
waters. The experience with the IWASR is a living exemple of this
statement.

Lelystad july 1988

THE ASHLAND OIL SPILL OF JANUARY 1988 - AN EPA PERSPECTIVE

Stanley L. Laskowski, Deputy Regional Administrator
Thomas C. Voltaggio, Chief, Superfund Branch

United States Environmental Protection Agency, Region III
Philadelphia, Pennsylvania USA

The Ashland Oil Spill of January, 1988 near Pittsburgh, Pennsylvania was one of the largest inland releases of oil in U.S. history. The response to this release lasted over a month and involved the combined resources of government, industry and the public. This paper discusses the spill incident, the response, and the causes of the release. The inquiry into the regulatory aspects of this incident is continuing and is not a focus of this paper.

P. Bockholts and I. Heidebrink (eds.), Chemical Spills and Emergency Management at Sea, 421—434.
© 1988 by Kluwer Academic Publishers.

Description of the Spill

On Saturday, January 2, 1988 at approximately 5:10 p.m., 3,881,841 gallons of No. 2 diesel fuel were released from a collapsing oil storage tank at the Ashland Oil Company terminal in Floreffe, Pennsylvania. The Ashland terminal is a small (35-tank) oil and tar tank farm located between State Highway 837 and the Duquesne Power and Light Company, approximately 200 yards from the Monongahela River. The catastrophic release created a wavelike surge of oil which passed over the banks of the facility's containment berms and into a nearby storm drain. Initial estimates were that 1,000,000 gallons of the diesel fuel entered the Monongahela via a storm sewer running beneath the Duquesne property. These estimates were later revised to approximately 750,000 gallons.

Most of the 750,000 gallons of oil that entered the river did so within the first two hours following the rupture of the tank. Even under the most perfect conditions, it would have been difficult to obtain the personnel and equipment necessary to block off the drains and the outfall to the river in such a short time. (See Figure 1). Several factors contributed to making conditions at the site less than ideal for responding to a major oil spill.

The rupture of the storage tank occurred after nightfall. Darkness prevented a thorough assessment of the extent of the spill which in turn delayed initial response actions. In response to reports of a gasoline leak, all electrical and telecommunication lines were disconnected at the Ashland terminal to diminish the threat of fire and explosion. Approximately 250 nearby residents were evacuated that evening. The resulting confusion made site access difficult and impeded the efforts of first responders to effectively contain the spill.

The Monongahela River current was initially estimated to be moving at 2 mph (or 3.2 kilometers/hr); however, the actual rate was 1.1 mph (or 1.76 kilometer/hr), and the flow rate continued to decrease as temperatures fell in the days that followed. Predictions of plume movement were difficult to make due to the changing river flow rate. The spill occurred in the Lock and Dam No. 3 pool at river mile 25. The plume quickly reached Lock and Dam No. 3 which is a little over one mile downstream at river mile 23.8. Passing over this dam and the series of navigational locks and dams on the Monongahela and Ohio Rivers, the oil was emulsified and mixed until it was dispersed throughout the water column. Once the oil was emulsified, it largely escaped containment by booms. Cleanup crews could find few suitable locations for oil collection downstream from the spill site, and river access was impossible at many points.

The Pittsburgh area experienced extremely cold weather during the first weeks of 1988. Very low air and water temperatures affected all aspects of response, monitoring, and cleanup activities. The frigid conditions on the river increased the risk of hypothermia for cleanup crews which led to the decision by the On-Scene Coordinator (OSC) to demobilize all personnel from the river on the fourth day after the spill. For several days, the ice cover on the Monongahela and Ohio Rivers was between 50 and 90 percent. The ice, which contained some of the oil, interfered with the placement of booms and sorbent materials and complicated estimations of the movement of the plume based on river velocity. Cold temperatures may have stabilized the oil emulsion in the river and contributed to the difficulty of oil recovery operations and increased the threat of contamination to subsurface water utility intakes.

Description of the Response

Notification Of Government Agencies

Ashland employees provided prompt and thorough notification within 15 minutes of the tank rupture and spill to the National Response Center (NRC) and to local emergency response agencies. In turn, the NRC notified the U.S. Coast Guard of the spill and at 1810 hours, or one hour after the spill the Coast Guard notified U.S. EPA Region III. Initial evaluation of the spill took place in darkness and as a result, early reports of the spill greatly underestimated its magnitude. Although all agencies received timely notification, important information about the spill was not known as the agencies prepared their initial response.

Initial On-Site Response

Local authorities were the first on-scene responders and immediately rerouted traffic away from the site and instituted site safety measures. Local mutual aid agreements were exercised and a temporary command post was established at the Floreffe Fire Hall. Early efforts by local response agencies and Ashland's cleanup contractor, O.H. Materials, Inc., were directed toward preventing the spilled diesel fuel from entering the Monongahela River by blocking off storm drains and by creating temporary containment dikes. These efforts were hampered by darkness and extreme cold. The response was further hampered because communication and power lines at the Ashland terminal were shut off as a precautionary measure at the time of the spill. Despite the outstanding efforts of local responders, the plugging, patching, booming, and building of dikes had minimal impact on the flow of oil; there was simply too much moving too quickly.

The Coast Guard acted as the first federal official on scene and closed the River to private and commercial traffic. They also mobilized their National Strike Force which is a specially trained unit of the Coast Guard organized to deal with environmnental emergencies on water ways. After inspecting the site, the Coast Guard provided information to EPA. EPA advised the Coast Guard that the position of Federal On-Scene Coordinator (OSC) would be assumed by EPA and that the OSC would arrive on site at first light. The EPA Region III Technical Assistance Team (TAT) was immediately dispatched to the site. The TAT is a contractor on-call 24 hours a day to provide engineering and scientific assistance to EPA during emergency responses.

Oversight of Ashland's Response Efforts

The Coast Guard monitored Ashland's efforts throughout the initial response phase and directed the river cleanup by providing advice to Ashland. The Coast Guard made the initial determination that Ashland's efforts were proper and thorough.

At 0740 hours the EPA OSC arrived on site and delivered a verbal "Notice to Suspected Discharger" to Ashland officials. The OSC advised Ashland that EPA would direct and monitor all phases of cleanup operations. Ashland agreed to EPA's direct control and also agreed to assume full responsibility for all cleanup costs incurred.

The Incident-Specific Regional Response Team (RRT) was not formally activated until Monday, January 4, 1988, although many of the Federal and State member agencies were actively involved in the initial incident response. The following agencies participated in RRT activities during the incident:

- U.S. Coast Guard, Second District, Fifth District, MSO
   Pittsburgh, National Strike Force LANTAREA Strike Team
- U.S. Environmental Protection Agency (Regions III, IV, V)
- U.S. Army Corps of Engineers (Cincinnati, Huntington,
   Louisville)
- U.S. Department of Interior (Philadelphia, Chicago)
- National Oceanic and Atmospheric Administration
   (Rockville, Seattle)
- Occupational Safety and Health Administration
- Federal Emergency Management Agency
- Pennsylvania Department of Environmental Resources
- Pennsylvania Emergency Management Agency
- West Virginia Department of Natural Resources
- West Virginia Department of Health
- Ohio Environmental Protection Agency
- Ohio River Valley Water Sanitation Commission
- Kentucky Department for Environmental Protection

The RRT was activated in order to provide advice and guidance to the OSC. RRT members acted as conduits of information to and from their respective agencies. During RRT teleconferences, the political, programmatic, and statutory implications of the actions of the responding agencies were discussed, and recommendations for appropriate response actions were made.

425

Delegation of Tasks

During the initial hours of the response, none of the agencies on site assumed the role of "lead response agency." Once on-scene the following morning, the EPA's OSC assumed the lead role. Specific tasks and responsibilities were delegated by the OSC to the agency best qualified to perform them.

State authorities immediately directed their efforts toward concerns over water quality, which by noon on January 3, 1988, became a separate, significant response phase managed by State and county authorities.

The Pennsylvania Department of Environmental Resources (PADER) and the Pennsylvania Fish Commission (PFC) began assessing the impact on water quality and downstream water intakes.

The Coast Guard, having recognized that Ashland and local authorities were controlling the site response, focused immediately on cleanup and recovery of oil in the river.

Effects on Water Supplies

The Monongahela River drains into the Allegheny River at milepost 0 with the confluence of the two rivers creating the Ohio River. Figure 3 (River map) shows the location of potentially affected downstream water users. The provision of information by emergency response agencies enabled downstream water suppliers to implement treatment procedures and increase storage volume before the spill affected their intakes. Public confidence in the water purveyors and in the government agencies responding to the spill was maintained because the quality of the water supplied never deteriorated.

Coordination of raw and finished water quality testing and reporting was critical because of the need to use that information in making decisions with respect to water plant operations. Use of private laboratories for quick turnaround time of sample results was vital for determining the effectiveness of carbon treatment and for verifying the quality of drinking water during the incident.

Where needed, emergency water stations were set up quickly and refill operations were well organized in spite of cold temperatures which necessitated the staging of water tankers indoors. Dairies and breweries provided bottled water and were given guidelines for protection of the quality of the bottled water.

The Commonwealth of Pennsylvania issued a Water Conservation Order for three counties during the incident. This order and the cooperation of the public allowed three of the four affected water suppliers in Pennsylvania to avoid water loss to their customers.

Water supply plants downstream from the Ashland terminal were promptly notified of the January 2, 1988 spill; however, the presence of oil throughout the water column and the corresponding impact on water supplies was not recognized until the following morning.

The following summarizes the experiences of the major water supply facilities during the spill incident:

Mile Point*	People Served	Water Company	Intake Closed	Intake Opened	Days Closed
-24.5	530,000	West Penn, Pa	1/3/88	1/6/88	4
4.5	160,000	West View, Pa	1/3/88	1/10/88	8
8.6	9,000	Robinson, Pa	1/3/88	1/10/88	8
36	9,000	Midland, Pa	1/4/88	1/7/88	4
40	17,000	E. Liverpool, Oh	1/4/88	1/7/88	4
59	1,000	Toronto, Oh	1/6/88	1/9/88	4
65	40,000	Steubenville, Oh	1/6/88	1/10/88	5
86	68,000	Wheeling, WV	1/8/88	1/10/88	3

[*Note: - means milepoint was on Monongahela River; all others were on the Ohio River].

Below Wheeling, the plume became difficult to find, yet several downstream users shut their river intakes down while the plume passed them by. Among these were:

137	Sistersville, WV
308	Huntington, WV
470	Cincinnati, Oh

River Monitoring

Three types of monitoring took place during the event: monitoring
of the river to define the spill mass and track its movements, monitoring
the effects on fish and wildlife, and monitoring at intakes to protect
water supplies. Flow and velocity forecasts by the National Weather
Service were initially utilized to predict progress of the oil plume.

Attempts at tracking the plume were initiated on January 4, 1988
and included the culmination of overflight observations and taste and
odor reports from treatment plant operators and lock and dam workers.
Flow and velocity data and weather forecasts were obtained from regional
experts. On January 5, PADER and the PFC began sampling on the
Monongahela and Ohio Rivers at three depths in the water column and
sediment on the river banks. Analyses performed were Oil and Grease
(O&G) and Total Organic Carbon (TOC). PADER, in cooperation with the
Allegheny County Health Department, also initiated sampling of surface
and ground water intakes along the rivers in Pennsylvania, with analyses
for volatile organics, TOC, and fuel oil performed by the PADER laboratory in Harrisburg.

A group at the Ashland command post consisting of representatives
of U.S. EPA, the National Oceanic and Atmospheric Administration (NOAA),
PADER, and the Ohio River Valley Water Sanitation Commission (ORSANCO)
set up a program of sampling river water at the first eight water supply
intakes downstream of the spill site. The sampling was initially set
up as three samples per day at each intake, but had to be cut back to
two per day due to logistical consideration. Analyses were first performed at a local contract laboratory (NUS) until a field laboratory
could be set up by the cleanup contractor, O.H. Materials, Inc.
Analyses included volatile organics, base neutrals, and No. 2 fuel oil.

On January 6, personnel from the U.S. EPA, West Virginia Department
of Natural Resources (WVDNR), and ORSANCO initiated an effort to track
the spill from tow boats. One boat equipped with a fluorometer moved
upstream from Wheeling, while a second boat equipped with a TOC analyzer
moved downstream from Wheeling. After the first day, it became apparent
that the fluorometer provided useful results for characterizing the
plume, while the TOC analyzer did not. Thereafter, the effort continued
utilizing one tow boat and a fluorometer with personnel from Ohio EPA,
WVDNR, and ORSANCO. Samples were collected at the point where the
fluorometer indicated the leading edge and the peak concentration of the
spill and were shipped to the WVDNR's Guthrie Laboratory for analysis of
volatile organics, base neutrals, and fuel oil.

Ohio EPA set up a monitoring system at eight sites on the Ohio River and commenced sampling on January 7. Analyses included TOC, O&G, and organics. Also, on on January 6, the Army Corps of Engineers (ACE)-Pittsburgh collected samples at New Cumberland Pike Island dams.

On January 13, personnel from the Corps of Engineers Huntington District took over the downstream tracking of the spill. They continued to perform the tracking, utilizing their own boat and a flow-through fluorometer, through January 23 when the spill left the limits of their jurisdiction (Meldahl Dam, river mile 436). The effort was then taken over by the Corps' Louisville District, which followed the spill until February 2, at which time the spill was no longer detectable by the boat-mounted fluorometer.

From January 19 through 23, a tow boat was again employed to assist in the tracking and to provide samples requested by downstream water users. Five-gallon samples were collected at the indicated peak for use by the utilities performing treatability studies.

On January 22, the spill reached Kentucky. Monitoring by the Kentucky Division of Water consisted primarily of fluorometers at water intakes. Fluorometers were successfully utilized at Maysville and Louisville.

Fluorometers were moved downstream to provide monitoring at water intakes in Evansville, Indiana and Cairo, Illinois. By that time, the spill had dispersed to the point where results were inconclusive. Monitoring at Cairo, just above the mouth of the Ohio River, was concluded on February 12.

In general, cold weather hampered the river monitoring efforts. Poor navigation conditions forced the monitoring crews to use slow moving tug boats which slowed the provision of monitoring information to water utilities.

River Cleanup

It is now known that virtually all the oil entered the river via a single route; a 24" drainage pipe on the adjacent Duquesne Power and Light Company property that discharged into the river at Duquesne's cooling water discharge. Once this route was identified, the fire department installed an underflow dam in front of the drain opening to prevent the oil from flowing into the river. However, due to delays caused by darkness, loss of power and communications, and an evacuation of the area, the dam was not installed until 2100 hours on January 2, 1988. By that time most of the oil had already reached the River. This first containment was replaced by an inflatable plug that failed early the next morning. At that time, an additional 50,000 gallons of oil were released into the river.

The oil spill from the Ashland facility quickly reached the first in a series of locks and dams along the Monongahela and Ohio Rivers. As the oil passed through the locks and dams, it apparently became emulsified and dispersed throughout the water column. As a result, the containment booms were only able to retain a percentage of oil that was not emulsified.

Traffic in the Monongahela River was prohibited for 58 hours following the Ashland spill and was restricted for seven days, enabling cleanup crews to move about freely in the river. Daily evaluation of the restrictions on river traffic were made by the Marine Safety Office (MSO) in Pittsburgh.

Efforts to contain and recover oil at this enormous spill included the utilization of over 150 people, eleven vacuum trucks, three cranes, and 20,000 feet of river boom. The very extreme weather conditions prevented extensive use of recovery methods such as the application of sorbent materials because ice cover prevented the sorbent materials from contacting the oil. In addition, the ice cover and high risk of injury for the work crews prevented the placing of booms in some otherwise strategic locations. Recovery efforts were centered around diversion and removal of the oil from natural or man-made pockets.

River cleanup operations spanned 30 river miles, and were conducted at times in extremely cold conditions. The U.S. Coast Guard provided personnel from both MSO Pittsburgh and National Strike Force LANTAREA Strike Team to oversee cleanup operations by contractors hired by Ashland. Cleanup monitoring consisted of visits to the cleanup sites by survey teams, who reported cleanup progress to the On-Scene Coordinator at the command post.

Providers of alternative oil recovery technology from all over the world contacted the EPA during the emergency. The following alternative technologies were proposed: biodegradation, viscoelastomers, degreasers, demulsifiers, solvent extractors, gelatinizing agents, water purifiers, boat-mounted oil/water separators, robots, bird feathers, and placement of pantyhose in the spillway.

All individuals and companies who contacted EPA concerning the use of their products at the spill were referred to the Ashland Oil Company.

At the request of Ashland, a field test of the effectiveness of a viscoelastomeric product was performed on the Ohio River during the cleanup. The product neither promoted nor inhibited the rate of oil recovery from the river.

Monitoring Effects on Wildlife

Dozens of individuals, under the guidance and direction of the Audubon Society and the Pennsylvania Game Commission, worked for days following the spill retrieving and cleaning oilsoaked waterfowl from along the rivers. Efforts to save waterfowl were hampered by low temperatures and by ice on the rivers which kept rescue workers on the shore. Although many birds were saved, estimates of waterfowl mortality range from 2,000 to 4,000 ducks, loons, cormorants, and Canada geese, among others.

Biologists from the WVDNR conducted shoreline counts along 120 miles of the Ohio River to determine the number of fish killed. In the week following the spill, several censuses of the dead and stressed fish were taken in the dam pools along the river.

WVDNR aquatic toxicologists designed oil-impact studies on species of mussels by taking censuses and samples from wellestablished mussel beds on the Ohio before and after the arrival of the oil slug. In separate studies, mussels and catfish were placed in cages in the river downstream from the spill and were collected after the plume had passed. The organisms will be analyzed to determine the adverse effects of the diesel fuel.

Summary of Some Problems During Spill Response - Lessons Learned

During the initial eighteen hours of the response, local, state, and federal agencies worked independently, and no single agency assumed an overall coordinating role. In the future, the responding agency should ensure that each member of the response organization understands his/her responsibilities and authority. It was also concluded that either a single command post or better liaison among multiple response sites would have enhanced response coordination.

The Regional Response Team was not activated until Monday January 4. It was concluded that the RRT could have provided valuable assistance to the responding agencies had it been involved in the first 24 hours after the spill.

Communication problems were encountered during the Ashland spill due to insufficient communications equipment at the command post and the large number of responding agencies.

Lack of available containment and monitoring equipment hindered the response. Delays were caused by the need to locate and transport essential equipment. It was recommended that inventories of locally available equipment be prepared to assist emergency responders in quickly locating needed equipment.

The Ashland Major Oil Spill could have been far more devastating
had public water supplies been contaminated or water shortages more
severe. It was recommended that emergency planning agencies and water
suppliers plan for the availability of contingency water supplies and
equipment.

Cause of the Tank Failure

An engineering firm was hired by Ashland Oil to determine the cause
of the tank failure. A report was released in late May which concluded
that the tank's failure began inside a ground-level plate and spread
vertically. The defect, described as a "clamlike" opening on the
interior wall, was below a T-joint, where horizontal and vertical welds
join a bottom plate to one above.

According to the report, the flaw had been in the plate since the
time it was made. It apparently went unnoticed when the tank first was
installed at a terminal near Cleveland. The flaw was missed again when
the tank was taken down and rebuilt at Ashland's Jefferson tank farm two
years ago.

Various experts had speculated that cold weather, brittle steel,
old welds or inadequate load testing caused the tank collapse. The
company report stated that all of those factors contributed to the
collapse, however the one factor which had largely been discounted was a
flaw in the plate itself, probably caused by an errant blowtorch.

The report stated the flaw that triggered the spill was merely one
of a set of circumstances:

- Rewelding. When the tank was rebuilt, the welding near the flaw
caused "hoop stress" weakening - a strain caused by the steel plate
contracting after the area had cooled.

- Old, brittle steel. The metal of the 1940's had a lowcarbon,
high-phosphorus composition that made it inherently less tough than
modern-day steel.

- Cool weather. The temperature of the steel in the cold, January
air was 38 degrees, which made it more susceptible to brittle fracture -
like "a piece of taffy breaking off."

- Load stress. The 4 million-gallon tank was filled to near the
top when it collapsed. The 3.8 million gallons was more than it had
held before.

Had it not been for the combination of all these factors, it was speculated that the oil might have drained out like a slow leak, not a sudden, catastrophic failure.

Specifically, company representatives took responsibility for several problems:

- Not building a new tank. New steel could have withstood temperature and load stresses better.

- The tank's ability to hold its capacity in oil was not tested properly by filling it with water.

- Proper construction permits had not been obtained from county officials.

Follow-up Activities to Ashland

In the wake of the Ashland release a number of follow-up investigations, assessments and legal actions were initiated which required a significant amount of effort and time by various regulatory agencies. A brief list of some of these activities includes:

- Governor's Task Force in Pennsylvania to
  investigate the causes of the spill
- State Senate hearings
- U.S. Congressional hearings and briefings
- Long-term environmental assessment by ORSANCO
- RRT evaluation of emergency response actions
- Multi-media facility compliance inspection by
  EPA and DER
- Spill prevention control and countermeasures (SPCC)
  inspection by EPA and DER
- National SPCC Task Force to review SPCC program
  and regulations
- Enforcement negotiations with Ashland for long-term
  soil and groundwater cleanup and start-up of
  facility
- Citizen lawsuits for damages

CONCLUSION

The willing cooperation between all regional elements was the outstanding factor which resulted in the successful protection of the public health and the environment during the Ashland Major Oil Spill. All responding agencies, groups, and individuals are to be commended for their performance throughout the emergency. Despite the magnitude of the spill, the rapid entry of oil into the river, and adverse weather conditions, all public water supplies were protected.

GLOSSARY OF ABBREVIATIONS

COE	United States Army Corps of Engineers
DOI	United States Department of Interior
EPA	United States Environmental Protection Agency
FEMA	Federal Emergency Management Agency
IDEM	Indiana Department of Environmental Management
KDW	Kentucky Division of Water
NOAA	National Oceanic and Atmospheric Administration
NRC	National Response Center
OEPA	Ohio Environmental Protection Agency
ORSANCO	Ohio River Valley Water Sanitation Commission
OSC	On-Scene Coordinator
PADER	Pennsylvania Department of Environmental Resources
PEMA	Pennsylvania Emergency Management Agency
PFC	Pennsylvania Fish Commission
RRT	EPA Regional Response Team
TATM	EPA Technical Assistance Team Member
USCG	United States Coast Guard
WVDNR	West Virginia Department of Natural Resources

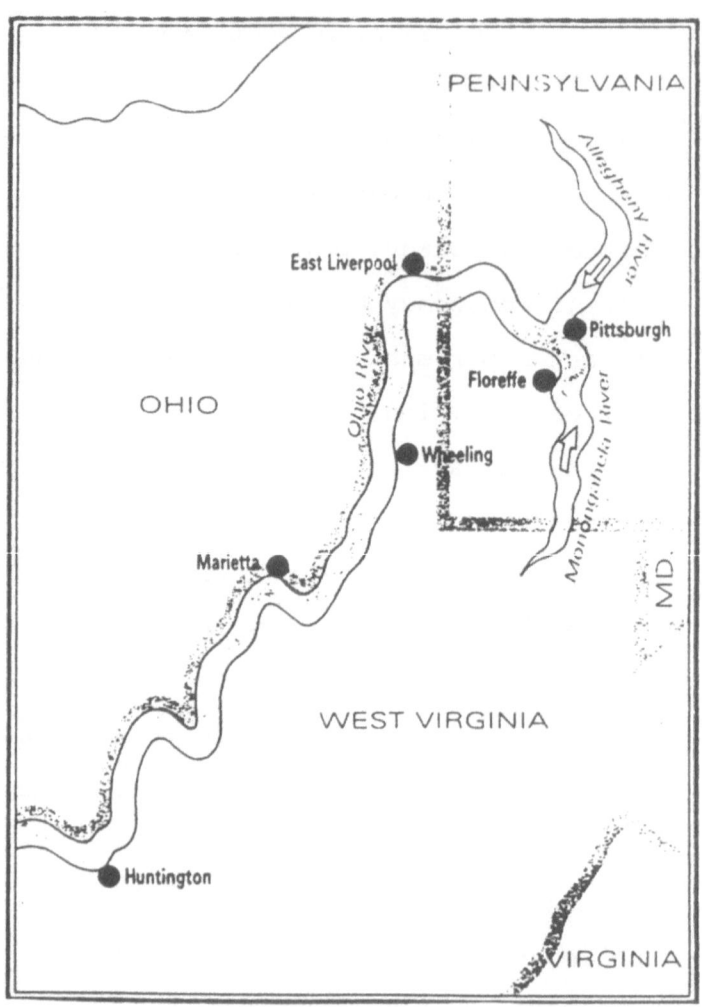

FIGURE 1

ENVIRONMENTAL EFFECTS OF THE JANUARY 2, 1988, DIESEL OIL SPILL INTO THE MONONGAHELA RIVER—PROGRESS REPORT

Edgar Berkey, Samuel M. Creeger, and Roger L. Price
Center for Hazardous Materials Research
University of Pittsburgh Applied Research Center
320 William Pitt Way
Pittsburgh, Pennsylvania 15238, U.S.A.

ABSTRACT. An investigation is being conducted into the environmental effects resulting from the January 2, 1988, storage tank collapse that spilled over 700,000 gallons of diesel oil into the Monongahela River near Pittsburgh. The study includes an analysis of the physical, chemical, and biological aspects of the spill to assess short and long-term environmental impacts. Work has involved analyzing the spill event to determine the rate of oil discharge into the river, compiling available water quality data from the spill to determine the progress and fate of the oil during the event, a mass balance, and gathering of field data to assess potential impacts on sediment, fish and wildlife. Results indicate the spill created only small to moderate environmental effects over the first 185 miles downriver. The majority of the spill was initially deposited over this region. Subsequently, heavy rains resuspended and carried away much of the oil, except in localized areas. The possibility of more subtle environmental effects is being considered for further studies.

1. Introduction

On January 2, 1988, a storage tank located at Ashland Oil's Floreffe, Pennsylvania, terminal suddenly ruptured and released nearly 3.9 million gallons of Diesel Fuel Oil No. 2. The sudden collapse sent a tidal-like wave of diesel fuel surging over the terminal's containment dike and resulted in a spill of diesel fuel into the Monongahela River approximately 25 miles upstream from its confluence with the Allegheny and Ohio Rivers at Pittsburgh, Pennsylvania. The path of the spill ultimately affected the water supply of numerous downstream communities.

For the first several weeks after the incident, Ashland Oil, its contractors, and federal, state, and local officials concentrated on emergency response to clean up and mitigate the effects of the spill on public water supplies. On January 20, 1988, Ashland awarded a grant to the University of Pittsburgh's Center for Hazardous Materials

P. Bockholts and I. Heidebrink (eds.), Chemical Spills and Emergency Management at Sea, 435—444.

Research (CHMR) for an independent study of the environmental effects of the spill. One of the goals of CHMR is to develop and provide useful technical information on key hazardous materials issues. The University of Pittsburgh is the largest research university near the spill site. CHMR was assisted in this work by the Civil Engineering Department, Graduate School of Public Health, and Center for Social and Urban Research. In addition, the Netherlands Organization for Applied Scientific Research (TNO), with which CHMR has a cooperative working agreement, participated on the project.

2. Approach

CHMR used a scientifically-based approach to research and gather information needed to study effects from the spill. Efforts focused on clarifying the route of diesel oil flow from the collapsed tank to the river and the rate of discharge; defining the conditions the oil was exposed to in the river; determining what happened to the oil in the river during and after the event; and assessing the short and longterm environmental effects on water quality, people, and wildlife.

The following activities were carried out.

1. Available information on the spill was gathered.
2. Relevant technical information and other data related to the spill were collected and studied.
3. The site of the tank collapse was inspected to determine the spill's route and collect field information.
4. A public information and education program has been conducted to help people understand various technical issues associated with the spill's effects.
5. During the period from June 22 through July 1, 1988, CHMR collected a total of 41 sediment samples along a 220 mile reach of the river system from above the spill site on the Monongahela River to Parkersburg on the Ohio River.
6. The following support studies are being coordinated and integrated with the overall project:
   o Evaluation of the Physical/Chemical Aspects--Department of Civil Engineering.
   o Mass Balance Analyses--Department of Civil Engineering;
   o Investigation into the Potential Bioaccumulation of Diesel Oil No. 2 in Wild Birds--Graduate School of Public Health;
   o Biological Effects Analysis--TNO, Den Helder.
   o Causes and Effects of Oil Spills--TNO, Apeldoorn; and
   o Risk Perception Survey--Center for Urban and Social Research;

3. Results

Results achieved to date are summarized below, as well as some of the remaining questions to be answered by on-going research and study.

3.1. HAZARDOUS MATERIAL RELEASED BY THE SPILL

The tank collapse released 3,882,000 gallons of Diesel Fuel Oil No. 2 into the Monongahela River. The particular batch of oil was a mixture of aliphatic and aromatic compounds having about twice the normal concentration (10%) of naphthalene compounds. Figure 1 portrays a portion of the Ohio River system and shows the elevation and location of locks and dams (L/D) near Pittsburgh. There are two additional dams on the Monongahela.

As the river water moves over each dam, it drops many feet in distance--a circumstance which adversely affected oil recovery efforts because it caused the oil, water, and suspended sediment to become increasingly mixed as each dam was passed. This mixing action caused the oil to contact and coat sediment particles suspended throughout the water column, which prevented the oil from floating back to the surface. During the spill event, detectable oil concentrations were found deep (i.e., 16-18 feet) in the river.

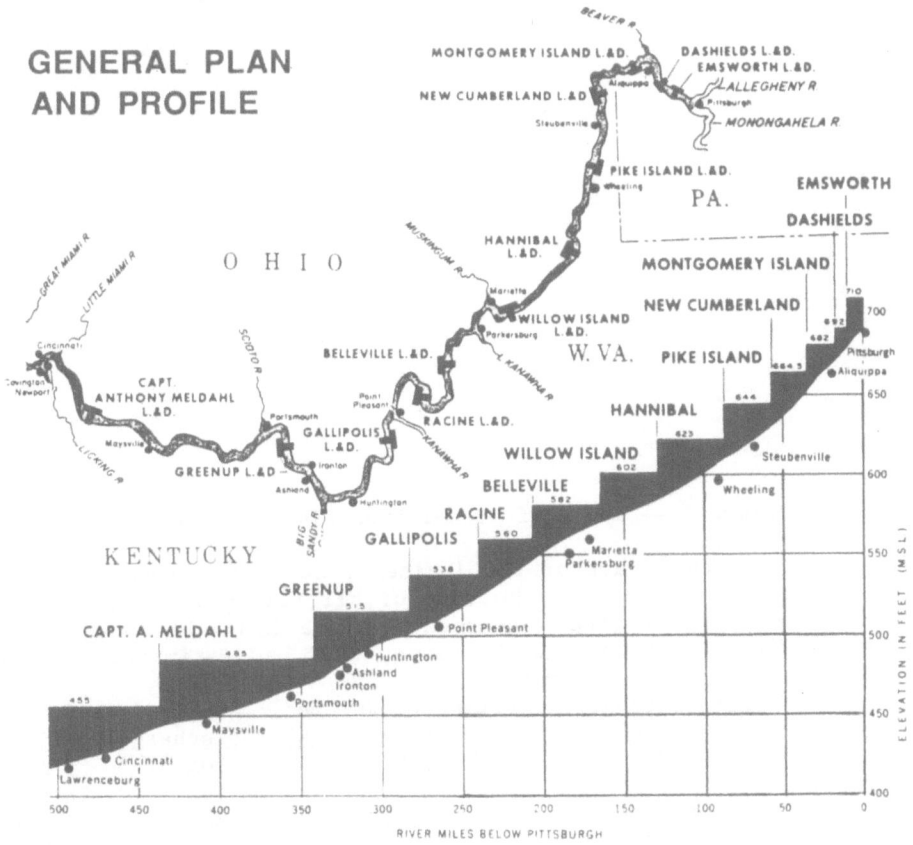

Figure 1 -- Ohio River System Near Pittsburgh.

3.2. RELEASE OF OIL TO RIVER

The route taken by the diesel oil from the ruptured tank into the Monongahela River is shown in Figure 2. The catastrophic release caused
diesel fuel to surge over the containment dike, flow overland across
an employee parking lot, and enter a storm water sewer system serving
an adjacent electric power station. This sewer system discharges
directly to the Monongahela River. It is estimated that 705,000 gallons of diesel oil were released based on the following:

3,882,000 Gallons	Released From Tank	(100.0%)
−2,967,000	Recovered at Terminal	(76.4%)
− 210,000	Remaining at Terminal	(5.4%)
705,000 Gallons	Released to River	(18.2%)

Of the oil released to the river, approximately 205,000 gallons
(29%) was recovered through skimming operations at three locations
downstream by cleanup contractors.

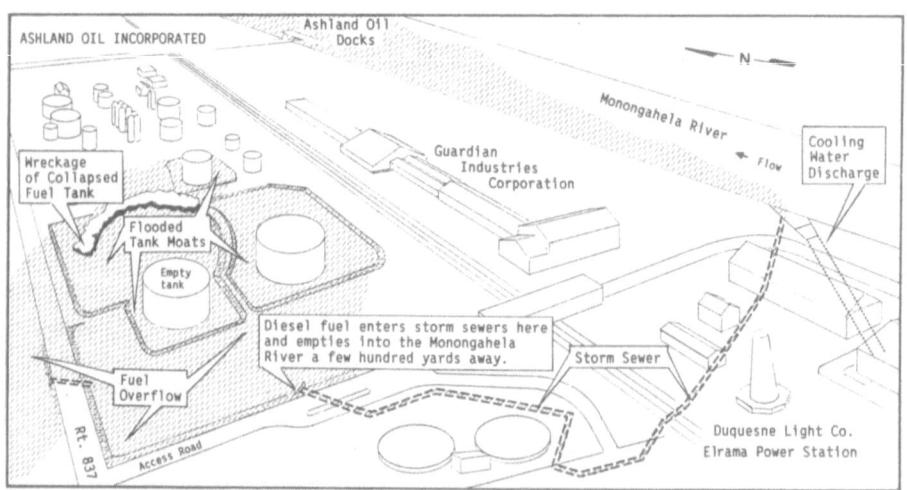

Figure 2 -- Route of Diesel Fuel Flow to River.

Due to the sudden nature of the incident, no measurements were made
during the event to determine the rate of oil release. Based on subsequent interviews, on-site assessments and a hydraulic analysis of
the pathway taken by the oil to the river, CHMR has estimated the rate
at which the oil entered the river:

Time Period	Est. Discharge Rate (gpm)	Estimated Volume Discharged (gals)	Cumul. Discharge	Cumul. Time
First 5 min	0–13,000	60,000	10%	5 min
Next 10 min	13,000	130,000	25%	15 min
Next 15 min	13,000–2000	150,000	50%	30 min
Next 1.5 hr	2,000–800	140,000	70%	2 hr
Last 9.5 hr	800–400	210,000	98%	11.5 hr
Plug Release	–	15,000	100%	
	TOTAL RELEASED	705,000		

Thus, about 50% of the oil entered the river in the first 30 minutes, while the remainder entered over the next 11 hours.

3.3. CONDITIONS IN THE RIVER

The flow rate of the Monongahela River at the time of the spill was high (18 cubic feet/second – CFS) compared to the typical average flow of 10 CFS. The river velocity at center channel was estimated to be a rapid 3 feet/second (FPS). The water temperature was 35–40°F, and the turbidity was high--a result of the significant stormwater runoff from rain events which took place several days before the spill.

Diesel fuel reached the first dam (located about 1 mile away) within minutes after the tank collapsed. By daybreak the following day, January 3, it was evident that the diesel fuel had spread across the full width of the river and was approximately 10 miles downriver. The second dam is located approximately 14 miles downriver and served to further disperse diesel fuel through the water column. On January 3, the fuel spill flowed into the Ohio River at Pittsburgh located 25 miles downriver.

Due to the extremely dry weather conditions, river flow rates dropped significantly for the next 15 days (through January 18) as shown in Figure 3, which gives river flow rate versus time at Wheeling, West Virginia, and Pomeroy, Ohio. The spill velocity declined from over 30 miles per day on the Monongahela to as low as 5 miles per day on January 8 in the vicinity of Wheeling. The velocity of the spill remained very low for the next 11 days until it reached an area between the Willow Island L/D and Parkersburg.

The Ohio River basin experienced heavy rains during the period from January 17 through January 19. After January 18, these heavy rains caused a substantial increase in river flows, as seen in Figure 3. Increased river flows from precipitation combined with contributions from major river tributaries served to further substantially dilute and spread the diesel fuel spill downriver.

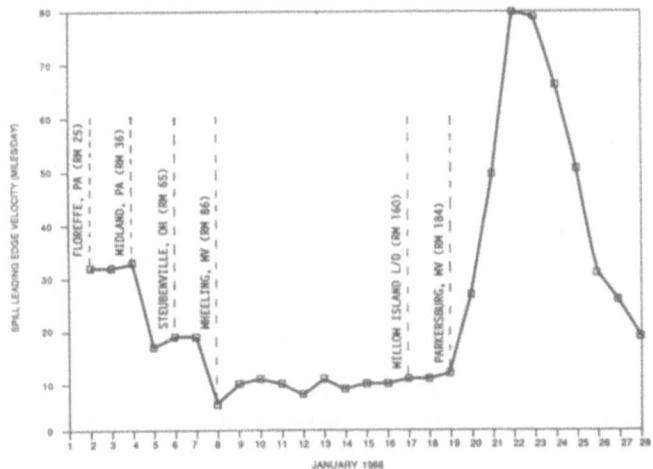

Figure 3 -- Spill Velocity During January.

Turbidity, a measure of suspended sediment in the water column, also decreased for the first few days as the spill progressed downriver. These decreases reflect increased removal of suspended sediment from the water column which can be expected when river velocities drop.

3.4. INITIAL FATE OF OIL IN RIVER

The transport and fate of diesel fuel in the water column of a river system typically depends on the actions of the following removal mechanisms: physical removal by cleanup contractors; volatilization; and adsorption onto particles and subsequent sedimentation.

The sampling and analysis conducted during the spill was performed principally for the purpose of responding to the immediate emergency and providing downriver water users with information regarding the spill's location. Some preliminary estimates regarding the transport and removal of diesel fuel from the water column can be derived from results of chemical analysis performed on river samples collected during the event. However, the quantity of samples collected and the type of analyses performed were totally inadequate for the purpose of subsequent accurate scientific determinations of contaminant fate.

Therefore, to scientifically predict the fate of diesel fuel in the river system from this incident requires significant modeling combined with results of laboratory simulation experiments. CHMR identified the need for such studies and initiated research in these areas soon after the incident.

The following provides preliminary estimates of the transport and removal of diesel fuel from the water column. These preliminary estimates were determined by CHMR using engineering and scientific judgment applied to the limited water quality data collected during the incident.

3.4.1. *Removal By Cleanup Contractors* A total of approximately 205,000 gallons of diesel fuel (29% of the 705,000 gallons released to the river) was recovered from the river through skimming operations conducted by cleanup contractors. The time frame over which diesel fuel was recovered from the river system follows.

Date	Time Frame	Approximate Quantity Recovered	Cumulative Pct. Total Recovered
1/7	First 5 days	140,000 gallons	68%
1/11	Next 4 days	25,000	80%
1/21	Next 10 days	20,000	90%
3/31	Next 2 months	20,000	100%
		205,000 gallons	

The diesel fuel was recovered at three locations which were respectively about 13 miles (where 70% was recovered), 32 miles (20%), and 37 miles (10%) downstream from the spill site.

3.4.2. *Volatilization* Volatilization resulted in the loss of some of the oil from the river to the atmosphere, particulary the lighter fractions. Volatilization is a complex process affected by many parameters. Conditions during the spill are believed to have been

such (low temperature, suspended material in the river, icing, low
flowrate between dams, etc.) that only the most volatile species
were removed from the water column by volatilization. Efforts to
quantify the contribution of volatilization as a transport mechanism
is the subject of on-going research at the University of Pittsburgh.

3.4.3. *Adsorption Onto Sediment Particles* Considerable suspended
sediment was present to adsorb diesel fuel as it entered and mixed
into the water column during the first few miles downriver from
Floreffe. The mixing action at each dam caused the oil to contact
and coat sediment particles suspended throughout the water column.
These oil coated sediment particles continued to remain suspended
throughout the depth of the river which prevented a significant percentage of the oil from floating back to the surface and becoming
available for removal by skimming operations.

3.5. INITIAL SETTLEMENT OF OIL

As the water/sediment/diesel fuel mixture continued downriver, its
velocity and turbidity declined significantly by the time it reached
Wheeling. These low velocities persisted as the spill moved further
until it reached an area between the Willow Island L/D and Parkersburg. Reduced river velocities cause suspended sediment to be removed from the water column and settle on the river bottom.

Analyses of water samples collected at several stations along the
river during the incident were used to provide a preliminary estimate of the total quantity of diesel fuel remaining in the water
column. Figure 4 shows diesel fuel concentration profiles estimated
at several locations, as well as the one determined at Wheeling.

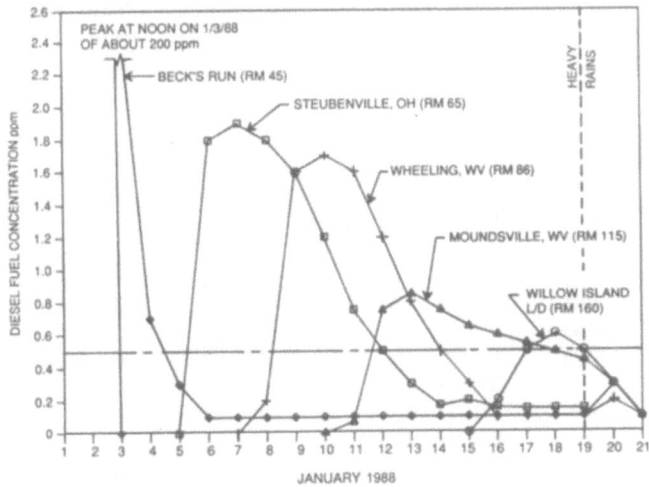

Figure 4 -- Diesel Oil Concentrations in River During Spill Event.

The total volume of diesel fuel within the water column which passed Wheeling can be estimated by combining this concentration profile with river flow rates at Wheeling for the same time period. This analysis indicated that only 125,000 gallons of diesel oil remained in the river as it passed Wheeling. Of this amount, CHMR has estimated that about 75,000 gallons deposited out between Wheeling and Willow Island L/D, leaving only about 50,000 gallons left in the water column after that point.

3.6. RESUSPENSION OF SEDIMENT

It is now evident that a substantial quantity of the oil-coated sediment initially deposited on the river bottom was resuspended during subsequent high river flow events (such as those which occurred during the third week in January--see Figure 3). This resuspended sediment was spread downstream over the Ohio River and possibly as far as the Mississippi River.

The upper 600 miles of the Ohio River is characterized by high average river velocities. The river bottom is principally inorganic rock and cobble with some sand and gravel. Sediment deposits in this part of the Ohio River are limited to a few localized areas of consistently low river velocities. Such areas are found near the shoreline (however, even much of the shoreline is sand and gravel) and a few locations where the river channel widens far beyond the norm. Even back channels are generally characterized by cobble, sand, and gravel. Oil is preferentially adsorbed to sediment with high organic content. Inorganic rocks, cobble, sand, and gravel will adsorb very small quantities of oil, but this is rapidly removed by the washing action of the moving river water column.

Between June 22 and July 1, 1988, CHMR collected 41 sediment samples along a 220 mile reach of the river system from above the spill site to above Parkersburg on the Ohio River. The results of this sampling survey combined with other on-going studies will further define and quantify the extent of residual contamination in this reach of the river system.

4. Conclusions

4.1. SHORT-TERM ENVIRONMENTAL EFFECTS

4.1.1. *Effects on Water Quality* Water quality results relevant to determining environmental impacts are summarized below:

Date	River Mile	Location	Peak Oil Concentration	Days Above 0.5 ppm
1/3	-4.5	Becks Run	100-200 ppm	1-2
1/9	86.0	Wheeling	2.0 ppm	5-6
1/12	110	Moundsville	0.70 ppm	6-7
1/19	160	Willow Island	0.60 ppm	1-2
1/21	265	Above Kanawha R.	0.25 ppm	NA
1/28	600	Louisville	0.05 ppm	NA

Water supply systems which depend on the Monongahela and Ohio Rivers for their source of supply were impacted. Diesel fuel contributes significant taste and odor to drinking water at low (ppb) concentrations. However, the combination of closing intakes, providing alternate supplies, and implementing a successful treatment process all served to protect the impacted systems from contamination. Drinking water systems supplied by wells were not impacted. When the spill passed Cincinnati, oil levels in the river had dropped to the point where immediate concern with drinking water had subsided.

4.1.2. *Effects on Wildlife* Short-term impacts on fish from the spill were most pronounced in the pool prior to Willow Island L/D where the largest number of fish kills was reported. The combination of appreciable diesel oil concentrations in the water (peak level of 0.60-.70 parts per million), relatively long exposure time for the fish (several days) prior to this point because of the slow water velocity in this pool, and lack of major diluting tributaries in this stretch of the river yielded conditions that would be expected to result in significant fish kills (i.e., but less than 10% of the population) based on toxicity data for diesel oil on fresh-water fish.

While the peak concentration of diesel oil in water was higher in the upper reaches of the Ohio River within Pennsylvania, the time over which the fish were exposed to the oil was much lower, and the presence of major tributaries (i.e., Allegheny River and Beaver River) provided opportunities for fish to take evasive action and enter cleaner water.

Short-term impacts on waterfowl from the spill were restricted to those stretches of the river system where there were significant amounts of floating oil on the water that could coat bird feathers. Because of the ability of the diesel oil to emulsify, these stretches are located entirely in Pennsylvania and include the Monongahela River and upper reaches of the Ohio River. An accurate assessment of the extent of the waterfowl kills has not been possible to complete because of the anecdotal nature of much of the information. The fact that the spill occurred during winter when the waterfowl and other birds present on the Monongahela and Ohio Rivers were primarily those that have become domesticated means that short-term impacts on more wild waterfowl species wintering in other parts of the U.S. were small.

4.2. POSSIBLE LONG-TERM ENVIRONMENTAL EFFECTS

4.2.1. *Effects on Water Quality* Oil that is sedimented will not significantly breakdown but will be slowly released from the sediment over a long period of time, acting as a long-term source into the water. However, it is expected that most of the oil originally deposited on the river bottom between Floreffe and Parkersburg has been removed by river flows. The remaining oil is believed to be limited to a few localized areas of consistently low river velocities. Slow release of oil from sediment is not expected to cause detectable diesel fuel concentrations in the river system.

4.2.2. *Effects on Wildlife and Fish* The most important oil components in the sediment that pose a potential source for toxic effects to aquatic species are naphthalene and naphthalene-related compounds.

While there is no potential for long-term accumulation of naphthalenes in fish, birds, mammals, and other vertebrates because of rapid metabolization, there is the possibility of chronic effects to species (i.e., bottom-dwelling organisms and species that feed on them) that are directly exposed to long-term, low-level oil released from the sediment.

Fish tissue analyses from ORSANCO and Ashland, which were performed on fish caught in March and April, 1988, at various locations from upstream of the spill site to 365 miles downstream, were found to contain certain naphthalene-related compounds at total levels from non-detectable to 9.4 parts per million. A statistical analysis of this data indicated that fish caught within 200 miles of the spill site contain more naphthalene compounds than the fish caught elsewhere. Nevertheless, the concentration levels are not believed to pose a risk to people eating the fish.

Any long-term impacts on fish and wildlife from Floreffe to Willow Island are expected to be be limited to a few localized areas of consistently low river velocities where elevated levels of oil-coated sediment may remain.

4.2.3. *Overall* Significant short or long-term environmental impacts resulting from the oil spill into the river are not expected beyond the location of the Willow Island L/D (Ohio River Mile 161.7). Heavy rains began within a day after the spill arrived at this location, greatly diluting the remaining oil concentration in the river and spreading oil-coated sediment over many river miles downstream. Localized areas remain where concentrations of diesel oil can be detected.

5. Future Work

Ongoing studies related to this spill are expected to focus on further defining the nature and extent of residual contamination in localized areas of the river system, defining any significant longterm effects on aquatic species and other wildlife inhabiting this stretch of the river system, and monitoring the situation.

RESEARCH AND DEVELOPMENT

LATEST RESULTS ON OIL SPILL COMBAT WITH ELASTOL$^{T.M.}$

R. Hingmann
BASF Aktiengesellschaft, Kunststofflaboratorium,
D-6700 Ludwigshafen, Federal Republic of Germany
F. Merlin, G. Peigné
Centre de documentation de recherche et d'expérimentation
sur les pollutions accidentelles des eaux, Pointe du Diable,
29263 Plouzané, France

1. INTRODUCTION

ELASTOL$^{T.M.}$* was developed by GTA with the assistance of BASF. It is manufactured in the form of a white powder with particle sizes between 100 μm and 1000 μm. The powder contains about 50% by weight of Polyisobutylene (PIB) of extremely high molar mass ($M_W \approx 6 \cdot 10^6$ g/mol). PIB is a non-toxic polymer that only consists of carbon and hydrogen atoms. The PIB-component is soluble in oil. When dissolved the macromolecules give rise to a distinct viscoelasticity as well as to a drastic increase in the elongational viscosity of the solution even at PIB concentrations of only a few 100 ppm. These properties make ELASTOL$^{T.M.}$ a very interesting oil spill treating agent [1]. In this work we report on laboratory tests as well as field tests with ELASTOL$^{T.M.}$ applied to 13 different types of oil at various concentrations [2].

2. RHEOLOGICAL PROPERTIES OF ELASTOL$^{T.M.}$ OIL SOLUTIONS

2.1. Viscoelasticity of ELASTOL$^{T.M.}$ solutions

In order to characterize the viscoelastic properties of ELASTOL$^{T.M.}$ oil solutions the storage modulus G' and the loss modulus G" were measured as functions of the angular frequency ω in small amplitude oscillatory shear experiments in a rotational rheometer. Figure 1 shows the concentration dependence of the moduli in diesel oil. The pure oil (full circles) is Newtonian and does not have a measurable storage modulus. With increasing ELASTOL$^{T.M.}$ concentration the storage modulus grows stronger than the loss modulus. For $\omega = 1$ s^{-1} (0.13 Hz) a fourfold increase in concentration (4 000 ppm to 16 000 ppm) yields a G" growth

---

* in USA registered trademark of General Technology Applications (GTA)
  Inc., Manassas, Virginia; in other countries of Elastogran
  Polyurethane GmbH, a company of BASF group

P. Bockholts and I. Heidebrink (eds.), Chemical Spills and Emergency Management at Sea, 447—456.
© *1988 by Kluwer Academic Publishers.*

448

of less than a factor of 20 whereas the storage modulus increases
approximately 100-fold.

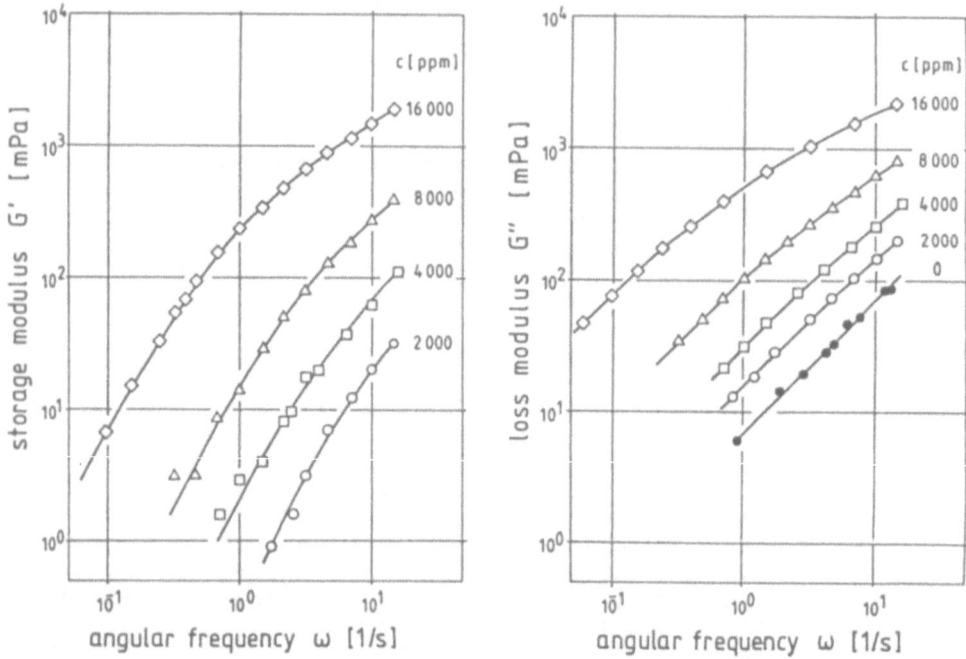

Figure 1. Storage modulus (left diagram) and loss modulus (right diagram) versus angular frequency of ELASTOL[T.M.] solutions in diesel oil
at various concentrations and 24 °C.

The behaviour of PIB macromolecules dissolved in oil can be understood
as follows: at rest the molecules have a random coil like equilibrium

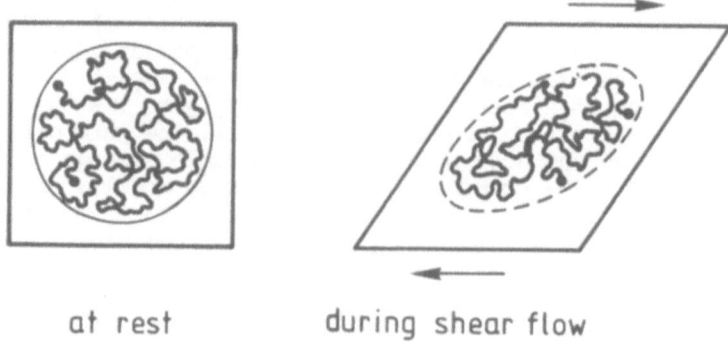

Figure 2. Equilibrium conformation (random coil) at rest and deformed
state during flow of a macromolecule in solution (schematic).

conformation (left side of Fig.2). In a flow field the coils immobilize part of the solvent giving rise to a viscosity increase with increasing polymer concentration. During flow the molecule attains an ellipsoidal conformation (right side of Fig.2) and will partially be oriented parallel to the direction of flow. This deformation of the coil, however, is reversible. When the flow is stopped the molecule will re-establish its equilibrium conformation.

2.2. Drawability (elongational viscosity) of ELASTOL[T.M.] solutions

One of the most striking effects of ELASTOL[T.M.] dissolved in oil is the dramatic increase in resistance to elongational flows. When the solution is stretched the PIB molecules are oriented which yields an elongational viscosity greater than three times the shear viscosity [3]. This behaviour is best demonstrated in the ductless siphon test [4].

A pipe being connected to a vacuum pump carries a nozzle at its other end. The solution is contained in a beaker. When the nozzle is brought into contact with the meniscus of the solution the latter is sucked out of the beaker. When the distance h between nozzle and solution meniscus is increased the solution will still flow upward until at a maximum ductless siphon height h_{max} the filament breaks.

For an untreated oil, in general, the maximum ductless siphon height h_{max} will be in the order of a few millimeters. With ELASTOL[T.M.] h_{max} is much greater and may reach values of half a meter or more (see Fig.3 and Table I). This behaviour is of tremendous importance for the performance of collecting devices used for oil spill combat.

The physical reason for the high elongational viscosity is schematically depicted in Fig.4. At rest the molecule has a coil like conformation, but in an extensional flow field, the molecule is stretched.

TABLE I Ductless siphon heights h_{max} at filament break for various oils at an ELASTOL[T.M.] concentration of 2 000 ppm. n_0 is the zero shear rate viscosity measured at 25 °C.

oil	n_0 [mPa·s]	h_{max} [mm]
petroleum	1.42	100
fuel oil	3.8	187
spindle oil	7.0	380
transformer oil	13.0	471
fine mechanics lubricant	34.3	805

The maximum ductless siphon height strongly increases with increasing oil viscosity. Therefore, the higher the oil viscosity, the lower the concentration of ELASTOL[T.M.] required to obtain a given value of h_{max}.

450

Figure 3. Ductless siphon test on a 4 000 ppm solution in fine mechanics lubricant. Nozzle diameter 1 mm. Distance between the nozzle and the solution 20 cm.

2.3. Consequences of ELASTOL[T.M.] on droplet formation

In an elongational flow field oil droplets can be broken up into two smaller droplets. With dissolved macromolecules, however, the elongational viscosity is increased and the break up process comes to an end at a bigger droplet size.

As a consequence the phase separation of oil/water emulsions formed with ELASTOL[T.M.] treated oils is much faster than for untreated oils. Table II summarizes some results of shaking tests with fuel oil/water mixtures. After a rest period of 15 min the water phase obtained from an ELASTOL[T.M.] treated mixture contains only 52 mg/l of organic carbon, compared to 20 000 mg/l for the untreated mixture.

at rest

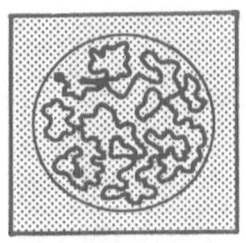

during elongational flow

$\Leftarrow$ 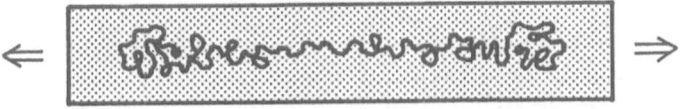 $\Rightarrow$

Figure 4. Stretching of the polymer chain when the solution is subjected to an elongational flow field (schematic).

TABLE II Oil content in water after shaking fuel oil/water (1/4) mixtures. 15 min shaking, 15 min for separation of oil and water phases. (TOC = total organic carbon).

	concentration [ppm]	TOC [mg/l]
pure fuel oil (η_0 = 3.65 mPa·s)	–	20 000
ELASTOL[T.M.] oil solution	2 000 4 000 10 000	23 22 23
ELASTOL[T.M.] powder added after 15min preshaking oil/water mix.	4 000	52

3. DISSOLVING BEHAVIOUR OF ELASTOL[T.M.] POWDER

The efficiency of ELASTOL[T.M.] in treating oil spills is strongly dependent on the dissolving speed of the powder after being spread onto oil layers floating on water. As a measure of the effective solution concentration the oil solution viscosity was determined as a function of time.

452

In Table III the time necessary to dissolve 2 000 ppm ELASTOL[T.M.] out of a total powder concentration of 10 000 ppm under different agitation conditions are listed. For fuel oil, the time to dissolve 20% of the applied ELASTOL[T.M.] is reduced from 90 h (no agitation) to 0.36 h by gentle agitation. With waves, the dissolving speed is even higher. Therefore, in practical applications under realistic sea conditions wave motion will drastically increase the dissolving speed of the powder. Furthermore, Tabel III shows that the dissolving time decreases with decreasing oil viscocity.

TABLE III   Time to dissolve 20% or applied ELASTOL[T.M.] powder.

	oil viscosity [mPa·s]	dissolving time [hours]		
		without agitation	gentle stirring	with waves*
fuel oil	3,65	90	0,36	–
Marcol 82	21,6	–	2,2	2,0
Primol 352	130	–	13	2,7

* taken from [5], wave length 1.5 m, wave height 0.13 m, 16 °C.

4. WATER/CRUDE OIL EMULSIONS

The formation of extremely stable water-in-crude oil emulsions, often called "chocolate mousse", is a major problem in combatting oil spills at sea. In order to simulate the formation of "chocolate mousse" from crude oil floating on water the apparatus shown in Fig.6 was used. Here the fluid was repeatedly picked up by a bucket-wheel and subsequently poured out onto the surface (1 revolution per second). In this arrangement the formation of stable emulsions was observed for Arabmed crude oil after less than 4 hours. The "chocolate mousse" formed with untreated Arabmed crude oil had a viscosity of 2000 mPa·s. In Table IV

TABLE IV   Viscosities for "chocolate mousse" formed with Arabmed in the bucket-wheel arrangement (shear rate 2 s^{-1}).

stirring time [h]	residence time [h]	η_0 [mPa·s] untreated crude oil	η_0 [mPa·s] with 2 000 ppm ELASTOL[T.M.]
4	4	2 000	1 400
8	32	7 800	2 000
12	100	12 200	2 700
16	104	5 400*	

* Addition of 2 000 ppm ELASTOL[T.M.] to the "chocolate mousse" obtained from untreated Arabmed after 100 hours .

the emulsion viscosities measured after various stirring and evaporation periods are listed.

When 2 000 ppm ELASTOL[T.M.] powder were spread on the "mousse" a drop in the emulsion viscosity from 12 200 mPa·s to 5 400 mPa·s was observed after another stirring period of 4 h. This means that the pumpability of the emulsion is improved and the advantage of elasticity is obtained by application of ELASTOL[T.M.] even after the formation of a stiff "mousse". When ELASTOL[T.M.] was spread on the oil layer before the formation of a "chocolate mousse", the effect of viscosity reduction was even more pronounced.

Figure 6. "Chocolate mousse" obtained with Arabmed in the bucket-wheel arrangement.

5. CONTAINMENT TESTS AT THE BERLIN MODEL BASIN

At the Versuchsanstalt für Wasserbau und Schiffbau (VWS) H.U. Oebius [5] performed containment tests in a 60 m long and 2.5 m wide basin with a boom which could be moved by a towing carriage at variable speeds. The quantity of interest was the boom velocity at which the first droplets appear on the backside of the boom (drainage failure).

In the case of calm water a concentration of only 1 000 ppm ELASTOL[T.M.] was sufficient to increase the critical velocity from 0.24 m/s to 0.48 m/s for an oil of η_0 = 22 mPa·s viscosity (MACROL[T.M.] 82).

With waves of wave length 1.5 m and 0.13 m wave height a critical boom velocity of about 0.30 m/s was measured on MARCOL$^{T.M.}$ 82. The application of 1 000 ppm and 3 000 ppm ELASTOL$^{T.M.}$ increased the critical

TABLE V   Critical boom speed v_c for Marcol 82 (η_0 = 22 mPa·s) at various ELASTOL$^{T.M.}$ concentrations with and without waves (wave length 1.5 m, wave height 0.13 m).

| | critical boom speed v_c [m/s] | |
c [ppm]	without waves	with waves
0	0.23	0,32 (0.28)
1 000	0.48	0.36
3 000	0.50	0.42 (0.48)
5 000	0.50	
10 000	0.48	

velocity up to 0.36 and 0.42 m/s, respectively. These numbers are summarized in Table V.

6. FIELD TESTS AT CEDRE

In April 88, tests were carried out on a pond at CEDRE testing facilities to assess the effect of ELASTOL$^{T.M.}$ on the efficiency of 3 different skimmers: an oleophilic rope skimmer, an oleophilic drum skimmer and a vacuum skimmer.

Each unit was tested with pure oil and oil treated with ELASTOL$^{T.M.}$. The oil was a mixture of diesel and heavy fuel oil with a viscosity of about 50 mPa·s. Between the treatment of the oil and the beginning of the recovery one hour delay was left to ELASTOL$^{T.M.}$ in order to dissolve.

These ease of the recovery, the flow rate of collected oil and the selectivity of the skimming were the main criteria for the assessment.

The oleophilic rope did not seem convenient for ELASTOL$^{T.M.}$; 2 000 ppm ELASTOL$^{T.M.}$ increases the adherence of the oil to the rope and makes its squeezing more difficult.

In return, with the oleophilic drum skimmer the results were surprising: with 3 000 ppm ELASTOL$^{T.M.}$, the drum picked up much more oil and it was possible to double the rotaing speed of the drum and to collect between 5 to 10 times more oil (see Table VI).

With the vacuum skimmer, ELASTOL$^{T.M.}$ improved mainly the selectivity of the recovery: the oil treated with 4 000 ppm ELASTOL$^{T.M.}$ became viscoelastic enough to raise the succion head few centimeters over the slick and to prevent the collecting of water. However, the delay needed to reach the desired degree of a viscoelasticity was longer than one hour.

TABLE VI   Oil recovery with the oleophilic drum.

	pure oil			oil treated with 3 000 ppm ELASTOL[T.M.]		
drum r.p.m.	17	21	21	21	40	40
Total flow-rate l/min	0,2	1,3	2	4,4	10	10
Oil flow-rate l/min	0,1	0,7	0,7	3,2	6	4

All the treated slicks showed a much better cohesiveness: the film of oil became more resistent and had lower tendency to break when it was drawn by the skimmer. It was like a draining effekt towards the skimming device which improved the recovery even at the end when the oil thickness became very thin.

However, in some cases, it will be necessary to adapt the skimming devices (i.e. increase of the rotating speed or of the pumping flow rate ...) to obtain the maximum advantage of ELASTOL[T.M.].

Finally these results are promissing enough for planning to carry out further experiments at sea with the French Navy.

7. SUMMARY AND CONCLUSION

In this work rheological properties of ELASTOL[T.M.] oil solutions were investigated. The most striking rheological effects are the dramatic increase of elasticity and elongational viscosity of the solutions compared to pure oils. This behaviour can be explained on the basis of elastic recoil and orientation by stretching of the PIB macromolecules in the solution. As a practical consequence for oil spill combat the performance of collecting devices is considerably improved. Furthermore, the formation of small oil droplets in agitated oil water mixtures is suppressed, leading to an accelerated phase separation. The dissolving time of ELASTOL[T.M.] becomes shorter with increasing agitation. Therefore, in practical applications wave motion will improve the dissolving behaviour of the powder. Finally, chocolate mousse formed with ELASTOL[T.M.] treated crude oils tends to be unstable, has a reduced water content, and is easier to pump.

Mesoscale testing [6] as well as field tests at CEDRE have demonstrated that these properties can successfully be transformed to a larger scale. Off-shore tests in Canada [7] have shown, that the effectiveness

456

of ELASTOL$^{T.M.}$ unter realistic sea conditions is even greater than expected from previous laboratory studies. Another off-shore test was performed during the NOFO oil on water exercise in June 1988. Preliminary results from this test are promissing.

REFERENCES

[1] Bobra, M.A., Kawamura, P.I., Fingas, M., Velicogna D., Laboratory and Tanktest Evaluation of Elastol$^{T.M.}$, Proceedings of the 10th artic and marine oilspill program technical seminar, June 9-11, 1987, Edmonton, p. 223

[2] H.M. Laun and R. Hingmann, Laboratory tests of ELASTOL$^{T.M.}$ (Oil Spill Combat Agent), submitted to 'Oil and Petrochemical Pollution', 1987

[3] Jones, D.M., Walters, K. and Williams, P.R., Rheol. Acta 26 (1987) 20

[4] Chao, K.C., Child, C.A., Grens II, E.A., and Williams, M.C.: Anti-Misting Action of Polymeric Additives in Jet Fuels, AICh.J. 30 (1984) 111

[5] H.U. Oebius, Versuchsanstalt für Wasserbau und Schiffbau, Berlin, Report No. 1097/87

[6] Bobra M., Kawamara P., Fingas M., Velicogna D., Laboratory and Mesoscale Testing of ELASTOL$^{T.M.}$ and Brand M Demoussifier, Proceedings of the 11th artic and marine oil spill program technical seminar, June 7-9, 1988, Vancouver, p. 391

[7] Gershey R., Batstone B., Field Tests of ELASTOL$^{T.M.}$ and Demoussifier, Proceedings of the 11th artic and marine oil spill program technical seminar, June 7-9, 1988, Vancouver, p. 443

RESEARCH AND DEVELOPMENT WITHIN CEDRE

Roger KANTIN
CEDRE (Centre de Documentation, de Recherche et d'Expérimen-
tations sur les pollutions accidentelles des eaux)
P.O. Box 72
29263 PLOUZANE
FRANCE

ABSTRACT. Research and development activities within CEDRE regarding accidental spills of hazardous susbtances in water have been concerned with the following :

- an evaluation of the amount of pollution along the French coastline,
- protection of the antipollution personnel working in hazardous areas,
- approaching a vessel in order to use equipment for identifying the substance or for measuring toxic vapors,
- workable methods of responding to a spill (lightening operations, confining and/or recovering toxic spills, treatment products, recovering toxic drums).

These and several points will be described in the following pages. One of the more important objectives for the R+D work is the preparation and up-dating of intervention guides.

INTRODUCTION

As for all scientific research, Research and Development within the framework of "Chemical Spills and Emergency Management at Sea" must be adapted to the necessities of the moment. With the increase of maritime transportation of hazardous substances other than oil (approximately 30 million tons per year in the English Channel), there is an increasing risk of an accidental spill. The CEDRE research and development program has been implemented in order to be better prepared in case of emergency.

During the past four years, CEDRE (Center for Documentation, Research and Experimentation on Accidental Water pollution), placed under the tutorship of the French Ministry of the Environment, has

457

P. Bockholts and I. Heidebrink (eds.), Chemical Spills and Emergency Management at Sea, 457—466.

458

enlarged its activities to include accidental pollution by hazardous substances. CEDRE work has been especially concerned with the following points :

- transportation of chemicals (bulk or packaged),
- accidents and incidents at sea (marine casualties),
- a study of chemicals' behaviours,
- a compilation of information sources and setting up specialized information service systems,
- an inventory of the various means of response,
- an evaluation of equipment used to detect and identify chemicals,
- protection of the pollution response personnel,
- adapting certain methods of responding to oil spills for use in responding to floating chemical spills (lightening operations, boom and skimmer operations, treatment),
- a procedural definition of the ways and means of recovering toxic drums,
- intervention guides.

This report is presented in two parts :

a) Finding and using appropriate information : it is necessary to collect and evaluate data on the maritime transportation of chemicals, on the chemicals themselves, and on other fundamental knowledge such as described in the previous presentation (Hazard Assessment/Decision Making : collection of relevant information).

b) The research and development tests and experiments which are aimed at improving a ship board intervention and at improving pollution response activities.

In order to clarify our discussion we will present concrete examples and some of our conclusions.

I. FINDING AND USING INFORMATION

These activities are summarized in the following table.

Finding Information	Using Information
Transportation statistics Accidents Priority List	Risk assessment of pollution along the French coastline
Inventory of information source Setting up information Study of chemical behaviour	Study of Accident Scenarios. What kind, how long, how great the risk.
Listing the means for responding to an accident Evaluating equipment	Intervention procedures

. A study of the maritime transportation of bulk or packaged hazardous substances along the French coastline has been especially concerned with Channel traffic and the packages which were loaded and unloaded in Marseilles.

. Incidents and marine casualties involving hazardous substances were analyzed (damaged vessels, floating or washed up drums).

. The classification and preliminary studies of the behaviour of spilled chemicals was undertaken in collaboration with the Swedish Coast Guard and Rijkswaterstaat.

. An inventory of available information sources has been prepared, and the data were analyzed for a limited number of chemicals.

. Specialized information service systems have been set up:

- SIMPA (Système d'Information sur les Moyens de lutte Anti-pollution: Information System for Antipollution Response Means) is useful for an up-to-date listing of the type and amount of pollution response equipment and products which are available in French Storehouses (for pollution by both petroleum products as well as other hazardous substances).

- H.A.C.S. (Hazard Assessment Computer System), an American system, has been adapted for use in French.

A listing of the ways and means of responding to pollution by various categories of hazardous substances (primarily based on Canadian and American documents).

. An evaluation of the various portable equipments which are on the market and which could be used for measuring or identifying gases or vapors in air or substances in water. Laboratory tests enabled an evaluation of the performance and procedures to be followed in a hazardous area for the following equipment : WARNEX EXPLOSIMETER, HNU photoionizer, DRAEGER test tubes (polytest and different reacting tubes), sampling tubes (MSA or DRAEGER) equipped with a GILIAN pump, COMPUR MONITOX individual toxic gas detectors (for detecting H_2S, Cl_2, NO_2, SO_2, CO, HCN).

II. NEW RESEARCH PROJECTS

1. RESEARCH AND DEVELOPMENT FOR PROTECTIVE EQUIPMENT

In collaboration with CEPPOL (Commission d'Etudes Pratiques de Lutte Antipollution - French Navy) and the private company MATISEC, CEDRE has perfected some protective clothing for use aboard a damaged vessel transporting hazardous substances.

The suit is made of NOMEX ($150g/m2$) which is coated on the outside with viton A ($300g/m2$) and with Butyl ($200g/m2$) on the interior; it is both chemically resistant and physically flexible. The face mask

460

is composed of polycarbonate coated with a thin teflon film.
The zipper and boots are coated with viton.

The KO2 closed circuit breathing apparatus is furnished
with an air supply kit which is designed to avoid depressions and an
alarm system. Extra air is released through pressure valves.

A system of communication is located in the suit's helmet
in order that the personnel can communicate with one other and with
other personnel beyond the polluted zone.

The personnel are equipped with a heart-beat monitor made
up of a sender attached to the wearer's chest and a watch-like receiver
worn on his wrist.

The physiological demands of wearing all the equipment have
limited the length of time that the protective clothing can be worn
safely to one hour. Beyond two hours, the wearer may experience heat
prostration, and for this reason new research will study another type
of breathing apparatus which is less restrictive, as well as the possibility of a liquid cooling vest.

2. RESEARCH AND DEVELOPMENT FOR CARGO TRANSFER.

In case of emergency, cargo can be transferred :

- from tank to tank,
- from container to container,
- from vessel to vessel.

Tests regarding the compatibility of various chemicals and
the components of pumps and hoses have been done in the laboratory.
This study has enabled us to recommend the FRAMO TK5 and TK8 systems
when used in conjunction with stainless steel 316 L hoses in the event
of a lightening operation.

3. RESEARCH AND DEVELOPMENT ON CONFINING AND RECOVERY DEVICES.

Materials which make-up the BALEAR (323, 333), POLLUGUARD
(130 MANCHE) and VIKOMA (SEABOOM) booms as well as the components of
the SIRENE 20, high sea ESCA, DESTROIL, CYCLONET and NETPOL 25 skimmers,
the SYKES, GUINARD, MOUVEX and STORK motorized pumps, and the DOP 10
and WILDEN M8 pneumatic pumps have all been tested.

Among the conclusions, we observed :

- stainless steel (AISI 316 L) is corroded at certain temperatures and for certain dilutions by caustic soda and phosphoric acid.

- Neoprene hoses (hydraulic alimentation or back flow) get
problems with most of the aromatic solvents.

- The pump propellors composed of an aluminium alloy are corroded by caustic soda and certain acids.

- The hypalon used in the booms is severely damaged by most chemicals.

- The tubes coated with polyamide are damaged by acids.

4. RESEARCH AND DEVELOPMENT ON DISPERSANTS.

The third generation dispersant FINASOL OSR 5 was first tested on a wide range of chemicals in the laboratory (using the flow through system) and then in a controlled natural environment using floating tanks.

Use of this dispersant was effective for certain chemicals (n-butyl acrylate, dodecylbenzene, ethylbenzene, ethyl-2-hexanol, styrene, xylene). The self-dispersing nature of some chemicals was also evaluated : a serie of tests has shown that n-butyl acetate, aniline, n-butanol and methyl-isobutyl-ketone have a self-dispersibility greater than 50% (flow-through system).

5. RESEARCH AND DEVELOPMENT ON SORBENTS.

Laboratory tests have been made using two bulk sorbents: one is polyurethane (BULTHANE) and the other is polypropylene (3M). The amounts of sorbents necessary for absorbing the chemicals studied are (for BULTHANE) between the optimal ratio needed for a heavy fuel oil (1/30 wt) and a light fuel oil (1/15 wt). However, the resulting agglomerate will not have a thick enough consistency, so in situ collection and towing will be impossible.

6. RESEARCH AND DEVELOPMENT ON GELLING AGENTS.

The products known as RIGIDOIL, ELASTOL and NORSOREX were tested on twenty different chemicals (floaters, dissolvers and sinkers) in their pure state as well as when they were spilled in water.

The alcohols (butanol, ethyl-2-hexanol, ethylene glycol, hexanol, methanol, ortho-cresol) were not able to be gelled in ratios recommended for oil spills ; in fact we needed to have a 1/1 ratio of the gelling agent and the chemicals in order to see any result on the pure products.

On the other hand, other chemicals such as benzene, chloroform, cyclohexane, carbon tetrachloride, trichlorethylene and xylene were able to be treated effectively at a much lower ratio of gelling agent/chemical both in their pure or diluted state. However, when sinkers were treated in this way, a strong agitation was necessary before the gelling agent was effective.

7. RESEARCH AND DEVELOPMENT FOR RECOVERING FLOATING OR WASHED-UP TOXIC DRUMS.

CEDRE has designed and developed an overpack drum to be used when recovering toxic drums both at sea and on land (the maximum deformations of the drums to be recovered were measured after numerous drop tests. The overpack was designed on the basis on this data).

On shore, a procedure for approaching and repacking a damaged drum (using a crow-bar and a dolly) has been defined and tested in real cases.

At sea, a recovery system using a fish net pocket suspended from an onboard hydraulic crane has been developed. Recovery procedures (with or without repacking in an overpack) have also been defined.

III. FUTURE RESEARCH

CEDRE plans to keep of the technological advances of 1989/90 regarding hazardous susbtances in testing any new products which are available on the market and in dealing with certain promising ideas such as the use of chemical foams to limit the evaporation of volatile substances.

An important aspect of future research will involve a controlled testing at sea in order to :

- observe the behaviour of spilled chemicals and verify the computer models,
- set up operational conditions for an intervention (protective clothing, measuring equipment),
- specific means of responding to chemical pollution.

Finally, the SEABEL system will be set up and will be adapted to national needs. Eventually the system will be enlarged to include other existing models.

IV. CONCLUSION

The R + D studies are aimed at being better prepared for an eventual response to an accidental spill whether it is a small emergency (a drum on the coast) or a larger pollution (such as the rupture of a wing tank on a chemical tanker).

The studies take into consideration the work of our international colleagues (such as the research of the Warren Spring Laboratory on dissolvents spilled at sea for determining valid models).

Two points seem vital for giving this knowledge its full value :

- a constant exchange of information regarding the progress of each of us in our research and development work,

   - all the knowledge we have should be integrated in the
intervention plans and in decision making systems in order that all
of us will be able to make use of our common know-how.

BIBLIOGRAPHY

A. CEDRE REPORTS ON CONTRACTS MADE WITH THE FRENCH NAVY AND THE FRENCH
   MINISTRY OF THE ENVIRONMENT.

- Evaluation des risques de pollution que courent les côtes françaises
  (par les méthodes d'analyse de la sécurité des systèmes), 1983.

- Transbordement en mer d'hydrocarbures et de substances dangereuses.
  Rapport R.83.686.R.

- Evaluation du risque de pollution accidentelle que courent les côtes
  françaises. Rapport R.85.80.E.

- Etude de vêtements et d'équipements de protection des personnels
  chargés d'intervenir dans une zone polluée par un produit chimique.
  Rapport R.85.87.R.

- Le risque de pollution par les épaves dangereuses. Le cas des produits
  toxiques autres que les hydrocarbures. Rapport R.86.138.R.

- Absorption et Agglomération de produits chimiques. Rapport R.86.193.R.

- Adaptation des équipements existants à l'allègement des transporteurs
  de produits chimiques. Utilisation des moyens propres aux navires.
  Recensement des navires stockeurs. Rapport R.86.210.C.

- Confinement et récupération des polluants flottants. Adaptation des
  équipements existants à une lutte contre les pollutions par produits
  chimiques. Rapport R.87.245.R.

- Transport Maritime des produits chimiques au large des côtes
  françaises: routes suivies, nature des produits, types d'accidents
  - une synthèse. Rapport R.87.262.C.

- Etude du risque de pollution lié au transport maritime de substances
  dangereuses. Rapport R.87.264.E.

- Lutte contre les pollutions accidentelles des eaux par substances dangereuses : équipements de mesure et de détection. Modèles d'évaluation des risques. Rapport R.87.265.E.

- Préparation à la lutte contre les pollutions accidentelles par substances dangereuses. Programme 1986-87. Rapport R.87.272.R.

- Mesure de la toxicité et de l'explosivité. Système d'échantillonnage de l'eau et de l'air ambiant. Rapport R.87.285.C.

- Méthodes de lutte contre les pollutions accidentelles des eaux par produits chimiques (dispersion au stade laboratoire). Etat d'avancement R.87.300.C.

- Equipements de protection des personnels chargés d'intervenir en mer dans une zone polluée par des produits dangereux. Rapport R.87.304.C.

- Guide d'allègement des navires en difficulté. Rapport R.87.316.C.

- Récupération de fûts. Rapport R.87.322.C.

- Recherche et expérimentations sur les pollutions relatives au transport maritime de substances dangereuses (formation et préparation à la lutte). Rapport R.88.330.C.

- Etude de l'adaptation aux besoins de la Marine Nationale du programme informatique H.A.C.S. d'évaluation des risques lors de pollutions accidentelles par substances dangereuses. Rapport R.88.341.C.

- Etablissement d'une liste prioritaire de substances dangereuses. Révision Février 1988. Rapport R.88.343.C.

- Evaluation de la dispersibilité des produits chimiques. Rapport R.88.359.C.

- Guide pour les interventions en mer dans les zones polluées par substances dangereuses. Rapport R.88.371.C.

- Approche d'un navire accidenté. Rapport R.88.392.C.

- Limites d'utilisation des équipements de protection des personnels chargés d'intervenir en mer dans une zone polluée : améliorations proposées. Rapport R.88.393.C.

- Miniguides d'intervention et de lutte face au risque chimique, concernant les produits suivants : acide phosphorique, acrylonitrile, ammoniac, benzène, butane, chlorate de sodium, chlore, chlorure de vinyle, isoprène, méthanol, nitrate d'ammonium, oxyde de propylène, phénol, soufre, styrène, urée. Rapport R.88.399.C.

B. REPORTS FOR THE UNITED NATIONS AND THE EUROPEAN ECONOMIC COMMUNITY.

- Risk assessment of marine casualties involving hazardous substances in the Mediterranean Sea. An overview. Report for IMO. R.87.232.C.

- Guide pour la collecte de données de l'inventaire des moyens de lutte contre les pollutions accidentelles par substances dangereuses. Report for EEC. R.88.344.C.

C. DOCUMENTS PREPARED FOR TRAINING COURSES AND CONFERENCES.

- Research on protective equipment for an intervention at sea in a zone polluted by hazardous substances. Emergency response 86, Vancouver, September 1986.

- Intervention sur subtances dangereuses. Conduite à tenir. Sécurité individuelle et collective. Stage CMIRT, December 1986.

- L'intervention sur un navire présentant des risques de pollution par substances dangereuses. Stage de formation INFOPOL, June 1986.

- Evaluation du risque de pollution accidentelle lié au transport maritime de produits chimiques. Stage de formation INFOPOL, June 1987.

- La règlementation du transport maritime et les moyens d'intervention contre les pollutions chimiques. Stage de formation INFOPOL, June 1987.

- Le trafic maritime de substances dangereuses en Méditerranée. Séminaire OTAN/CDSM BREST, March 1987.

- Comportement de substances dangereuses déversées en mer : étude basée sur cinq produits types. Séminaire OTAN/CDSM BREST, March 1987.

- Les risques de pollution accidentelle liés au transport maritime en Méditerranée. Conférence Maritime Régionale, Toulon, October 1987.

- Classification of hazardous chemicals in maritime trade. Workshop on maritime transportation of hazardous chemicals, Bahrain, June 1987 and MEDIPOL, March 1988.

- Effects of spilled hazardous chemicals and the fate of these spills. Workshop on maritime transportation of hazardous chemicals, Bahrain, June 1987.

- Les risques de pollution liés au transport maritime des produits chimiques autres que les hydrocarbures. Séminaire sur le Devenir des Polluants Chimiques, Brest, February 1988.

- Description of the problems associated with combating marine pollution by hazardous substances. An overview. (English and French). MEDIPOL, Malta, March 1988.

- Les appareils de contrôle et de mesure de produits chimiques en cas d'intervention. MEDIPOL, Malta, March 1988.

- Presentation of the results of the NATO-CCMS workshop, Brest, March 1987. Seminar on Risk analysis and damage assessment for spills of oil and other harmful substances, Copenhagen, 14-18th, March 1988.

- Behaviour of spilt bulk chemicals : floaters, sinkers and dissolvers. Behaviour of packaged goods. RISC training course, June 1988.

- Response to floaters, sinkers and dissolvers. Response to packaged goods. RISC training course, June 1988.

Status Report for the Nordic Research and Development Co-operation
concerning Response to Chemical Accidents.

C. G. Ulf Bjurman
Deputy Assistant Under-Secretary
Ministry of Defence
103 33   STOCKHOLM
Sweden

ABSTRACT

The Nordic States are within the framework of the Nordic Council of
Ministers in co-operation carrying out a research and development
programme concerning response to chemical accidents both on land and
at sea. In the first phase a mutual strategy and level of ambition
for emergency response to such accidents will be presented. Different
studies have been and are being conduced to make this possible, inter
alia studies of the risks and emergency response in certain geographical areas. Work has already been started on certain practical issues such as sources of information, equipment for indication of gases
and the establishment of a practical testing centre. The proposal for
a strategy will be presented in the beginning of 1989 and the programme is planned to be concluded in 1990, when the work will continue in normal Nordic fora for collaboration between the countries.

PREFACE

The Nordic research and development co-operation concerning response
to chemical accidents both on land and at sea was started in the
latter part of 1985 by the Environmental Committee of the Nordic
Council of Ministers. The background for the research and development
programme is the very serious chemical accidents that have happened
in recent years. It has been found that the ability to deal with such
accidents is very limited.

The work is carried out in mutual Nordic projects or as partly financed projects. Some national work has also been started on the initiative of the Nordic steering committee. As for chemical spills at sea,
work is also being carried out within the framework of the different
regional agreements which the Nordic States are parties to. All this

P. Bockholts and I. Heidebrink (eds.), Chemical Spills and Emergency Management at Sea, 467—478.

work will be integrated into the Nordic research and development
co-operation. The aim is to present at the beginning of 1989 to the
Nordic Council of Ministers a proposal for a Nordic strategy and
level of ambition for emergency response to chemical accidents. The
strategy and level of ambition will form a base for the development
of the technique for dealing with chemical accidents both in the form
of systems of equipment and of methods. The technique will be subject
to practical trials and tests. This part of the work has already
started and is estimated to be completed in 1990. After a pilot study
a number of mutual projects have been carried out. A small symposium
was arranged in 1986 and a larger symposium in 1987. A new symposium
is planned for the beginning of 1989.

The co-operation is lead by a steering committee under the chairman-
ship of Mr. Ulf Bjurman, Sweden. Members of the committee and con-
tacts for their respective countries are Mr. Leif Palle, the National
Board Environment Protection, Denmark, Mr. Olli Pahkala, the Ministry
of the Environment, Finland, Mr. Kjell Kolstad, the State Pollution
Control Administration, Norway, and Mr. Nils-Olof Sandberg, the
National Rescue Services Board, Sweden. Secretary of the committee is
Mr. Morten Hauge, Norway. Mr. Torkild Barnkob, the National Fire
Inspection, Denmark, Mr. Arne Kjaer Sörensen, the Analytic-Chemical
Laboratory of the Civil Defence, Denmark, Mr. Jukka Metso, the
Ministry of the Interior, Finland, Mr. Risto Lautkaski, the State
Tecnical Research Institute, Finland, Mr. Gunnar Hem, the National
Directorate for Fire- and Explosionprotection, Norway, Ingunn
Valvatne, the State Pollution Control Administration, Norway, Mr. Bo
Zetterström, the National Rescue Services Board, Sweden, and Mr.
Björn Looström, the Swedish Coast Guard, have also taken part in the
work of the steering committee.

1   INTRODUCTION

The chemical accidents or environmental accidents are of great concern due to a number of serious accidents that have happened in
recent years. As the ability to deal with such accidents is very
limited, it has been decided to carry out a co-ordinated Nordic research and development programme. The aim of the programme is to produce a mutual policy for questions concerning emergency response and
to develop the technique for dealing with chemical accidents both in
the form of systems of equipment and of methods. The Nordic co-operation has already proved to be important both nationally and internationally, where co-ordinated nordic action has been possible.

The co-operation programme is based on a pilot study which showed
that it is important to increase the preparedness in Nordic countries
for chemical accidents. The programme is conducted in the form of
mutual Nordic projects and partly finansed projects. Due to very
limited appropriation, much of the work has to be carried out on the
national level. The Nordic steering committee has taken the initiative to such national work. International co-operation in the field
of combating chemical spills at sea is also being conducted within
the regional agreements which the Nordic countries are parties to.
The steering committee co-ordinates its work as far as possible with
this co-operation. To the greatest possible extent there is collaboration with the chemical industry. Different chemical industries
have also been studied in connection with the meetings of the
steering committee.

Different authorities have the responsibility for emergency response
in case of chemical accidents in the Nordic countries. The organisation of the national co-operation has therefore been developed in
different forms. In Denmark the National Boards of Civil Defence,
Fire Inspection and Environment Protection have formed a joint group.
The Ministry of the Interior and the Environment work together in the
programme in Finland. A special group has been formed in Norway which
collaborates with the committee for hazardous activities. In Sweden
the National Rescue Services Board leads a steering committee in
which there are representatives from the chemical industry, the
National Board of Occupational Safety and Health, the National

National Environment Protection Board, the Swedish Association of
Local Governments, the Boards of Chemical Inspection, Shipping and
Technical Development, the Defence Institute and the Swedish Coast
Guard.

In the following there will to start be an account of the steering
committee's work-plan. This will be followed by the result ot a pilot
study and records of a number of projects which have been carried out
or are going on at present. Some projects are being considered and
these are described. Several practical combating questions, which are
being dealt to symposia which have been held or are planned. Finally,
a summary on the planned future work is included.

2   WORK-PLAN

The steering committee has adopted the following work-plan for the
programme. A pilot study was carried out to get sufficient material
for a decision on how to conduct the work. The pilot study also included a general analasis of the risks in the Nordic countries. The
programme is divided into the sub-programmes A-E.

- A consists of resarch and development in order to build up knowledge for evaluation and choice of strategy for dealing with chemical accidents and of a reasonable level of ambition for the emergency services.

- B consists of a study of the present preparedness and of the resources and organization in the Nordic countries.

- C consists of studies of the efficiency of the emergency services.

- D consists of the development of the technique for dealing with chemical accidents in the form of systems of equipment and method.

- E consists of trials and tests of the technique and of equipment.

Besides these sub-programmes there is a special sub-programme for international contacts which aims at building up knowledge and avoiding duplication of work. Information about chemical accidents that have happened and expereience from these accidents is exchanged within the steering committee.

Under 1988 work will be continued on the sub-programmes A,B and C and with international contacts. The work will successively be more and more concentrated on sub-programmes D and E in 1989 and 1990. The programm is planned to be completed in 1990.

Besides the pilot study several projects have been carried out, others are going on and more are planned. Some projects or studies dealing with more practical aspects of dealing with chemicals are already being conducted. Nordic research and development symposia have been held and another is planned to be held in the beginning of 1989. These symposia are intended for people from industry, consulting firms, research institutions and authorities.

3   PILOT STUDY

The aim of the Nordic pilot study, which was performed by the Norwegien consultant firm Petreco, was to produce the necessary basic for the planning of the work. It was also intended to form a base for future work on contingency planning, technique for dealing with chemical accidents and the level of ambition in the Nordic countries. A review was made of current Nordic literature in this field, of nordic and other international agreements of interest and of the present significant legislation in the Nordic States. The existing means for dealing with chemical accidents are described in the study under the following headlines:

- responsible authorities;
- existing resources;
- training;
- financing.

A list containing 106 chemicals, which are used in ingreater quantities in the Nordic States, has been compiled for us as an example in
the work. Consideration has then been taken to the following properties of the substances:

- acute toxicity for mammals;
- acute toxicity for aquatic life;
- degree of difficulty to clean them up;
- reduction after a spill of the possibilities of using the environment as a resource;
- bioaccumulation;
- the quantity of the substance used;
- danger for fire and explosion.

The ten substances from the list that are given the highest priority
in the pilot study are the following:

- carbon disulfide;
- sulfuric acid;
- hydrogen sulfide;
- formaldehyde solution;
- tetra ethyl lead;
- chlorine;
- acrylonitrile;
- nitric acid;
- arsenic compounds;
- ammonia.

The treat from acute spills of chemicals in the Nordic States has
been assessed on the grounds of:

- the concentration of chemical activities;
- a picture of the chemicals transported;
- the statistics for accidents.

Hypotetical incidents, which could happen, have also been used to
illustrate the consequences of acute chemical spills. These scenarios
indicate that the potential risk for accidents is considerable. The
pilot study showed that it is important to increase the ability to
deal with chemical accidents in the Nordic States.

472

The countinued work was proposed to be concentrated on:

- a more complete and detailed collection of data and analysis of consequences of accidents and ability to deal wit the accidents;
- a development of the Nordic co-operation in this field;
- a development and trials of practical technique and equipment.

4   PROJECTS

4.1 Two industrial centres in the Nordic countries

Petreco in Norway has carried out a study in which the risks and the emergency response in two industrial areas in the Nordic countries have been assessed. Information has been collected of the transports, the amounts of chemicals handled and the different forms of production from the Borregård factories, Sarpsborg, Norway, and from five different small chemical companies in the Hangö peninsular, Finland. The information has been used for the construction of nine scenarios which from the base for the assessment of the consequences of accidents at three different levels of preparedness. In both the areas it has been found necessary to improve the emergency preparedness for acute chemical spills, in Borregård mainly for spills of condensed gas and in the Hangö area by co-operation and a general increase of the resources.

4.2 A model for the assessment of the need for prepardness

The result of the project concerning two industrial centres in the Nordic countries will form the base for a new project in which the main aim is to develop a modell for dimensioning the emergency response to hazardous substances at a plant. The task will be to develop a method for collecting information about the risks for chemical accidents at a plant and assess the different risk criteria. A method which can give guidance for establishing a risk value for a plant and the sorrounding geographical area will also be elaborated. Further, the criteria for dimensioning the emergency response to different sorts of substances will be worked out. The modell will be applied on a large number of plants in the Nordic countries. An evaluation of the results that have been reached with the modell and the existing preparedness will be carried out. The modell is expected to be most appropriate for small and medium-sized plants. In an adjusted form the modell should also be suitable for contingency planning in a region.

473

4.3 Regional analyses of risks and consequences

A project concerning an analysis of risks and consequences in the County of Malmöhus, Sweden, is being carried out by the regional government in collaboration with other authorities, municipalities and other interesed bodies. The risk situation there has such a wide spectrum that it should be possible to use the results of the project also in other places in the Nordic countries. To begin with, the dimensioning risk cases for the rescue services will be clarified. Assessments of the consequences of these risk cases will be made in a second step. These assessments will include all consequences for society, such as consequences for human health and the environment and also the direct and indirect costs for society. The demands that should be put on the rescue services will be defined and considered from cost/benefit point of view. The project will result in an action plan with priority given to different options. Within the framework of the project a symposium will be held on 23 and 24 August 1988. Special attention will in the project be devoted to chemical indu- stries, transports and storage of substances which are dangerous to the environment and human health but also other risks will be taken into consideration.

4.4 Registration of transports by road and rail of chemicals

A number of serious accidents and near-accidents have happened in connection with the transport of hazardous gods. Chemical products are transported nationally but also in the Nordic countries to a great extent in transit, especially through Denmark. In a project the quantity and type of chemicals which are transported in bulk by road and rail will be registred. Oilchem Recovery Denmark will in a pilot study investigate how a geographical registration can be organized in the most appropriate way. Information will be gathered from authori- ties and industry of how they manage the transports. Proposals for systems of registration will be elaborated. The need for control of the transports will be considered. The pilot study will also produce proposals for reporting formats.

4.5 Chemical accidents at sea

Veritec, Norway, has in a project made an analasis of chemical acci- dents at sea. In Nordic waters the probability for a serious accident causing chemical pollution is 0,1-0,2 per annum. Case-studies, which have dealt with constructed accidents with ammonia in the Sound and the Frier Fjord in Norway, show that the consequences can be 1-3 dead persons and 50-200 persons who need treatment in hospital. The con- sequences in the sea will be large numbers of dead fish but no long- term effects. The findings in the rapport are that, for parts where dangerous gas i shipped and where many people are concentrated in the neighbourhood, there is a need for a preparedness which can give hospital treatment to 50-200 wounded within 2-3 hours.

474

4.6 Sources of information

The State Technical Research Institute, Finland, has made an investigation of which sources of information the rescue services in the Nordic countries use in connection with chemical accidents. These sources have been evaluated and recommendations have been given for the development of the systems and for Nordic co-operation. The information systems can be divided into three groups: the first phase sources of information, complementary sources of information and expert assistance. The first phase sources of information help to indentify the substance and the immediate risks. They also give information about what protective equipment should be used and which measures should be undertaken first. Surch information is most important in connection with transport of hazardous substances. The complementary sources give further information about substances and methods for combating the spills. Such information can be obtained from sources which are at the disposal of the rescue services or manuals etc. Expert assistance is nearly always needed in connection with major accidents or if the substance is unusual or if the risks for human health and the environment are serious. Such help can be obtained from a central centre or local experts. The report describes the different sources of information that are used in Denmark, Finland, Norway and Sweden. Some systems that are used in other countries are also presented in order to show the alternatives and ideas for the development of the systems. The needs are considered to be very similar in the Nordic countries and the existing systems, which have been developed almost completely independently of each other, are similar.

4.7 Spreading in water and environmental consequences

A study of the spreading of chemicals in water and the consequences of such spills for the marine environment is being carried out by the Swedish Meteorological and Hydrological Institute and the Institute for Water and Air Protection. The US Coast Guard HACS-system has been studied. The Canadian TIPS-system has also been studied partly. Descriptions of the systems in the European Communities have also been taken into consideration. It has, however, been found that these systems have limited capacity. Certain parts of HACS, especially the newest programmes, can be used for creating a system. There is a proposal for the continuation of the work. The part dealing with the environmental consequences i still going on.

4.8 Equipment for the indication of gases

The Analytic- Chemical Laboratory of the Civil Defence, Denmark, and the National Rescue Services Board, Sweden, are together conducting a pilot study concerning equipment which can be used in operations for measuring and identifiering gases in connection with chemical spills.

The aim of the study is to see which existing equipment can fulfil
requirements which have been set up. Three different levels have been
set up in the requipments. If the required equipment does not exist
on the market, then efforts will be made to develop such equipment.
To get efficient equipment is considered to be most important for the
safety of the rescue personnel, for defining which areas that are
dangerous for the public and in order to be able to conduct an efficient rescue operation.

5   PLANNED PROJECTS

5.1 Information systems

The responsible authorities in the Nordic countries are at present
considering the results and proposals of the investigation which was
carried out by the State Technical Research Institute in Finland
concerning sources of information. Further information is also being
gathered about the problems and faults in the present systems. The
intention is to carry on with co-operation in the field of information systems.

5.2 Registration of transports of chemicals by road and rail

The pilot study which is being carried out is expected to be continued.

5.3 Information which can be used in an operation

In a study it is planned to collect information about equipment and
other resources for combating chemical spills. This work will be
co-ordinated with work which is being carried out by the European
Communities.

5.4 Spreading in water and environmental consequences

When the results of the on-going study are presented, the possibility
of carrying on in order to develop practical and operationel systems
will be considered.

5.5 A testing centre for chemical spills

The intention is to produce a general plan for a Nordic Test Centre
for chemical spills. The requirements for such a centre will be
studied. The different testing facilities will be defined and the
costs for them determined. In the study the results that can be
reached at such a centre will be described. The aim is to get more
efficient methods and equipment for response to spills but also to be
able to study the spreading and behaviour of the chemicals.

5.6 Mutual strategy etc.

Based on the results of the different projects and also the work
which is carried out nationally, work will be conducted to produce a
mutual Nordic strategy and level of ambition for the emergency services for response in case of chemical accidents. The result of this
study will be presented to the Nordic Council of Ministers.

6   PRACTICAL EMERGENCY RESPONSE

6.1 Training

The training of the rescue personnel in the Nordic countries for
response to chemical spills has been studied. Standard requirements
and recommendations for the training are now being considered. Mutual
aids for training, i.e. video films, will be produced.

6.2 Mutual use of resources

Mutual use of special resources for combating chemical spills is
being considered. This includes strik teams, special equipment and
experts etc. The work will be continued within the frame-work of the
Nordic Agreement on Assistance in Case of Accidents, which is expected to be signed within short.

6.3 Response vehicles

The possibility of adopting mutual recommendations for the equipment
for chemical response in vehicles or containers is being studied. The
work is based on the different levels that are in use in the Danish
rescue services.

6.4 Protective equipment

The possibilities of reaching a Nordic standard for protective equipment are being studied. Investigations are also being carried out to
define where further development of the equipment is needed. This is
most probably in the field of flamable and poisonous gases.

6.5 Equipment for indication of gases

The project which is going on has been described earlier.

6.6 Investigation within industry

Certain methods and equipment have been been developed within industry. An investigation is planned to discover what exists and spread
information about it.

6.7 Statistics

The possibilities of getting truly comparable statistics are being studied.

6.8 The Seveso Directive

Within Finland, Norway and Sweden the Seveso Directive is being studied in order to see if all the requirements are covered by the national legislation.

7 SYMPOSIA

A small symposium was held in 1986 which delt with research work going on concerning combating of chemical spills. The work within the Swedish Chemical industry in order to improve preventive and response measures at plants dealing with hazardous substances was also discussed at this symposium.

A Nordic symposium with more than 100 participants took place during two days in May 1987. The object of the symposium was to inform about the work within the Nordic research and development programme and to discuss the future work. The symposium was divided into four sessions, dealing with case studies, the threats, the consequences which the threats lead to and response measures. In all 24 experts from the different Nordic countries made brief introductions and there was a discussion at the end of each session. Some of the results of the discussions are the following. There was a general feeling that the Seveso Directive should be adopted in all the Nordic countries. The transit traffic with hazardous substances should be better identified. Research institutions in the Nordic countries should collaborate more with one another. The present spreading modells are not reliable. The use of risk analises has decreased and instead analises of consequences (scenarios) are implemented. Information concerning preventive and response measures should be improved in general. An investigation was carried out at the end of the symposium to find out what was the experts' opinion on the level of knowledge and on the possibilities of reaching results in different aspects of emergency response to chemical accidents. The result of this investigation is used in the research and development work.

The Nordic symposium was very successful and a new symposium is therefore planned in the beginning of 1989.

8   FUTURE WORK

The work, which is being conducted at present and which will be completed in 1988-1989, is expected to be sufficient to make it possible
to present a Nordic strategy and level of ambition for the preparedness for chemical accidents. The results will be analyzed and the
proposal for strategy and level of ambition presented in the beginning of 1989.

Work will continue with the practical combating questions. It is
important to develop information systems, which are practical and
easy to use in an operation to make the response efficient and for
the safety of the rescue personnel. Practical and efficient equipment
for combating chemical spills is also essential. This also applies to
the protective clothing of the personnel. The equipment has so far
not been given much attention. Inter alia the possibilities of improving the measures which are on site at a plant for immediate

mitigation should be subject to extensive efforts. The spreading of
chemicals in the air and in water and also the consequences for the
environment is a subject which needs to be dealt with. The testing
centre is planned and is considered to be important for the development of method and equipment and also to be able to study the spreading and behavions of the chemicals. The work, which will be carried
out in these fields, will be concluded in 1990. After this work will
be continued in normal Nordic fora.

SOURCES OF POLLUTION IN INDIAN OCEAN - RISK & MANAGEMENT

TAYYAB SAIFY AND S.A. CHAGHTAI
DEPARTMENT OF BOTANY
SAIFIA COLLEGE, BHOPAL 462 001 INDIA

ABSTRACT. Indian Ocean, though smaller than Pacific and Atlantic Oceans in area, volume of water and depth; its waters have a great significance as compared to all other oceans due to notable spreading of the bottom water and intermediate water towards the equator. Amounts of oil and Petroleum discharged into the Indian Ocean have been estimated to be around 3-5 x 10^6 tonnes annually which is approximately 40% of the total petroleum spill of the world oceans. The total recorded incidence of accidental oil spill in the Indian Ocean is 38 tanker and non-tanker disasters and 22 blow outs during the period of 1975 to the mid of 1988, resulting in a general tendency to ascribe "Tar lumps" or "Oil Slicks" affloat everywhere. Such incidents increase albedo values paralysing the entire ecosystem of the spilled region below the water surface. The pollution caused by heavy metals and their salts, continental wash and river runoff increase the turbidity load by 24 x 10^8 tonnes annually. In addition to it municipal and domestic sewage and a large number of chemicals of agricultural importance are continuously coming to the Indian Ocean. Photochemical and biological degradation of pollutants may be enhanced many folds by usin g additional reflectors affixed on the rescue vessels. These reflectors can minimize the rate of albedo values by increasing euphotic zone and can keep the natural ecosystem intact.

1. INTRODUCTION

In accordance with the geological history, due to large accumulations of organic matter tremendous petroleum deposition zone developed in the Persian Gulf, hence Indian Ocean is considered as the biggest petroleum producing region of the globe. As data reveals, the offshore petroleum production rate of Saudi Arabia alone in 1979 was 1.03 x 10 bbl/year or 147 x 10^6 mta which became approximately double by the end of 1987. For this reason majority of Gulf countries exploit Indian Ocean without considering its managerial problems and risks which may likely come to the fore. Sen Gupta and Qasim (1985) and British Petroleum (1986) reported that the global marine transport of oil in 1986 was 1264 MT; of which 447 MT or 35.4% of the total was shipped from Gulf countries alone.

P. Bockholts and I. Heidebrink (eds.), Chemical Spills and Emergency Management at Sea, 479—488.

Majority of coastal countries export their crude petroleum to other countries only for the reason that they are lacking advance technology and sophistication. India is playing a leading role to monitor the Indian Ocean through its department of Ocean Development and National Institute of Oceanography In this country a decade of marine pollution investigations has been completed since the commissioning of the first oceanographic research vessel Gaveshani in February 1976. The ocean research vessel Sagar Kanya of India commissioned in June 1983, has since provided a wealth of information which can be accepted as the state of art documentation on the health of the seas around India (Qasim & Sen Gupta, 1983).

Although the extent of pollution in the seas around India is not very severe, it is the high time we recognise that it needs sustained scientific efforts aimed towards its improvement. The present state of affairs though not so serious a situation is largely due to the nature and geographical orientation of the northern Indian ocean, which is subjected to the impacts of semidiurnal tides associated with the biannual reversal of the direction of Monsoon and the resulting sea currents. These natural phenomena promote enough flushing and provide adequate exchange of water masses resulting into good dispersal of all the incoming pollution loads. Unlike the Baltic and Mediterranean seas which are closed and therefore heavily contaminated due to increasing of land based pollutants from countries around them, Indian Ocean has the advantage of open seas that ensures good dispersal of pollutant species (Qasim et al., 1988).

2. INDIAN OCEAN IN GEOLOGIC PERSPECTIVE

Indian Ocean which extends between the meridian of Cape Augulhas (20°E) to the meridian of the South-East Cape of Tasmania (147°E) is smaller than Pacific and Atlantic oceans with an area of 74.12×10^6 Km^2, volume of water being 284.61 km^3. Its mean depth is 3840 m, while maximum depth is 7455 m. Major water bodies connected with the Indian Ocean are Bay of Bengal, Gulf of Arabian Sea, Gulf of Suez and Gulf of Aqaba alongwith Red Sea through Gulf of Aden, Persian Gulf through Gulf of Oman, Laccadive Sea, Andman Sea, Mozambique Channel and Great Australian Bight. The bottom topography of the "Deep Sea Sills" plays an important role in the physical oceanography because it influences the distribution of the quality of deep water at the bottom in the area of Kerguelen, Maldive and East Indian Ridges in the Indian Ocean.

As per the geological history the Persian Gulf is very old. It has existed ever since the placeozoic period and belongs to a geosyncline, along which the Arabian plate is moved beneath the Eurasiatic plate. The deepening of the gulf by sinking motion is balanced by sedimentation. The delta of Shatt Al Arab is extended seaward into the Gulf by upto 50 m every year. Consequently, sediments of more than 15 Km. thickness have been accumulated. Owing to the fact that during the earth's history conditions in the area of the primary Persian Gulf often were suitable for production and accumulation of organic matter, large petroleum deposits were formed which are among the most productive ones on the globe.

3. SOURCE OF POLLUTION IN INDIAN OCEAN

Extesnive search of oil and petroleum areas in the coastal region, vigorous development of industries, widespread use of chemicals in agriculture and forestry, input of river runoff and continental wash, dumping of sewage and municipal waste are some of the prominent sources of pollution in the Indian Ocean.

3.1 OIL AND PETROLEUM INPUT

Crude oil and fuel oil petroleum hydrocarbons consists of alkanes, cyclo-alkanes and aromatic compounds containing alteast one benezene ring. These substances have been recorded in the Indian Ocean, in the form of "Tar lumps" or "Oil Slicks". The alkanes or aliphatic hydrocarbons available in the Indian Ocean consist of the fully saturated normal alkalenes (also called paraffins) and branched alkanes of the general molecular formula $CnH2n + 2$, where the value of 'n' ranges from C_1 to usually around C_{40}, although compounds with 60 carbons have also been reported. Most important group of branched compounds is the isoprenoid hydrocarbons (above C_{13}) consisting of isoprene building blocks, mainly pristance (C_{19}) and phytane (C_{20}) (Albaiges, 1980).

Various types of cycloalkanes also called naphthenes or cycloparaffins, are available in the Indian Ocean alongwith some specific plant and animal precursors (e.g. steranes, diterpanes, triterpanes etc) which serve as important molecular markers in oil spill and geochemical studies. Aromatic hydrocarbons contain one or more aromatic (benzene) rings connected as fused rings (i.e. naphthalene) or lined rings (i.e. biphenyls). Such hydrocarbons are less abundant in Indian Ocean than saturated ones.

All the six types of nonhydrocarbon petroleum as classified by Posthuma (1977) have been recorded from Indian Ocean which comprise of sulphur compounds nitrogen compounds, prophyrins, oxygen compounds, asphaltenes and trace metals.

Sulphur compounds form the most important group of hydrocarbons their constituents are organically bound as heterocyclic derivatives. The organosulphur compounds consists of thiols, sulphides, disulphides, cyclic sulphides and thiophenes. Nitrogen is present in all crude oils in the form of pyridines, quinolines, benzoquinolines, acridines, pyrroles, indoles, carbazoles and benzacarbazoles (Clark and Brown 1977). The porphyrins occur as organometallic complexes of Vanadium and Nickel and are nitrogen containing compounds derived from chlorophyll which consists of four pyrrile rings linked together. Oxygen compounds present in crude oil, are less than 2% and are found primarily in distillation fractions above 400°C in the form of phenols, carboxylic acids, ketones, easters lactones and few other compounds. Asphaltenes are the NSO compounds of higher molecular weight (1000-10,000) containing 10-20 fused rings with aliphatic and nephthenic side chains and contribute significantly to the properties of petroleum in geochemical formations and in spill situation in relation to emulsification behaviour. In petroleum, trace metals are mainly Vanadium and Nickel some times reaching thousands of parts per million. They are present mainly in porphyrin complexes and other organic compounds (Clark and Brown, 1977).

Refined petroleum products introduced to the marine environment of Indian Ocean include gasoline, Kerosene, Jet fuels, fuel oils (No. 2, No. 4, No. 5 & No. 6) or Bunker oil and lubricating oils.

Estimates of oil and petroleum hydrocarbons discharged to the Indian Ocean have varied from 3 to 5 x 10^6 tonnes annually, which is comparatively 40% of the total oil spill of the world oceans (Table I), following the discharge of oil into water a slick is formed due to low water solubility of most of the oil components. Currents, waves and winds act to spread the oil slick into thin films or they get accomulated and form the larger lived lumps that float on the sea until they sink, dissolved in sea water or are beached. Tar lumps are the residual hydrocarbons remaining un-dergraded after the various physical and biological processes have acted on the spilled oils for varying lengths of time. The physical appearance of tar balls or lumps varies in size ranging from a few millimeters to several meters in diameter, Some oil residues are soft and others are hard. Some even have living organisms on their surface, attesting to the fact that they have been in the sea for a longer period.

Table - 1

Approximately annual input rate of petroleum hydrocarbons from various sources in the Indian Ocean.

Source	Probable values in X 10^6 tonnes		
Offshore Production	0.04	–	0.08
Costal refineries	0.15	–	0.35
River runoff	0.80	–	1.25
Natural seeps	0.45	–	0.63
Urban runoff	0.10	–	0.65
Atmosphere	0.42	–	0.58
Transportation			
Tankers	0.60	–	1.10
Accidents	0.40	–.	0.80
Loading and Unloading	0.10	–	0.50
Thermal Operations	0.01	–	0.30
Total	3.07	–	5.74

3.2 FATE OF SPILLED OIL

The fate of oil spilled at sea depends upon the composition of the oil, and on such external factors as light and temperature. Free-grade and Hatchett (1970) used aritficial light sources simulating the action of the sun. and found that the total rate of decomposition corresponded to a 2.5 um slick in 100 hr. Saify and Nasir (1987) have proposed a theory

"Saify's Photon Impact" and proved that if the "reflectors" are used than the source of light i.e. the sun can be brought very close to the oil spill. By utilizing such reflectors not only intensity of light can be enhanced but the rate of light penetration can also be increased i.e. the euphotic zone will also increase which naturally helps in the steep rise of the temperature of the entire region. These reflectors alongwith the activation of photochemical oxidation, can also enhance biological degradation of spilled oil through aerobic microbes. By this way a 2.5 μm slick of oil can be decomposed within 20 hours only. The first step of microbial degradation is to convert the hydrocarbon molecules to fatty acids. This results in the formation of chocolate mousse and a colloidal effect is produced that acts to further the rate of microbial degradation and disperse the oil in the sea. Essential number of reflectors can be fixed upon the rescue vessels at different angles (Saify 1987).

Approximately 1 sq. ft. size of reflector may induce the phtochemical reaction sufficient for the area of 100 sq. meter. The rate of photochemical reaction will increase many times depending upon the composition of spilled oil. Micro-organism are found universally in the water strata as well as at the bottom sediments of the ocean. However, it has been recorded that they are more abundant in areas of anamolous concentration of hydrocarbons. These organisms constitute an important link in the food chain and in the productivity cycle. Because natural microflora is actively oridising hydrocarbons in the ocean, petroleum may therefore, be considered as another source of carbon for use in this cycle. Thus when reflectors are used entire food chain can be activated and the limiting factor equation will behave as follows.

$$Y = y \text{ opt} \left(\frac{1}{1-2^b} \right) (1-2^x)(1-2^y)(1-2^z)$$

Where

Y   =   Photosynthetic yield per m^2 of production area.
Y opt   Yield obtained, if all the factors are present at optimal levels.
x,y,z   Are the factors influencing the photosynthetic process.
b   =   Is the increased euphotic zone when reflectors are used.

3.3 OPTICAL PROPERTIES

In Indian Ocean the intensity and the spectral composition of electromagnetic radiation are influenced by the optical properties of sea water. It plays an important role in the heat budget of Indian Ocean and atmosphere because they determine the proportion of the radiative energy that is taken up by the sea and converted into heat. Knowledge of the ratio of outgoing to incoming radiation is more important than the knowledge of reflectively for practical problem i.e. the albedo If Ea (above the sea surface) and Eb (below the sea surface) are the radiations that arrive at the sea than the precentage of albedo will be -

$$\% \text{ Albedo (A)} \propto \frac{Eb}{Ea}$$

When oil pollution takes place than the Eb decreases and sometimes it becomes zero. In such conditions the entire biological activities below sea surface decline sharply which creates adverse impact on the entire ecosystem. In such conditions if the reflectors are used they will increase the Eb values and can nullify the harmful impacts of spill oil on the existing ecosystem as

$$\% \text{ Albedo } (A') = \frac{Eb \times S}{Ea}$$

$$S = \text{Solar constant} = 1.94 \text{ Cal cm}^{-2} \text{ min}^{-1}$$

A' = Albedo value after the use of reflectors.

3.4 TOTAL CONTINENTAL RUNOFF AND SALINITY

Potential changes pertaining to the ecological characteristics in the water column of the Indian Ocean as well as in the bottom are miniscule as compared with the changes occuring daily throughout geological times. Except four major rivers i.e. Brahmaputra ($20 \times 10^3 \text{ m}^3/\text{sec}$) Irrawadi ($14 \times 10^3 \text{ m}^3/\text{sec.}$) Indus ($3.9 \times 10^3 \text{ m}^3/\text{sec.}$) and Zambessi ($2.5 \times 10^3 \text{ m}^3/\text{sec.}$) all other rivers are smaller with respect to their runoff capacity hence hardly 7 to 8% total continental runoff is added to Indian Ocean with respect to entire world oceans. Thus approximately 74000 m^3/sec or $2.336 \times 10^3 \text{km}^3$ per year total continental water is added to the Indian Ocean which brings about 24×10^8 tonnes of sediments annually. Visible turbidity can be recorded at the point of major river enteries. The surface water of the inner Persian Gulf reaches high salinity values i.e. 40 to 45% owing to the large excess of evaporation over precipitation. The continental runoff from the Shatt Al Arab obviously is too small to disturb this salt enrichment. This water flows towards the exit of the Gulf in the strait of Hormuz and with a still high salinity of 39.5% it enters the Gulf of Oman, where it sinks to a depth of 200 to 300 m, forming high salinity intermediate water.

3.5 HEAVY METALS, RADIONUCLIDES AND PESTICIDES

Heavy metals and their salts constitute the most widely distributed group of highly toxic and long retained substances. This group of pollutants is commonly found in coastal region. It enters through the sewer waters of commercial plants, other mining products, metal processing factories chemical and galvanizing plants and other industries. Metal pollution in the Indian Ocean has not yet reached alarming levels (Qasim et al, 1988). Detailed studies on the heavy metals in Indian Ocean have been done by Ganeshan et al (1980), Kureishy (1985), Naqvi (1987), Parulekar et al (1985), Sanzgiri & Braganca (1981), Topping (1969).

By this time there is no record of radionuclides from Indian Ocean and to maintain this status it is essential to keep a close watch that Indian Ocean should not be used as dumping ground for radioactive nuclear wastes However, the chemicals which are extensively used for the agricultural production have become serious threat for Indian Ocean which are continuously entering along coastal region through continental runoff. At present more than 200 poisonous chemicals of various constituents have been recorded from Indian Ocean in different dilutions such as the chemicals containing arsenic (Sodium arsenite, Paris green etc.), organochlorine compound (DDT, hexachlorohexane, aldrin, chlordane, polychloropinene, polychlorocamphene, DDD, DGE, DFDF, etc.) Organophosphorous compounds (Methylcaptophos, Chlorophos, Phosphamide, Carbophos, thiophos, metaphos, methylnitrophos, trichlorometaphos etc) sulphur organic compound (Captan, tetramethylthiuroamide sulphite etc.), mercuric organic compounds (Granosan mercusan etc) derivatives of carbonic acids (Sevin, Avadex etc) derivatives of phenoxyacetic acid (2,4-D, 2, 4, 5 - T etc) derivatives of nitrophenol (dinitro-orthocresol, nitrophenol, etc) salts of heavy metals (barium chloride, copper sulphate) alkaloids (nicotine sulphate, anabasine sulphate etc) alongwith various injurious chemicals.

A recent study (Sarkar and Sen Gupta 1988) of pesticides and other chemical residues in the sediments along the east coast of India recorded that these chemicals are of land origin, as high concentration of such chemicals can be recorded mainly at the river mouth.

3.6 DOMESTIC SEWAGE

All along the coastal region specially in the vicinity of shore areas of all metropolitan cities with large industrial complexes, significantly higher quantity of sewage input can be recorded. Sen Gupta and Sankaranarayanan (1975) observed an increase of about 40% (0.82 - 1.13 μ mol/litre) of phosphate phosphorous concentration during 1959-1974 in the nearshore water of Bombay. More recent observations reveal a concentration of approximately 2 μ mol/litre (Zingde 1985). Sabnis (1984) estimated 365 million tonnes (MT) as a total volume of all discharges coming from the Bombay city alone every year. A similar estimate of about 396 MT was given by Ghosh et al (1973) from the environs of Calcutta city. Similar type of understanding is required for entire coastal region of Indian Ocean to quench the thirst of mystery about the status of its eutrophication.

4. FISH & FISHING

Due to the enormous quantity of hydrocarbons, domestic sewage and some of the essential metals are being introduced in the Indian Ocean which activate the entire food chain and productivity cycle, cattle breeding yields 300 to 600 Kg. of meat per hectare per year but respective catches of sea fish amount 10 to 50 Kg. It is not surprising that the oceanic yield compares rather unfavourable as we do not attentively care for it. If instead of utilizing the ocean for various destructive activities they are properly managed, than it can serve the mankind more effectively. Keeping in view the peaceful approach India has launched a programme of ocean development through which she will monitor and search the Indian Ocean thoroughly upto the depth of 5000-6000 ft., within its own territory.

5. RISK & MANAGEMENT

The main responsibility to monitor Indian Ocean lies upon oil and petrol exporting countries as in the course of transportation of their products sometimes they release large quantity of Petroleum in small areas which become highly damaging and poisonous causing neuroparalytic, hemolytic, enzymatic, protoplasmic or the localised damage.   The second draw back with Indian Ocean is that it has not been so extensively monitored as Atlantic and Pacafic Oceans.   Since the oceans are unified life supporting system hence the management of entire world ocean is a most essential task which can be achieved by the following means and measures.

1.   SOFAR Channel (Sound Fixing and Ranging) through which the position of an airplane or a ship in distress, or a point of a impact of a missle can be determind even from a great distance when an explosive charge is fixed in the SOFAR Chennel, should be commissioned all over the coastal countries.

2.   Every coastal country should equip itself with an advanced research vessel compulsorily.

3.   Every coastal country should have its own rescue vessel so that an immediate helping hand may be extended to the spot of distress, in emergency.

4.   Such rescue vessels should be equiped with suitable number of reflectors which may be used as and when needed.

5.   The programme of Integrated Global Ocean Station System (IGOSS) should be honestly implemented all over the world oceans, to record planetary Waves or Rossby waves so that the monsoon current and tidal forecast can be possible.

6.   Like International Council for the Exploration of Sea (ICES) of European Oceanographers, regional associations should be framed so that the regional emergency management may be made possible.

7.   Every rescue vessel should also be equiped with high power water filtration plant so that the oil layer of the spill region may be removed upto a possible extent on spot filtration.

8.   United Nations World Ocean Monitoring & Protection Agency(UNWOMPA) should be established under UNO Leadership for which maximum funds should be provided by the petrol exporting countries.

9.   Fixed Research Towers (FRT) and Underwater Laboratories (UWL) should be commissioned all over the world oceans.

6. CONCLUSION

Due to various reasons the pollution of the oceans can't be totally ruled out. The harmful impacts of such unavoidable processes can be minimised by photochemical and biological degradations. It will be in the fitness of things if natural resources could be utilised more efficiently to induce such phenomena. In this regard alongwith other measures if the reflectors are used as proposed by Saify and Nasir (1987) in their theory "Saify's Photon Impact" than photochemical and biological degradations of various pollutants can be enhanced manifolds.

7. ACKNOWLEDGEMENTS

The authors gratefully acknowledged CSIR New Delhi for the award of Project No. 37(15)/84-EMR II and MP Council of Science and Technology Bhopal Project No. 3783/CST/85/SCO(B1). We are also thankful to Dr. S.Z. Qasim, Deptt. of Ocean Development for the guidance in preparing the manuscript.

REFERENCES

Albaiges, J. 1980 Fingerprinting petroleum pollutants in the Mediterranean Sea pp 68-81. In J. Albaiges, (ed) Analytical techniques in environmental chemistry. Pergamon Press New York.

British Petroleum 1986 BP Statistical Review of world energy London.

Clark, R.C. Jr and Brown, D.W. 1977 Petroleum: Properties and analyses in biotic and abiotic systems, pp. 1-89. In D.C. Malins (ed) Effect of petroleum on Arctic and Subarctic marine environments and organisms 1 Nature and Fate of Petroleum. Academic Press, New York.

Freegarde, M and Hatchett, C.G. 1970. Coprography in marine animals. Limnol. Oceanogr., 12 443-450.

Ganeshan, R.; Shah, P:K.; Turel, Z.R and Haldar, B.C. 1980 Study of heavy metals (Cd, Hg and Se) in the environment around Bombay by radiochemical neutron activation analysis, Transactions 4th International Conferences on Nuclear methods in Environment and Environmental Research ; University of Missouri, U.S.A.

Ghosh, B.B.; Ray and Gopalakrishnan, V. 1973 Survey and characterisation of waste waters discharged into Hooghly estuary, J. Inland Fish Soc. India. 5 82-101.

Kureishy, T.W. 1985 Studies on Mercury, Cadmium and Lead in marine organisms in relation to marine pollution from the seas around India. Ph.D. Thesis Aligarh Muslim Univeristy, Aligarh.

488

Naqvi, S.W.A. 1987 Relationships between nutrients and dissolved oxygen and nitrate reduction in the Arabian Sea, Doctoral dissertation, Poona University, Poona.

Parulekar, A.H.; Ansari, Z.A.; Harkanta, S.N. and Rodrigues, C. 1985. Long term variation in benthic macrovertebrates of Bombay India; in Biology of marine benethic organisms; techniques and methods as applied to the Indian Ocean (eds) M.F. Thomson, R. Sarojini and R. Nagabhushanam Oxford and IBH Publishing Co. New Delhi pp 485-494.

Posthuma, J. 1977 The composition of petroleum. Rapp. P-V Reun Cons. Int. Explor. Mer. 171 7-16.

Qasim, S.Z. and Sen Gupta, R. 1983 Environmental Characteristics of Ocean; in Encyclopaedi-a of Environmental Science and Enginee-ring (ed) J.R. Pfafflin and E.N. Ziegler (New York; Gordon and Breach Science Publishers) 1 pp 294-309.

Qasim, S.Z.; Sen Gupta, R. and Kureishy, T.W. 1988 Pollution of Seas around India. Proc. Indian Acad. Sci. 97 (2) 117-131.

Sabnis, M.M. 1984 Studies on some major and minor elements in the pollu-tted Mahim River Estuary, Doctoral Dissertation, University of Bombay, Bombay.

Saify, T. 1987 Fresh water resources in India - Management and allied problems. Ind. J. App. & Pure Biology 2 (1) 1-14.

Saify, T. and Nasir, S.A. 1987. Management of Surface Water Quality in India. Wat. Sci. Tech. 19 (9) 87-96.

Sanzgiri, S. and Braganca, A. 1981. Trace metal in Andaman Sea. Indian J. Mar. Sci. 8 254-256.

Sarkar, A. and Sen Gupta, R. 1988 Studies on organochlorine pesticide residus in marine sediments along the east cost of India. Mar. Pollut. Bull. (in press).

Sen Gupta, R. and Sankaranarayanan 1975 Pollution studies of Bombay, Mahasagar Bull Natl. Inst. Oceanogr. 7 73-78.

Sen Gupta, R. and Qasim, S.Z. 1985 The Indian Ocean - an environmental Overview, in The Oceans - Realities and Prospectus (ed) R.C. Sharma (New Delhi; Rajesh Publications) 7-40.

Topping, G. 1969 Concentrations of Mn, Co, Cu and Fe in the Northern Indian Ocean and Arabian Sea. J. Mar. Res. 27 318-326.

Zingde, M.D. 1985. Wastewater effluents and coastal marine environment of Bombay, Proc. Semin. Sea Water Qual. Demands, Bombay 20 1-20.